Abelian Varieties and Number Theory

CONTEMPORARY MATHEMATICS

767

Abelian Varieties and Number Theory

Moshe Jarden
Tony Shaska
Editors

AMERICAN MATHEMATICAL SOCIETY

2020 *Mathematics Subject Classification.* Primary 14K05, 14A10, 14H05, 14H40, 14K15, 14H52, 11G05, 11G09, 14H55.

Library of Congress Cataloging-in-Publication Data

Names: Jarden, Moshe, 1942– editor. | Shaska, Tanush, editor.
Title: Abelian varieties and number theory / Moshe Jarden, Tony Shaska, editors.
Description: Providence, Rhode Island : American Mathematical Society, [2021] | Series: Contemporary mathematics, 0271-4132 ; volume 767 | Includes bibliographical references.
Identifiers: LCCN 2020043007 | ISBN 9781470452070 (paperback) | ISBN 9781470464233 (ebook)
Subjects: LCSH: Abelian varieties. | Number theory. | Geometry, Algebraic. | AMS: Algebraic geometry – Abelian varieties and schemes – Algebraic theory. | Algebraic geometry – Foundations – Varieties and morphisms. | Algebraic geometry – Curves – Algebraic functions; function fields [See also 11R58]. | Algebraic geometry – Curves – Jacobians, Prym varieties [See also 32G20]. | Algebraic geometry – Abelian varieties and schemes – Arithmetic ground fields [See also 11Dxx, 11Fxx, 11G10, 14Gxx]. | Algebraic geometry – Curves – Elliptic curves [See also 11G05, 11G07, 14Kxx]. | Number theory – Arithmetic algebraic geometry (Diophantine geometry) [See also 11Dxx, 14Gxx, 14Kxx] – Elliptic curves over global fields [See also 14H52]. | Number theory – Arithmetic algebraic geometry (Diophantine geometry) [See also 11Dxx, 14Gxx, 14Kxx] – Drinfeld modules; higher-dimensional motives, etc. [See also 14L05].
Classification: LCC QA564 .A212 2021 | DDC 516.3/53–dc23
LC record available at https://lccn.loc.gov/2020043007

Contemporary Mathematics ISSN: 0271-4132 (print); ISSN: 1098-3627 (online)
DOI: https://doi.org/10.1090/conm/767

---

10 9 8 7 6 5 4 3 2 1 26 25 24 23 22 21

Dedicated to Gerhard Frey on his 75th birthday.

Contents

Preface

This volume contains a collection of papers in honor of Gerhard Frey's 75th birthday. Gerhard Frey' s contributions in mathematics over the last five decades have been in number theory, arithmetic and algebraic geometry, field arithmetic, and cryptography. His contribution to Fermat's Last Theorem, by converting it to a problem about elliptic curves is well known. Such elliptic curves today are known as Frey curves. It was Frey's idea and then work of Ribet which connected Fermat's Last Theorem to the Taniyama-Shimura conjecture of elliptic curves which led to the proof of Fermat's Last Theorem by Andrew Wiles. Gerhard Frey also has major contributions to Abelian varieties and their isogenies, field arithmetic, rational points on curves, and many other topics. Gerhard is the single most important person on development of hyperelliptic curve cryptography. He was a director of the Institute of Experimental Mathematics in Essen for many years, editor of Manuscripta Math., and held positions in Erlangen, Saarland, and Essen. He had 19 PhD students until he retired. On the occasion of his 75th birthday we are organizing this volume recognizing his work and achievements. Below we briefly describe each contribution.

Following Gerhard Frey's article [**Fry73**], one calls a field K **PAC** if every nonempty geometrically integral algebraic variety V over K has a K-rational point. It turns out that infinite extensions of finite fields are PAC [**FrJ08**, p. 196, Cor. 11.2.4] (Y. Ershov), non-principal ultra-products of distinct finite fields are PAC, [**FrJ08**, p. 200, Cor. 11.3.4] (J. Ax). In addition, if K is a countable Hilbertian field, then for each positive integer e and for almost all $\sigma := (\sigma_1, \dots, \sigma_e) \in \mathrm{Gal}(K)^e$ the fixed field $K_{\mathrm{sep}}(\sigma)$ in K_{sep} of $\sigma_1, \dots, \sigma_e$ is PAC [**FrJ08**, p. 380, Thm. 18.6.1].

Here we denote the algebraic closure of a field K by $\tilde{K}$, the separable algebraic closure of K in $\tilde{K}$ by K_{sep}, the maximal purely inseparable extension of K in $\tilde{K}$ by K_{ins}, and the absolute Galois group $\mathrm{Gal}(K_{\mathrm{sep}}/K)$ of K by $\mathrm{Gal}(K)$. Also, "almost all" is used in the sense of the Haar measure of the profinite group $\mathrm{Gal}(K)^e$.

However, non-separably-closed Henselian fields are not PAC (a result of Frey-Prestel [**FrJ08**, p. 205, Cor. 11.5.5]). In view of this result, Florian Pop has introduced the notion of an ample field that combines both types of fields. A field K is **ample** if for every geometrically integral algebraic variety V over K with a simple K-rational point the set $V(K)$ is Zariski-dense in V (see [**Pop96**, p. 2] or [**Jar11**, p. 68, Def. 5.3.2]). Among others, every PAC field and every Henselian field is ample (a result of Pop, [**Jar11**, Section 5.6]). In addition, every power series field $K_0((X_1, \dots, X_n))$ over an arbitrary field K_0 is ample [**Pop96**, p. 93, Thm. 5.11.3].

The striking property of an ample field K that Pop proved is that every finite split regular embedding problem over K is solvable. In particular, the inverse

Galois problem over $K(t)$, with t transcendental over K, has a positive solution ([**Pop96**, Main Theorem A] or [**Jar11**, p. 88, Thm. 5.9.2]).

Returning to Frey's contribution to Field Arithmetic, [**FyJ74**] proves for a countable Hilbertian field K, for each $e \geq 1$, for almost all $\sigma \in \mathrm{Gal}(K)^e$, and for every non-zero abelian variety A over $K_{\mathrm{sep}}(\sigma)$, that $\mathrm{rank}(A(K_{\mathrm{sep}}(\sigma)))$ is infinite. This result was generalized for many classes of fields by several authors (e.g. Petersen [**Pet06**] and [**GJ06**]). In particular, Fehm-Petersen proved in [**FeP10**] that

(1) $\mathrm{rank}(A(M)) = \infty$ for every non-zero abelian variety A over an ample field M of characteristic zero.

The proof of (1) relies among others on the Lang-Mordell Conjecture proved by Faltings in [**Fal83**]. In an unpublished paper from 2010, Fehm and Petersen proved (1) also for ample fields K of positive characteristic. The proof relied on a generalization of a theorem of Grauert-Manin:

(2) Let K be a finitely generated extension of $\mathbb{F}_p$ and let C be a geometrically integral algebraic curve over K such that $\mathrm{genus}(C_{\tilde{K}}) \geq 2$ and $C_{\tilde{K}}$ is not birationally equivalent to a curve defined over $\tilde{\mathbb{F}}_p$. Then, $C(K_{\mathrm{ins}})$ is finite.

A proof of (2) was given in [**Kim97**]. However, at that time the proof was unsatisfactory, so Fehm and Petersen did not publish their result.

In the meantime, Damian Rössler has carried out Kim's program and even improved Kim's result by producing a quantitative version of it: Let F be a function field of one variable over an algebraically closed field of positive characteristic p. Let C be a smooth proper geometrically connected curve over F such that $C(F) \neq \emptyset$, $\mathrm{genus}(C_{\tilde{F}}) \geq 2$, and $C_{\tilde{F}}$ is not birationally equivalent to a curve defined over $\tilde{\mathbb{F}}_p$. Let n be the smallest integer with $p^n > 2\,\mathrm{genus}(C_{\tilde{K}}) - 2$ and n is greater than an explicitly given positive integer n_0. Then, $C(F_{\mathrm{ins}})C(F^{p^{-n+1}}) = \emptyset$. Using the result of Grauert-Manin [**Sam66**, p. 107, Thm. 1] about the finiteness of $C(F^{p^{-n+1}})$ (Grauer-Manin), Rössler concludes that $C(F_{\mathrm{ins}})$ is finite.

This has allowed Fehm-Petersen to safely prove (2) (see [**FeP21**, Thm. 3.1]) and to complete the proof of (1) also in positive characteristic under the assumption that M is not an algebraic extension of a finite field [**FeP21**, Thm. 1.1].

Still relating to the main result of [**FyJ74**], Bo-Hae Im and Michael Larsen obsereved that the absolute Galois group $\mathrm{Gal}(K_{\mathrm{sep}}(\sigma))$ is finitely generated (in the sense of profinite groups). That lead them to the following conjecture:

(3) If the absolute Galois group of a field M of characteristic 0 is finitely generated, then $\mathrm{rank}(A(M)) = \infty$ for every non-zero abelian variety A over M. The same holds in positive characteristic provided A is not isotrivial.

Another conjecture, this time of Markus Junker and Jochen Koenigsmann, says that if a field M is infinite and $\mathrm{Gal}(M)$ is finitely generated, then M is ample. It would follows that (3) is a special case of (1).

However, it seems that at this time, nobody has a clue to whether the Junker-Koenigsmann conjecture is true or not. So, in the meantime Im and Larsen have been proving their conjecture in many special cases. Their proofs invoke methods from various branches of Mathematics: Probability, Diophantine Geometry, Combinatorics, Arithmetic, and Analytic Number Theory. The paper [**ImL21**] surveys the special results obtained using those methods.

Akio Tamagawa proves in [**Tam21**] that for every pair (X, Y) of smooth affine curves over $\tilde{\mathbb{F}}_p$ there exists a curve Z admitting étale morphisms to both X and Y. This stands in sharp contrast to a result of Mochizuki saying that there are only finitely many pairs of "general curves" over $\mathbb{C}$ having a common étale covering.

On the other hand, Tamagawa proves that if X is a smooth connected affine curve of genus ≥ 2 over an algebraically closed field F of positive characteristic p and X is not-defined over $\tilde{\mathbb{F}}_p$, then there are only finitely many curves X' as above such that X and X' have a common étale overing. Among others, the proof applies the version of the Mordell-Lang conjecture for algebraically closed fields of positive characteristic proved in [**FeP21**, Prop. 4.2].

Lior Bary-Soroker and Alexei Entin prove the global function field analog of an explicit quantitative version of Hilbert's irreducibility theorem due to S. D. Cohen [**BSE21**]:

If $f(T_1, \ldots, T_n, X)$ is an irreducible polynomial over $\mathbb{F}_q(u)$, then the proportion of n-tuples $(t_1, \ldots, t_n)$ of monic polynomials in u of degree d for which $f(t_1, \ldots, t_n, X)$ is reducible out of all n-tuples of degree d monic polynomials is $O(dq^{-d/2})$. This result appears to fill a forty year old gap in the literature.

Gebhard Böckle and Florian Breuer focus in [**BoB21**] on a Galois representation associated to torsion points of a generic Drinfeld module of arbitrary rank in positive characteristic.

Specifically, the authors consider the ring of integers A of a global function field F of positive characteristic. Let $\mathfrak{p}$ be a maximal ideal of A and let $A_\mathfrak{p}$ be the $\mathfrak{p}$-adic completion of A at $\mathfrak{p}$. The group $\mathrm{Gal}(F)$ acts on the universal Tate Module associated to a certain Drinfeld module of rank r. This action gives rise to a continuous representation $\rho\colon \mathrm{Gal}(F) \to \mathrm{GL}_{r-1}(A_\mathfrak{p})$. The main result of [**BoB21**] is that ρ is surjective.

Arno Fehm, Dan Haran, and Elad Paran give a new proof to a well known theorem about the realization of every finite group G over $\mathbb{C}(z)$. Their proof uses the method of algebraic patching introduced in [**HV96**] and further presented in [**Jar11**]. In that method one considers the field $\hat{K} = K((t))$ of formal power series in t over a field K, constructs certain finitely many cyclic extensions of $\hat{K}(z)$, for another variable z, and "patches them together" to a G-extension of $\hat{K}(z)$. Having done so, one "pushes down" the latter G-extension to a G-extension of $K(z)$.

In [**FHP21**] the authors carry out the patching procedure directly over $\mathbb{C}(z)$, replacing the locally compact field $\hat{K}$ of the preceding paragraph by the locally compact field $\mathbb{C}$. The proofs of [**FHP21**] assume only knowledge of complex functions theory that an undergraduate student studies.

Anna Cadoret describes in [**Cad21**] an elementary strategy to study the locus where a finite family of linearly independent 1-cohomology classes for the étale fundamental group remains linearly independent under specialization. She applies that strategy to the injectivity of the specialization morphism on the second graded piece of the l-adic Abel-Jacobi filtration on Chow groups varying in the fibers of a smooth projective morphism. When the base scheme is a curve, this provides a generalization of a theorem of Silverman about the sparsity of the jumping locus of the rank in the fibers of an abelian scheme.

David Harbater, Julia Hartman, Valentijn Karemaker, and Florian Pop compare in [**HKP21**] a few local-global principles of smooth varieties over a function field of several variables F over a complete discrete valuation field. They prove that if the local-global principle holds for one family, then it holds for the others. The proof uses, among others, the Artin approximation theorem and resolution of singularities of arithmetic surfaces.

Barry Green and Peter Roquette represent in [**GrR21**] the Deuring theory of reductions of valued function fields of one variable. A function field F/K of one variable is said to have a good reduction with respect to a valuation v of F if there exists a transcendental element $t \in F$ over K such that $[F : K(t)] = [\bar{F} : \bar{K}(\bar{t})]$ where the bar denotes reduction with respect to v and $\bar{t}$ is transcendental over $\bar{K}$ and if, in addition, $\text{genus}(F/K) = \text{genus}(\bar{F}/\bar{K})$. Under this assumption, there is a homomorphism $\mathfrak{a} \to \bar{\mathfrak{a}}$ of the group of divisors of F/K onto the group of divisors of $\bar{F}/\bar{K}$ that maps principal divisors onto the corresponding principal divisors.

The authors give an elementary proof to the statement that an arbitrary function field F/K of one variable has a good reduction at "almost all" valuations of F.

Brendan Creutz and Josśe Phelipe Voloch consider in [**CrV21**] a finite field k, a curve F over k with function field K, and an elliptic curve E over K which is defined over k. It is known from the work of Tate and Milne that the Tate-Shafarevich group Sha(E/K) is finite, and Milne gave a formula for its order in terms of the Frobenius eigenvalues of E and F.

In this paper the authors use the structure of the set of l-isogenies of elliptic curves to compute the group structure of Sha(E/K) in the case where E and F are isogenous ordinary elliptic curves.

Yuri G. Zarhin considers in [**Zar21**] a field K of prime characteristic p, an integer $n \geq 5$, an irreducible polynomial $f(x)$ over K of degree n whose Galois group is either the full symmetric group S_n or the alternating group A_n. Let l be an odd prime different from p, $\mathbb{Z}[\zeta_l]$ the ring of integers in the lth cyclotomic field, $C_{f,l}\colon y^l = f(x)$ the corresponding superelliptic curve and $J(C_{f,l})$ its Jacobian. He proves that the ring of all $\tilde{K}$-endomorphisms of $J(C_{f,l})$ coincides with $\mathbb{Z}[\zeta_l]$ if $J(C_{f,l})$ is an ordinary abelian variety and $(l,n) \neq (5,5)$.

Aharon Razon considers in [**Raz21**] a finite dimensional central simple alternative algebra A over a field F of characteristic 0 and let the symmetric group S_n act on the tensor product $A^{\otimes n}$ of n copies of A. He proves that the fixed algebra $\text{Sym}^n A$ under that action is a direct sum of central simple algebras over F. This result generalizes the case where A is associative, which in a sense goes back to Hermann Weyl.

We hope this book will be helpful to all mathematicians working with Abelian varieties, especially those who are crossing over from other areas of mathematics. Our special thanks to all the authors who contributed papers, referees for all their work and effort, and the AMS production staff for their help in preparing this book.

Moshe Jarden

Tony Shaska

Bibliography

[BoB21] G. Böckle and F. Breuer, *The generic monodromy of Drinfeld modular varieties in special characteristic,* this volume.

[BSE21] L. Bary-Soroker and A. Entin, *Explicit Hilbert's irreducibility theorem in function fields,* this volume.

[Cad21] A. Cadoret, *Representations of étale fundamental groups and specialization of algebraic cycles,* this volume.

[CrV21] B. Creutz and J. F. Voloch, *Tate-Shafarevich groups of constant elliptic curves and isogeny,* this volume.

[Fal83] G. Faltings, *Endlichkeitssätze für abelsche Varietäten über Zahlkörpern* (German), Invent. Math. **73** (1983), no. 3, 349–366, DOI 10.1007/BF01388432. MR718935

[FHP21] A. Fehm, D. Haran, and E. Paran, *The inverse Galois problem over* $\mathbb{C}(z)$, this volume.

[FeP10] Arno Fehm and Sebastian Petersen, *On the rank of abelian varieties over ample fields*, Int. J. Number Theory **6** (2010), no. 3, 579–586, DOI 10.1142/S1793042110003071. MR2652895

[FeP21] A. Fehm and S. Petersen, *Ranks of abelian varieties and the full Mordell-Lang conjecture in dimension one,* this volume.

[FrJ08] Michael D. Fried and Moshe Jarden, *Field arithmetic*, 3rd ed., Ergebnisse der Mathematik und ihrer Grenzgebiete. 3. Folge. A Series of Modern Surveys in Mathematics [Results in Mathematics and Related Areas. 3rd Series. A Series of Modern Surveys in Mathematics], vol. 11, Springer-Verlag, Berlin, 2008. Revised by Jarden. MR2445111

[Fry73] Gerhard Frey, *Pseudo algebraically closed fields with non-Archimedean real valuations*, J. Algebra **26** (1973), 202–207, DOI 10.1016/0021-8693(73)90020-3. MR325584

[FyJ74] Gerhard Frey and Moshe Jarden, *Approximation theory and the rank of abelian varieties over large algebraic fields*, Proc. London Math. Soc. (3) **28** (1974), 112–128, DOI 10.1112/plms/s3-28.1.112. MR337997

[GJ06] Wulf-Dieter Geyer and Moshe Jarden, *The rank of abelian varieties over large algebraic fields*, Arch. Math. (Basel) **86** (2006), no. 3, 211–216, DOI 10.1007/s00013-005-1492-x. MR2215309

[GrR21] B. Green and P. Roquette, *An introduction to Deuring's theory of constant reduction,* this volume.

[HV96] Dan Haran and Helmut Völklein, *Galois groups over complete valued fields*, Israel J. Math. **93** (1996), 9–27, DOI 10.1007/BF02761092. MR1380632

[HKP21] D. Harbater, J. Hartmann, V. Karemaker, and F. Pop, *A comparison between obstructions to local-global principles over semi-global fields,* this volume.

[ImL21] B.-H. Im and M. Larsen, *Abelian varieties and finitely generated Galois groups,* this volume.

[Jar11] Moshe Jarden, *Algebraic patching*, Springer Monographs in Mathematics, Springer, Heidelberg, 2011. MR2768285

[Kim97] Minhyong Kim, *Purely inseparable points on curves of higher genus*, Math. Res. Lett. **4** (1997), no. 5, 663–666, DOI 10.4310/MRL.1997.v4.n5.a4. MR1484697

[Pet06] Sebastian Petersen, *On a question of Frey and Jarden about the rank of abelian varieties*, J. Number Theory **120** (2006), no. 2, 287–302, DOI 10.1016/j.jnt.2005.12.006. MR2257548

[Pop96] Florian Pop, *Embedding problems over large fields*, Ann. of Math. (2) **144** (1996), no. 1, 1–34, DOI 10.2307/2118581. MR1405941

[Raz21] A. Razon, *Symmetric tensor products of octonion algebras,* this volume.

[Roe21] D. Rössler, *Purely inseparable points on curves*, this volume.

[Sam66] P. Samuel, *Lectures on old and new results on algebraic curves*, Notes by S. Anantharaman. Tata Institute of Fundamental Research Lectures on Mathematics, No. 36, Tata Institute of Fundamental Research, Bombay, 1966. MR0222088

[Tam21] A. Tamagawa, *Correspondences on curves in positive characteristic*, this volume.

[Zar21] Y. G. Zarhin, *Endomorphisms of ordinary superelliptic Jacobians*, this volume.

Contemporary Mathematics
Volume **767**, 2021
https://doi.org/10.1090/conm/767/15393

Abelian varieties and finitely generated Galois groups

Bo-Hae Im and Michael Larsen

To Gerhard Frey on the occasion of his 75th birthday

ABSTRACT. This paper surveys the methods that have been used to attack the conjecture, still open, that an abelian variety over a characteristic 0 field with finitely generated Galois group is always of infinite rank.

1. Introduction

Let K be a field finitely generated over $\mathbb{Q}$. By Néron's extension [**Ln**, Chapter 6, Theorem 1] of the Mordell-Weil theorem, for every abelian variety A/K, the group $A(K)$ is finitely generated.

We say a field K is *anti-Mordell-Weil* (AMW) if for every non-trivial abelian variety A/K, the group $A(K)$ is of infinite rank.

In 1974, Frey and Jarden [**FyJ**] proved the following theorem:

THEOREM 1. *If K is finitely generated over $\mathbb{Q}$ and n is a positive integer, the set of $\boldsymbol{\sigma} := (\sigma_1, \ldots, \sigma_n) \in G_K^n$ such that $\bar{K}(\boldsymbol{\sigma})$ is AMW is of Haar measure* 1.

Here $G_K := \mathrm{Gal}(\bar{K}/K)$, and $\bar{K}(\boldsymbol{\sigma})$ denotes the fixed field of $\bar{K}$ under $\langle\boldsymbol{\sigma}\rangle := \langle\sigma_1, \ldots, \sigma_n\rangle$. Note that $\langle\sigma_1, \ldots, \sigma_n\rangle$ denotes the closure of the subgroup of G_K generated by the σ_i; a closed subgroup of G_K is understood to be *finitely generated* if it is of this form.

In [**Lr**], one of us conjectured that more is true: that in fact $\bar{K}(\boldsymbol{\sigma})$ is AMW for all $\boldsymbol{\sigma}$. This conjecture remains open. In this paper, we will present several different approaches to the problem, review what is known, and discuss some variants and related open problems.

This conjecture can be reformulated as follows:

CONJECTURE 1. *Every characteristic* 0 *field with finite transcendence degree and finitely generated absolute Galois group is AMW.*

Indeed, $\bar{K}(\boldsymbol{\sigma})$ has a finitely generated Galois group, and every characteristic 0 field with finite transcendence degree and finitely generated absolute Galois group can be written in this form.

This implies the following, apparently stronger, statement:

The first author was supported by Basic Science Research Program through the National Research Foundation of Korea(NRF) funded by the Ministry of Science, ICT & Future Planning(NRF-2017R1A2B4002619).

The second author was partially supported by NSF grant DMS-1702152.

CONJECTURE 2. *Every characteristic* 0 *field with finitely generated absolute Galois group is AMW.*

Indeed, let K be any characteristic 0 field and $K' := \bar{K}(\boldsymbol{\sigma})$ an algebraic extension of K with a finitely generated absolute Galois group. Any abelian variety over K' is obtained by extension of scalars from an abelian variety over a finitely generated subfield $k \subset K'$. Let k' and $\bar{k}$ denote the algebraic closure of k in K' and $\bar{K}$ respectively. Then $\bar{k}$ is an algebraic closure of k', and there is a natural continuous homomorphism $\pi\colon G_{K'} \to G_{k'}$. As $\bar{k}$ is linearly disjoint with K' over k' and algebraic over k', we have $K' \otimes_{k'} \bar{k} \cong K'\bar{k}$ is a subfield of $\bar{K}$, and every k'-automorphism of $\bar{k}$ extends uniquely to an automorphism of $K'\bar{k}$ which is trivial on K' and thus to an automorphism of $\bar{K}$ which is trivial on K'. We conclude that π is surjective, so $G_{k'}$ is finitely generated. Thus Conjecture 1 for k' implies Conjecture 2.

The following special case of Conjecture 1 remains open, even in the case that $K = \mathbb{Q}$ and A is an elliptic curve:

CONJECTURE 3. *Let A be an abelian variety over a number field K. Then the rank of A is infinite over $\bar{K}(\boldsymbol{\sigma})$ for all $\boldsymbol{\sigma}$.*

There is also a characteristic p version, but some care is needed because a subfield of $\bar{\mathbb{F}}_p$ cannot be AMW because every point in $A(\bar{\mathbb{F}}_p)$ lies in the finite group $A(\mathbb{F}_q)$ for some q and is therefore of finite order. For characteristic p fields, one can define a field K to be AMW if every abelian variety A/K *which is not isotrivial* has infinite rank over K. There are no such abelian varieties over subfields of $\bar{\mathbb{F}}_p$, so the condition is vacuous, and we can formulate:

CONJECTURE 4. *Every field with finitely generated absolute Galois group is AMW.*

Here the absolute Galois group of a field K means $\mathrm{Gal}(\bar{K}/K)$, where $\bar{K}$ is a *separable* closure of K.

The purpose of this paper is to survey what is known about these conjectures and related questions. The emphasis will be on the wide variety of different methods which have been brought to bear, with varying degrees of success, on special cases.

2. Probabilistic methods

Throughout this paper, a *variety* will be a separated, geometrically reduced scheme over a field; it need not be connected. A *curve* is a variety of dimension 1.

We recall that a field K is *PAC* if every geometrically irreducible variety V over K has a K-point. This implies that $V(K)$ is dense in V. We say K is *ample* (the terms *large* and *anti-mordellic* also appear in the literature) if every non-singular curve has either no K-points or infinitely many. Any K-point on a curve which is irreducible over K but not over some algebraic extension of K must lie in at least two geometric components and therefore must be a singular point. In other words, a K-point on any non-singular curve must lie on a geometrically connected irreducible component of that curve. Thus PAC implies ample.

The "PAC Nullstellensatz" [**FdJ**, Theorem 18.6] asserts that if K is a countable field satisfying the Hilbert irreducibility theorem (e.g., any global field [**Ln**, Chapter 9, Theorem 4.2]), then with probability 1, $\bar{K}(\boldsymbol{\sigma})$ is PAC. It follows that $\bar{K}(\boldsymbol{\sigma})$ is ample. Markus Junker and Jochen Koenigsmann [**JK**], made the following conjecture:

CONJECTURE 5. *Every infinite field with finitely generated absolute Galois group is ample.*

Arno Fehm and Sebastian Petersen proved [**FP**, Theorem 2.6(a)] that, at least in characteristic 0, every ample field is AMW. This shows that Conjecture 5 implies Conjecture 2 and also gives a new proof of Theorem 1.

The same approach can be used to give a probabilistic analysis of Galois representations coming from the non-torsion part of a non-trivial abelian variety A over a field K finitely generated over $\mathbb{Q}$. Let $V_A := A(\bar{K}) \otimes \mathbb{Q}$, regarded as a space with discrete topology, and consider the representation $\rho_A \colon G_K \to \operatorname{Aut}_{\mathbb{Q}} V_A$. This representation is continuous since

$$V_A := \sum_L A(L) \otimes \mathbb{Q} = \bigcup_L A(L) \otimes \mathbb{Q},$$

where the union (or sum) is taken over all of the (countably many) finite Galois extensions L/K. By Néron's theorem, each summand $A(L) \otimes \mathbb{Q}$ is finite-dimensional.

Every ordered n-tuple $\boldsymbol{\sigma} \in G_K^n$ defines a homomorphism $e_{\boldsymbol{\sigma}} \colon F_n \to G_K$, where F_n is the free group on n generators. The composition $\rho_A \circ e_{\boldsymbol{\sigma}}$ is then a countable sum of finite dimensional $\mathbb{Q}$-representations of F_n, each of which factors through a finite quotient of F_n. It is therefore a direct sum of irreducible $\mathbb{Q}$-representations of F_n which factor through finite quotients. We call such an irreducible representation an *atom*.

By the *generic representation* of F_n we mean the direct sum of a countably infinite number of copies of every atom. Thus, $\rho_A \circ e_{\boldsymbol{\sigma}}$ is always a subrepresentation of the generic representation of F_n.

THEOREM 2. *If K is finitely generated over $\mathbb{Q}$ and n is a positive integer, the set of $\boldsymbol{\sigma} \in G_K^n$ for which the G_K-representation $\rho_A \circ e_{\boldsymbol{\sigma}}$ is generic for all non-trivial A/K has measure* 1.

Note that the space of $G_{\bar{K}(\boldsymbol{\sigma})}$-invariants of V_A is exactly $A(\bar{K}(\boldsymbol{\sigma})) \otimes \mathbb{Q}$, so the infinite multiplicity of the trivial atom in $\rho_A \circ e_{\boldsymbol{\sigma}}$ is equivalent to the infinite rank of A over $\bar{K}(\boldsymbol{\sigma})$.

PROOF. Since there are countably many isomorphism classes of abelian varieties over K, it suffices to prove the statement for a single one. Likewise, it suffices to prove that each atom has infinite multiplicity in V_A with probability 1. We therefore fix an atom $\alpha \colon F_n \to \mathrm{GL}(W)$ which factors through some finite quotient G of F_n (so that W is the representation space of an irreducible rational representation). In particular, G is generated by some n-element subset.

We fix an embedding $\iota \colon G \hookrightarrow \mathsf{A}_N$, $N \geq 4$, such that the restriction of the (unique) irreducible $N-1$-dimensional $\mathbb{Q}$-representation of A_N to G contains W as a subrepresentation. This is possible for all $N \geq |G|+2$, by composing the regular permutation representation of G with an embedding $S_{|G|} \hookrightarrow \mathsf{A}_N$. Note that S_m embeds in A_N for all $N \geq m+2 \geq 3$, and the permutation representation of S_m is a subrepresentation of the restriction of the irreducible $N-1$-dimensional representation of A_N.

Since the K-points in any projective space are Zariski-dense, Bertini's theorem implies that there exists a non-singular curve X on A defined over K and passing through the identity 0. Let g be the genus of X. Since there are no rational curves on an abelian variety, we have $g \geq 1$. We fix a non-empty G_K-stable finite subset S of points in $X(\bar{K}) \smallsetminus \{0\}$.

For any $N \geq 2g + 2|S|$, we consider the space of meromorphic functions f on X which vanish to order ≥ 2 on every point of S, which are holomorphic except at 0, and which have a pole of order $\leq N$ at 0. By the Riemann-Roch theorem, the projective space of such functions has dimension $N - 2|S| - g$; imposing the condition that $\text{ord}_0 f \geq 1 - N$ or the condition that $\text{ord}_s f \geq 3$ for any specified $s \in S$ defines a projective subspace of codimension 1, while imposing the condition that $\text{ord}_t f \geq 2$ for some $t \in X \smallsetminus (S \cup \{0\})$ defines a projective subspace of codimension 2. The union over t of these codimension 2 subspaces is therefore a subvariety of codimension 1.

It follows that there is a dense open subvariety of functions f on X which have a pole of exact order N at 0, no other poles, zeroes of exact order 2 at each point in S, and no other multiple zeroes. As K-points are Zariski dense in projective space, we can take f to be defined over K. Thus, f defines a morphism $X \to \mathbb{P}^1$ of degree N, which gives a degree N extension of function fields $K(X)/K(\mathbb{P}^1)$.

The Galois group H of the minimal Galois extension of $K(\mathbb{P}^1)$ containing $K(X)$ is a permutation group of degree N, defined up to conjugation. Local monodromy considerations at ∞ and 0 show that H contains an N-cycle and an element of order 2 with $|S|$ 2-cycles (and therefore at least 2 fixed points, as $N \geq 2|S| + 2$.) If N is prime and sufficiently large, H cannot therefore be contained in the group of affine transformations on the field with N elements. If, in addition, $N > 23$ and N is not of the form $\frac{q^m-1}{q-1}$ for any prime power q and integer $m \geq 2$, then H must contain A_N [**Fe**, Theorem 4.2]. We conclude that there exists an n-tuple $(h_1, \ldots, h_n) \in H^n$ such that

$$\langle h_1, \ldots, h_n \rangle = \iota(G).$$

There is no difficulty choosing an arbitrarily large prime that is not of the form $\frac{q^m-1}{q-1}$. For the latter expression to be prime, either q must be a power of 2, or m must be odd. The number of expressions of the form $\frac{2^{km}-1}{2^k-1}$ less than M is $O(\log^2 M)$; the number of expressions $1 + q + \cdots + q^{2k} < M$ where k is a fixed positive integer and $q \geq 2$ is variable is less than $M^{1/2k} - 1$, so the number if k is allowed to vary is $O(M^{1/2})$, while the number of primes $< M$ grows like $M/\log M$. So we may fix a prime N with $\mathsf{A}_N \subset H \subset S_N$.

Suppose $c \in K \subset \mathbb{P}^1(K)$ such that the kernel G_L of the action of G_K on the N-element set $f^{-1}(c)$ satisfies $G_K/G_L \cong H$. Defining

$$V_c := \text{Span}_{\mathbb{Q}} f^{-1}(c) \subset A(L) \otimes \mathbb{Q},$$

the action of $\text{Gal}(L/K)$ on this space is a quotient of the N-dimensional permutation representation of H, which decomposes as an irreducible $N-1$-dimensional representation and a trivial representation.

If the quotient V_c does not contain the $N-1$-dimensional irreducible, it must have trivial H-action, so $x_1 - x_2 \in A(L)_{\text{tor}}$ for all $x_1, x_2 \in f^{-1}(c)$. By a theorem of Geyer and Jarden [**GJ**, Proposition 1.1], since $x_1 - x_2$ is defined over a bounded degree extension of a given finitely generated field, there are only finitely many possibilities for it, independent of the choice c.

For each $t \in A(\bar{K})_{\text{tor}}$ there are three possibilities for the translation map $\tau_t\colon A \to A$:

(1) $\tau_t(X) \neq X$;
(2) $\tau_t(X) = X$ but $f \circ \tau_t|_X$ and f are distinct;
(3) $f \circ \tau_t|_X = f$

In case (1), $f(x_1) = f(x_2)$ can only happen for finitely many pairs (x_1, x_2), so there are finitely many possibilities for $c = f(x_1)$. In case (2), there are finitely many c for which there exists x_1 with $x_1, x_1 + t \in f^{-1}(c)$.

The set of torsion points t for which case (3) occurs forms a finite subgroup T of A, and the morphism f factors through $X \to X/T$. Since $\deg f = N$ is prime, this means $|T| = 1$ or $|T| = N$. The case $|T| = 1$ means $x_1 - x_2 \in T$ implies $x_1 = x_2$, so this case can be disregarded. This leaves the case $|T| = N$, which implies f is the quotient map by translation by T. This cannot happen, since T acts freely on A and therefore on X, while every degree N morphism from an irreducible curve to $\mathbb{P}^1$ is ramified. Thus, at the cost of excluding finitely many values of c, we may assume that the action of H on V_c contains an irreducible factor of degree $N - 1$.

If the composition of $e_{\boldsymbol{\sigma}}$ with the quotient map $G_K \to H$ has image $\iota(G)$, then $\rho_A \circ e_{\boldsymbol{\sigma}}$ contains at least one copy of the atom α. The probability that this occurs for a single value c is at least $N!^{-n}$. However, by Hilbert irreducibility, for every finite sequence $K_1, \ldots, K_m$ of finite extensions of K there exists $c \in K \subset \mathbb{P}^1(K)$ such that the kernel G_L of the action of G_K on the N-element set $f^{-1}(c)$ satisfies $G_K/G_L \cong H$, and L is linearly disjoint from $K_1 \cdots K_m$ over K. We define K_{m+1} to be L, and iterate. By linear disjointness, the conditions on the compositions of $e_{\boldsymbol{\sigma}}$ with different maps $G_K \to \mathrm{Gal}(K_i/K)$ are independent. By the second Borel-Cantelli lemma, this implies that with probability 1, the atom α occurs infinitely many times in $\rho_A \circ e_{\boldsymbol{\sigma}}$. □

3. Diophantine geometry

Let K be any Hilbertian field. Given an abelian variety A/K, a finite group G, and a free $\mathbb{Z}$-module Λ of rank n and an integral representation $G \to \mathrm{Aut}(\Lambda)$, we define $B := \mathrm{Hom}(\Lambda, A) \cong A^n$. The action of G on Λ determines an action of G on B. Note that $\Lambda \subset \mathrm{Hom}(B, A)$, and this embedding is compatible with G-actions, where G acts on $\mathrm{Hom}(B, A)$ through its action on B.

Suppose that the quotient variety B/G contains a rational curve C. Let Y denote the inverse image of C in B. If Y is irreducible, we can apply Hilbert irreducibility to the morphism $f\colon Y \to Y/G = C$ to conclude that for "most" $c \in C(K)$, the inverse image of c in Y consists of a single point $y \in Y$ whose residue field L is a G-extension of K. The embedding of Y in B gives a point of $B(L)$, which we again denote y.

The action of $\mathrm{Gal}(L/K)$ on the Galois orbit of y is compatible with the action of G on B. Composing the embedding of Λ in $\mathrm{Hom}(B, A)$ with the evaluation map on y, we get a G-equivariant map from Λ to $A(L)$. The span of the image in $A(L) \otimes \mathbb{Q}$, regarded as $\mathrm{Gal}(L/K)$-representation is therefore a quotient representation of $W := \Lambda \otimes \mathbb{Q}$ as G-representation. Generically [**IL1**, Proposition 2.1], this span is in fact isomorphic to W as G-representation.

In favorable situations, this construction can be used to give unconditional proofs that $A(\bar{K}(\boldsymbol{\sigma}))$ has infinite rank. If $W^H \neq 0$ for all subgroups of G generated by n elements, then a copy of W in $A(L) \otimes \mathbb{Q}$ guarantees a non-zero element in $A(L \cap \bar{K}(\boldsymbol{\sigma})) \otimes \mathbb{Q}$. By linear disjointness arguments, one can construct an infinite linearly independent sequence of such elements.

In [**Lr**], this idea is implemented in the following concrete form. Suppose E is an elliptic curve over K and a_i and b_i are periodic sequences in K with period 3 such that for all i (or equivalently for $i = 1, 2, 3$),

$$y^2 = (x - a_i)(x - a_{i+1})(x - b_i)(x - b_{i+1})$$

is K-isomorphic to E. For all $c \in K$ distinct from all a_i and b_i, we have

$$\prod_{i=1}^{3} (c - a_i)(c - a_{i+1})(c - b_i)(c - b_{i+1}) \in (K^\times)^2.$$

Thus for all $\sigma \in \mathrm{Gal}(\bar{K}/K)$ there exists $i \in \{1, 2, 3\}$ such that

$$\sqrt{(c - a_i)(c - a_{i+1})(c - b_i)(c - b_{i+1})} \in \bar{K}(\sigma)$$

Finding such a_i and b_i for a given E amounts to realizing the c-line in E^3/G, where G is the Klein 4-group with each non-zero element inverting two coordinates of E^3.

Ideally, one would like to choose G and Λ such that $\Lambda^G = (0)$ but $\Lambda^H \neq (0)$ for all n-generated subgroups of G, where n is a fixed (possibly large) integer. For any n, such a pair exists; for instance, one can take $G = (\mathbb{Z}/2\mathbb{Z})^{n+1}$, and let Λ denote the quotient of the integral regular representation by its group of G-invariants. Unfortunately, for such pairs (G, Λ), for $n \geq 2$, we do not know if $\mathrm{Hom}(\Lambda, A)/G$ has any rational curves. (There is one exception: for the elliptic curve

$$y^2 = (x - 1)(x - \zeta_7)(x - \zeta_7^2)(x - \zeta_7^4),$$

and $n = 2$, the Hamming code gives [**Lr**, Theorem 6] a rational curve on $\mathrm{Hom}(\Lambda, A)/G$.)

There are example of pairs (G, Λ) for which $\mathrm{Hom}(\Lambda, A)/G$ has many rational curves. For instance, a theorem of Eduard Looijenga [**Lo**] shows that if G is a Weyl group and Λ is suitably chosen, this quotient is actually a weighted projective space. On the other hand, examples where most points in $\mathrm{Hom}(\Lambda, A)/G$ lie on rational curves are rare and never too far from the reflection group case [**KL**]. It may also happen that $\mathrm{Hom}(\Lambda, A)/G$ may fail to be uniruled and still have some rational curves. There are some interesting examples of this phenomenon, especially when A is an elliptic curve; for instance most of the smaller sporadic groups can be realized in this way [**IL1**].

On the other hand, there is some evidence that, especially for quotients of higher dimensional abelian varieties by finite groups, rational curves may be the exception rather than the rule. For instance, by a theorem of Gian Petro Pirola [**Pi**], most Kummer surfaces of dimension ≥ 3 have no rational curves. It would be interesting to have a criterion in terms of a positive integer d, a finite group G, and an integral representation Λ of G for whether $\mathrm{Hom}(\Lambda, A)/G$ has a genus 0 curve for all abelian varieties A of dimension d.

Note that if Λ_1 and Λ_2 are integral representations such that the rational representations $W_i := \Lambda_i \otimes \mathbb{Q}$ are isomorphic to one another, then the $\mathrm{Hom}(\Lambda_i, A)$ admit G-equivariant isogenies in both directions, so the existence of genus 0 curves depends only on the underlying rational representation. As a very preliminary step in this direction, in [**IL5**], we gave the following sufficient criterion, extending the result of [**KL**]:

THEOREM 3. *If G has an element g which acts on the Lie algebra of B with eigenvalues $\lambda_i = e^{2\pi i x_i}$, $0 \leq x_i < 1$, and $\sum_i x_i \leq 1$, then B/G has at least one rational curve.*

If G is an alternating group of even degree d and W is the $d-1$-dimensional irreducible representation, then $W^H \neq 0$ for all cyclic subgroups H of G. Indeed, H is generated by a single element, which cannot be a d-cycle and must therefore have at least two orbits in its action on $\{1, 2, \ldots, d\}$; it follows that W^H has dimension at least 1. Using this construction, in [**IL2**] we proved Conjecture 2 for $n = 1$:

THEOREM 4. *Let A be a non-trivial abelian variety over a field K which is not locally finite, which does not have characteristic* 2, *and which has (topologically) cyclic Galois group. Then the rank of A over K is infinite.*

4. Combinatorics

According to Conjecture 5, every pointed non-singular curve X/K has infinitely many points over $\bar{K}(\boldsymbol{\sigma})$ for all $\boldsymbol{\sigma}$. This obviously implies the following statement:

CONJECTURE 6. *If K is a finitely generated field over $\mathbb{Q}$ and X is a pointed non-singular curve over K, then for all $\boldsymbol{\sigma}$, there are infinitely many $\bar{K}(\boldsymbol{\sigma})$ points on X.*

For some curves, Conjecture 6 can be proved directly [**IL4**]:

THEOREM 5. *If $a_1, \ldots, a_{2g+2}$ are pairwise distinct elements of any infinite field $K' := \bar{K}(\boldsymbol{\sigma})$, not of characteristic* 2, *with finitely generated absolute Galois group, then the (non-singular) split hyperelliptic curve*

$$y^2 = (x - a_1) \cdots (x - a_{2g+2})$$

has infinitely many points over K'.

To explain the strategy of proof, we consider the case that the characteristic of K is 0, $g = 1$, and $a_i = 1 - i$ for $i = 1, \ldots, 4$. As $G_{K'} = \langle \boldsymbol{\sigma} \rangle$ is finitely generated, by Kummer theory, $(K')^\times \otimes \mathbb{F}_2$ is finite. Each positive integer n determines a class in this group, so by van der Waerden's theorem, there exist four positive integers in arithmetic progression, a, $a + d$, $a + 2d$, $a + 3d$, all of which are equivalent modulo squares in K'. Thus

$$(a/d)(a/d + 1)(a/d + 2)(a/d + 3) \in (F')^2,$$

so there is a rational point on the curve with $x = a/d$. In the general case, the proof uses the Hales-Jewett Theorem [**HJ**, Theorem 1].

This result implies that if K' is not locally finite, then for any abelian variety A which admits a non-constant K'-morphism from a split hyperelliptic curve X, the rank of A over K' is infinite. In particular, this is the case for all elliptic curves with all 2-torsion points rational.

The other class for which we are aware of a combinatorial proof of Conjecture 6 is projective curves X of the form $ax^n + by^n + cz^n = 0$. By [**BL**], if $a, b, c \in \mathbb{Q}$ then X has infinitely many points over $\bar{\mathbb{Q}}(\boldsymbol{\sigma})$ for all $\boldsymbol{\sigma}$ provided it has at least one point over $\mathbb{Q}$. In fact, less suffices; it is enough that X has points over every completion of $\mathbb{Q}$. The crucial point is to solve equations of the form $au + bv + cw = 0$, where u, v, w lie in chosen cosets of an arbitrary finite index subgroup of $\mathbb{Q}^\times$. This is done using the circle method.

5. Arithmetic

In this section and the next, we focus on Conjecture 3 in the case that $A = E$ is an elliptic curve. Here we consider methods from arithmetic geometry; in the following section, which is mainly conjectural, we consider what might be hoped for from analytic number theory.

We start with an elliptic curve $E/\mathbb{Q}$. By modularity, we have a good supply of algebraic points on E, namely, the Heegner points.

Let K be an imaginary quadratic field of $\mathbb{Q}$ with discriminant D and N be the conductor of $E/\mathbb{Q}$. For each positive integer c relatively prime to ND, let $\mathcal{O}_c$ be the order of index c in the ring of integers $\mathcal{O}_K$ of K. Then the elliptic curve E with $E(\mathbb{C}) = \mathbb{C}/\mathcal{O}_c$ defines a point on the modular curve $X_0(N)$, and its image P_c on E under the modular parametrization of $X_0(N)$ to E is called the Heegner point of conductor c; it is defined over the ring class field H_c of conductor c. If K satisfies the so-called *Heegner hypothesis*, i.e., all primes dividing the conductor N of E split in K, then there exists a non-torsion Heegner point in the collection of all Heegner points. (See [**Dar**].)

In [**I2**], it is proved that Heegner points span an infinite-dimensional subspace of the Mordell-Weil group $E(H)$ over the compositum H of all ring class fields with conductor prime to ND. In particular, since the ring class fields H_{rp^m}, where r and p are relatively prime to ND, have dihedral Galois group over $\mathbb{Q}$, if an automorphism $\sigma \in G_{\mathbb{Q}}$ does not fix K, then by the norm-compatibility relation among the Heegner points over H_{rp^k}, it can be shown that the rank of E is unbounded over the fixed subfields of H_{rp^m} under σ as m increases. If σ fixes all imaginary quadratic extensions, then automatically the rank of E over the fixed subfield under σ is infinite by an elementary argument. Either way, E has infinite rank over $\bar{\mathbb{Q}}(\boldsymbol{\sigma})$ for $n = 1$.

The same strategy has been applied [**BI**] to extend this result to elliptic curves over global function fields of odd characteristic parametrized by Drinfeld modular curves, and for elliptic curves over totally real fields parametrized by Shimura curves.

If k is a totally real number field and f is a new form on $\mathrm{GL}_2(\mathbb{A}_k)$ of weight 2 with level condition associated with $\mathfrak{c}$, where $\mathbb{A}_k$ denotes the ring of adèles of k and $\mathfrak{c}$ is a non-zero ideal of $\mathcal{O}_k$, then by [**Zh**] there exists an elliptic curve E'/k of conductor $\mathfrak{c}$ such that the L-functions of E' and f coincide up to factors at primes dividing $\mathfrak{c}$ and there exists a Shimura curve X/k and a surjective k-morphism from X to E'. So we say that E/k has a modular parametrization by a Shimura curve if E is k-isogenous to E' arising as above. If K is an imaginary quadratic extension of k such that the discriminant of K/k is prime to $\mathfrak{c}$ and satisfying a splitting or non-splitting property depending on the degree of k over $\mathbb{Q}$, which is the Heegner hypothesis in the case of totally real fields, and if E/k has a modular parametrization, then we can construct Heegner points on E via the isogeny as before.

If k is a global function field of odd characteristic, for any elliptic curve E/k, there exists a morphism from the Drinfeld modular curve $X_0(\mathfrak{c})$ to E, where the conductor of E is $\mathfrak{c} \cdot \infty$ for an ideal $\mathfrak{c}$ of $\mathcal{O}_k$ [**GR**]. We say K/k is an *imaginary quadratic extension* if the place ∞ does not split in K/k. In this case, the Heegner hypotheses for K is the condition that all primes dividing $\mathfrak{c}$ split in K/k. For K

satisfying the Heegner hypothesis, we can construct Heegner points via this Drinfeld modular curve parametrization as before.

Although these results are now encompassed by Theorem 4, Heegner point methods, possibly in conjunction with ideas from analytic number theory, remain a viable approach to Conjecture 3 for elliptic curves over suitable global ground fields.

Tim and Vladimir Dokchitser [**DD1**] gave an approach to Conjecture 3 for elliptic curves for general n. Assuming the Birch-Swinnerton-Dyer conjecture, they proved that the conjecture holds if K has a real place or E has non-integral j-invariant.

Their main idea was to show first that there is a quadratic extension M/K where the root number $w(E/M) = -1$ and the rank of $E(M)$ is odd. If a place v of K which is real or nonarchimedean with $v(j(E)) < 0$ is fixed, then they proved that by the weak approximation theorem, M can be taken as a quadratic extension such that the negative contribution of the local root number occurs only above v, so that the global root number $w(E/M)$, which is the product of local root numbers, is -1.

If $G \subset G_K$ is finitely generated, for an odd prime p, they constructed a Galois extension F/K containing M such that $\mathrm{Gal}(F/K) \cong \mathbb{F}_p^r \rtimes C_2$ where C_2 acts by -1, the image of G in $\mathrm{Gal}(F/K)$ has order at most 2, and the primes of M above bad reduction primes of E/K split completely in F. Then, by taking any index p subgroup V of $\mathbb{F}_p^r$, V is a normal subgroup of $\mathrm{Gal}(F/K)$ and $\mathrm{Gal}(F^V/K)$ is isomorphic to the dihedral group of order $2p$. As the image of of G in $\mathrm{Gal}(F/K)$ has order ≤ 2, they got a degree p-extension $L \subseteq F^V$ over K fixed by G and applied the congruence modulo 2-relation among the ranks of E over M, L and K ([**DD2**]) from which they concluded that by the p-parity conjecture,

$$\mathrm{rk}_L\, E > \mathrm{rk}_K\, E.$$

6. Analytic number theory

The method of §3 is based on the idea that where there are rational curves, there are rational points. However, many varieties without rational curves nevertheless have many rational points. The following conjecture seems to us likely to be true but unlikely to follow from the method of §3:

CONJECTURE 7. *Let $E/\mathbb{Q}$ be an elliptic curve. For all n there exist infinitely many linearly independent $n+1$-dimensional $\mathbb{F}_2$-subspaces $V_i \subset \mathbb{Q}^\times/(\mathbb{Q}^\times)^2$ such that for all i and all non-zero $v \in V_i$, the twist E_v has positive rank over $\mathbb{Q}$.*

This conjecture implies that Conjecture 3 holds for E. Indeed, let $\mathbb{Q}(\sqrt{V_i})$ denote the extension of $\mathbb{Q}$ obtained by taking square roots of coset representatives of all elements of V_i. By Kummer theory, $\mathrm{Gal}(\mathbb{Q}(\sqrt{V_i})/\mathbb{Q})$ is dual to V_i, and the image of $\langle \boldsymbol{\sigma} \rangle$ fixes a subspace of V_i of positive dimension. Thus there exists $v_i \in V_i$ such that $\mathbb{Q}(\sqrt{v_i}) \subset \bar{\mathbb{Q}}(\boldsymbol{\sigma})$. As E_{v_i} has positive rank, the non-trivial eigenspace of the action of $\mathrm{Gal}(\mathbb{Q}(\sqrt{v_i})/\mathbb{Q})$ on $E(\mathbb{Q}(\sqrt{v_i}))\otimes\mathbb{Q}$ has positive rank. By taking points of E in linearly disjoint fields $\mathbb{Q}(\sqrt{v_i})$, we obtain an infinite sequence of elements in $E(\bar{\mathbb{Q}}(\boldsymbol{\sigma})) \otimes \mathbb{Q}$, which can easily be seen to be linearly independent (see, e.g., [**Lr**, Theorem 5]).

It is known [**ILR**] that there are examples of curves E such that Conjecture 7 holds for $n = 1$. The conjecture would hold in general if we knew:

CONJECTURE 8. *Let $E/\mathbb{Q}$ be an elliptic curve. There exists an infinite dimensional $\mathbb{F}_2$-subspace $V_0 \subset \mathbb{Q}^\times/(\mathbb{Q}^\times)^2$ such that for all non-zero $v \in V_0$, the twist E_v has positive rank.*

Using analytic arguments and the known positive density of positive rank quadratic twists of certain elliptic curves over $\mathbb{Q}$ (see, e.g. [**Va**]), we proved [**IL3**] that there exist elliptic curves E over $\mathbb{Q}$ with arbitrarily large finite subspaces $V_0 \subset \mathbb{Q}^\times/(\mathbb{Q}^\times)^2$ satisfying the positive rank condition for E. On the other hand, Conjecture 8 would be an easy consequence of the following conjecture:

CONJECTURE 9. *Let $E_1, \ldots, E_n$ be elliptic curves over $\mathbb{Q}$. Then there exists $d \in \mathbb{Q}^\times/(\mathbb{Q}^\times)^2$ such that the twist of every E_i by d has positive rank.*

By estimating $\sum_{d<N} L'(1, E_d)$, Perelli-Pomykała [**PP**] showed that the proportion of twists of a fixed elliptic curve E by $d < N$ which have analytic rank 1 grows faster than $N^{-\epsilon}$ for all $\epsilon > 0$. The same is true for rank 1 by Kolyvagin's theorem. Choosing $\epsilon < 1/n$, this gives a heuristic argument for Conjecture 9. It would be interesting to try to bound

$$\sum_{d<N} L'(1, (E_1)_d) \cdots L'(1, (E_n)_d)$$

away from zero.

Acknowledgment

We would like to thank Moshe Jarden for his comments on an earlier version of this paper.

References

[BF] Lior Bary-Soroker and Arno Fehm, *Open problems in the theory of ample fields* (English, with English and French summaries), Geometric and differential Galois theories, Sémin. Congr., vol. 27, Soc. Math. France, Paris, 2013, pp. 1–11. MR3203546

[BL] Jean Bourgain and Michael Larsen, *A Finitary Hasse Principle for Diagonal Curves*, `arXiv:1404.2849`.

[BI] Florian Breuer and Bo-Hae Im, *Heegner points and the rank of elliptic curves over large extensions of global fields*, Canad. J. Math. **60** (2008), no. 3, 481–490, DOI 10.4153/CJM-2008-023-0. MR2414953

[DD1] Tim Dokchitser and Vladimir Dokchitser, *A note on Larsen's conjecture and ranks of elliptic curves*, Bull. Lond. Math. Soc. **41** (2009), no. 6, 1002–1008, DOI 10.1112/blms/bdp079. MR2575331

[DD2] Tim Dokchitser and Vladimir Dokchitser, *On the Birch-Swinnerton-Dyer quotients modulo squares*, Ann. of Math. (2) **172** (2010), no. 1, 567–596, DOI 10.4007/annals.2010.172.567. MR2680426

[Dar] Henri Darmon, *Rational points on modular elliptic curves*, CBMS Regional Conference Series in Mathematics, vol. 101, Published for the Conference Board of the Mathematical Sciences, Washington, DC; by the American Mathematical Society, Providence, RI, 2004. MR2020572

[FP] Arno Fehm and Sebastian Petersen, *On the rank of abelian varieties over ample fields*, Int. J. Number Theory **6** (2010), no. 3, 579–586, DOI 10.1142/S1793042110003071. MR2652895

[Fe] Walter Feit, *Some consequences of the classification of finite simple groups*, The Santa Cruz Conference on Finite Groups (Univ. California, Santa Cruz, Calif., 1979), Proc. Sympos. Pure Math., vol. 37, Amer. Math. Soc., Providence, R.I., 1980, pp. 175–181. MR604576

[FyJ] Gerhard Frey and Moshe Jarden, *Approximation theory and the rank of abelian varieties over large algebraic fields*, Proc. London Math. Soc. (3) **28** (1974), 112–128, DOI 10.1112/plms/s3-28.1.112. MR337997

[FdJ] Michael D. Fried and Moshe Jarden, *Field arithmetic*, 3rd ed., Ergebnisse der Mathematik und ihrer Grenzgebiete. 3. Folge. A Series of Modern Surveys in Mathematics [Results in Mathematics and Related Areas. 3rd Series. A Series of Modern Surveys in Mathematics], vol. 11, Springer-Verlag, Berlin, 2008. Revised by Jarden. MR2445111

[GR] E.-U. Gekeler and M. Reversat, *Jacobians of Drinfeld modular curves*, J. Reine Angew. Math. **476** (1996), 27–93, DOI 10.1515/crll.1996.476.27. MR1401696

[GJ] Wulf-Dieter Geyer and Moshe Jarden, *The rank of abelian varieties over large algebraic fields*, Arch. Math. (Basel) **86** (2006), no. 3, 211–216, DOI 10.1007/s00013-005-1492-x. MR2215309

[HJ] A. W. Hales and R. I. Jewett, *Regularity and positional games*, Trans. Amer. Math. Soc. **106** (1963), 222–229, DOI 10.2307/1993764. MR143712

[I1] Bo-Hae Im, *The rank of elliptic curves with rational 2-torsion points over large fields*, Proc. Amer. Math. Soc. **134** (2006), no. 6, 1623–1630, DOI 10.1090/S0002-9939-05-08494-7. MR2204272

[I2] Bo-Hae Im, *Heegner points and Mordell-Weil groups of elliptic curves over large fields*, Trans. Amer. Math. Soc. **359** (2007), no. 12, 6143–6154, DOI 10.1090/S0002-9947-07-04364-4. MR2336320

[IL1] Bo-Hae Im and Michael Larsen, *Realizing Rational Representations in Mordell-Weil Groups*, `arXiv:math/0401209`.

[IL2] Bo-Hae Im and Michael Larsen, *Abelian varieties over cyclic fields*, Amer. J. Math. **130** (2008), no. 5, 1195–1210, DOI 10.1353/ajm.0.0020. MR2450206

[IL3] Bo-Hae Im and Michael Larsen, *Parallelopipeds of positive rank twists of elliptic curves*, Indiana Univ. Math. J. **60** (2011), no. 1, 311–318, DOI 10.1512/iumj.2011.60.4398. MR2952420

[IL4] Bo-Hae Im and Michael Larsen, *Some applications of the Hales-Jewett theorem to field arithmetic*, Israel J. Math. **198** (2013), no. 1, 35–47, DOI 10.1007/s11856-013-0009-8. MR3096628

[IL5] Bo-Hae Im and Michael Larsen, *Rational curves on quotients of abelian varieties by finite groups*, Math. Res. Lett. **22** (2015), no. 4, 1145–1157, DOI 10.4310/MRL.2015.v22.n4.a9. MR3391880

[ILR] Bo-Hae Im and Álvaro Lozano-Robledo, *On products of quadratic twists and ranks of elliptic curves over large fields*, J. Lond. Math. Soc. (2) **79** (2009), no. 1, 1–14, DOI 10.1112/jlms/jdn048. MR2472130

[JK] Markus Junker and Jochen Koenigsmann, *Schlanke Körper (slim fields)*, J. Symbolic Logic **75** (2010), no. 2, 481–500, DOI 10.2178/jsl/1268917491. MR2648152

[KL] János Kollár and Michael Larsen, *Quotients of Calabi-Yau varieties*, Algebra, arithmetic, and geometry: in honor of Yu. I. Manin. Vol. II, Progr. Math., vol. 270, Birkhäuser Boston, Boston, MA, 2009, pp. 179–211, DOI 10.1007/978-0-8176-4747-6_6. MR2641190

[Ln] Serge Lang, *Fundamentals of Diophantine geometry*, Springer-Verlag, New York, 1983. MR715605

[Lr] Michael Larsen, *Rank of elliptic curves over almost separably closed fields*, Bull. London Math. Soc. **35** (2003), no. 6, 817–820, DOI 10.1112/S0024609303002431. MR2000029

[Lo] Eduard Looijenga, *Root systems and elliptic curves*, Invent. Math. **38** (1976/77), no. 1, 17–32, DOI 10.1007/BF01390167. MR466134

[PP] A. Perelli and J. Pomykała, *Averages of twisted elliptic L-functions*, Acta Arith. **80** (1997), no. 2, 149–163, DOI 10.4064/aa-80-2-149-163. MR1450922

[Pi] Gian Pietro Pirola, *Curves on generic Kummer varieties*, Duke Math. J. **59** (1989), no. 3, 701–708, DOI 10.1215/S0012-7094-89-05931-0. MR1046744

[Va] V. Vatsal, *Rank-one twists of a certain elliptic curve*, Math. Ann. **311** (1998), no. 4, 791–794, DOI 10.1007/s002080050209. MR1637976

[Zh] Shouwu Zhang, *Heights of Heegner points on Shimura curves*, Ann. of Math. (2) **153** (2001), no. 1, 27–147, DOI 10.2307/2661372. MR1826411

Department of Mathematical Sciences, KAIST, 291 Daehak-ro, Yuseong-gu, Daejeon, 34141, South Korea
Email address: bhim@kaist.ac.kr

Department of Mathematics, Indiana University, Bloomington, Indiana, 47405
Email address: mjlarsen@indiana.edu

Contemporary Mathematics
Volume **767**, 2021
https://doi.org/10.1090/conm/767/15394

Ranks of abelian varieties and the full Mordell-Lang conjecture in dimension one

Arno Fehm and Sebastian Petersen

ABSTRACT. Let A be a non-zero abelian variety over a field F that is not algebraic over a finite field. We prove that the rational rank of the abelian group $A(F)$ is infinite when F is large in the sense of Pop (also called ample). The main ingredient is a deduction of the 1-dimensional case of the relative Mordell-Lang conjecture from a result of Rössler.

1. Introduction

A considerable amount of work has been done to show that over certain fields, for example over certain infinite algebraic extensions of global fields, all abelian varieties have infinite rank. Here, the **rank** of an abelian variety A over a field F is the rank of the Mordell-Weil group $A(F)$, i.e. $\dim_{\mathbb{Q}}(A(F)\otimes\mathbb{Q})$. See [**Ros73, FJ74, RW02, Lar03, Im06, GJ06, Pet06, IL08, IL13**] to mention a few examples.

In all cases known to us where one actually succeeded in proving that over a field every abelian variety has infinite rank, these fields turn out to be ample or are conjectured to be ample, where a field F is called **ample** (or **large** following [**Pop96**]) if every smooth F-curve C satisfies $C(F) = \emptyset$ or $|C(F)| = \infty$. For example, every separably closed, real closed, or Henselian valued field is ample. Also so-called pseudo-algebraically closed fields (a term coined by Gerhard Frey) are ample. See [**Pop96, Jar03, Pop10, FP11**] for many more examples of ample fields, and [**BSF12, Pop14**] for surveys on the importance and ubiquity of ample fields in contemporary Galois theory and other areas.

In this work, we prove (see Section 5) our conjecture from [**FP10**]:

THEOREM 1.1. *Let F be an ample field that is not algebraic over a finite field and let A/F be a non-zero abelian variety. Then the rank of $A(F)$ is infinite.*

In [**FP10**] we proved several special cases of Theorem 1.1. In particular, we prove the case when F has characteristic zero. In this work, we complete the case

2020 *Mathematics Subject Classification.* Primary 14K05, 12E30, 14G05.

Key words and phrases. Abelian variety, rational rank, ample field, large field, Mordell-Lang conjecture.

This work was supported by the European Commission under contract MRTN-CT-2006-035495, and the Minkowski Center for Geometry at Tel Aviv University, established by the Minerva Foundation.

The second author was supported by the NCN grant OPUS16 UMO 2018/31/B/ST1/01474 of the Polish ministry of science.

of positive characteristic. In the proof of Theorem 1.1, we use and strengthen results of Ghioca and Moosa [**GM06**] on the Mordell-Lang conjecture in positive characteristic: The 1-dimensional case of the Mordell-Lang conjecture is reduced to a special case, which is then solved by recent work of Rössler [**Rös21**]. What we get (see Section 4) is the following:

THEOREM 1.2. *Let K be an algebraically closed field of positive characteristic p, A/K an abelian variety, C a closed subcurve of A, and $\Lambda \subseteq A(K)$ a subgroup of finite rank. If $\Lambda \cap C(K)$ is infinite, then C is elliptic or birational to a curve defined over the algebraic closure of $\mathbb{F}_p$.*

Theorem 1.1 reproves and generalizes several previous infinite rank results. In Section 6 we give a few concrete applications of Theorem 1.1.

Notation

Let K be a field. We denote by $\tilde{K}$ the algebraic closure of K, by K_{sep} the maximal separable, and by K_{ins} the maximal purely inseparable extension of K in $\tilde{K}$. If V is a K-scheme and F/K is a field extension, then $V_F := V \times_{\mathrm{Spec}(K)} \mathrm{Spec}(F)$. If V and W are K-schemes and $f\colon V \to W$ is a K-morphism, then $f_F\colon V_F \to W_F$ stands for the base extension of f. Throughout this paper, a K-variety is a separated algebraic K-scheme which is geometrically integral, and a K-curve is a K-variety of dimension 1. The absolute genus of a K-curve C is the genus $g(C')$ of a smooth projective model C' of $C_{\tilde{K}}$. We denote by $J_{C'}$ the Jacobian variety of C'. Subvarieties are always understood to be closed in the ambient variety. If $f\colon V \to W$ is a morphism and $X \subseteq W$, then we define $f^{-1}(X) := (V \times_W X)_{\mathrm{red}}$, the reduced induced scheme structure on the fiber product. If $f : G \to H$ is a homomorphism of group schemes over K, then $\ker(f) := G \times_{H,e} \mathrm{Spec}(K)$ is the scheme theoretic fibre of the unit section e.

2. Isotrivial and special curves

In this section we prove some preliminary results about isotrivial curves and special curves in the sense of Hrushovski, cf. [**Hru96**]. The main conclusion here, probably known to experts, is that a subcurve of an abelian variety is special if and only if it is elliptic or isotrivial.

Throughout this section, let K/K_0 be an extension of algebraically closed fields.

A K-variety V is **K/K_0-isotrivial** (or simply isotrivial) if there exists a K_0-variety V_0 and a K-birational map $V \dashrightarrow V_{0,K}$. Note that if V is a smooth projective curve, then V is K/K_0-isotrivial if and only if there exists a K_0-variety V_0 and a *K-isomorphism* $V \to V_{0,K}$, i.e. V descends to K_0. If F is a subfield of K, then we say that an F-variety V is K/K_0-isotrivial if V_K is K/K_0-isotrivial.

The following lemma also appears as [**BBP14**, Fact 2.15, Remark 2.17], but we include a full proof for the convenience of the reader:

LEMMA 2.1. *Let A/K be an abelian variety. Assume that there exists an abelian variety B_0/K_0 and an isogeny $\varphi : B_{0,K} \to A$. If $\dim(A) = 1$ or if φ is separable, then there exists an abelian variety A_0/K_0 with $A_{0,K} \cong A$.*

PROOF. The kernel $N := \ker(\varphi)$ is a finite subgroup scheme of $B_{0,K}$. If $\dim(A) = 1$, then $\dim(B_0) = 1$ and thus there exists a finite subgroup scheme N_0/K_0 of B_0 such that $N = N_{0,K}$, see [**Con06**, Theorem 2.3]. If φ is separable, then N/K is a finite *étale* subgroup scheme of $B_{0,K}[n]$ for a suitable integer n killing N, and thus there exists a finite subgroup scheme N_0/K_0 of $B_0[n]$ such that $N = N_{0,K}$, see [**Con06**, Lemma 3.11]. In both cases it follows that $A_0 := B_0/N_0$ is an abelian variety over K_0 (see [**Con06**, 3.7] for the existence of the quotient) and $A_{0,K} = B_{0,K}/N_{0,K} = B_{0,K}/N \cong A$, as desired. □

LEMMA 2.2. *Let A/K be an abelian variety. Assume that there exist an abelian variety B_0/K_0 and a surjective homomorphism $B_{0,K} \to A$. Then there exists an abelian variety A_0/K_0 and a purely inseparable isogeny $A \to A_{0,K}$.*

PROOF. By the Poincare reducibility theorem [**Mil86a**, 12.1] there exists an abelian subvariety B' of $B_{0,K}$ such that A is K-isogenous to B'. By [**Con06**, 3.21] there exits an abelian subvariety B_0' of B_0 such that $B' \cong B'_{0,K}$. After replacing B_0 by B_0' we can assume that there exists an isogeny $\varphi : B_{0,K} \to A$. We put $B := B_{0,K}$.

There exists $n \in \mathbb{N}$ such that $\ker(\varphi)$ is a subgroup scheme of $\ker([n]_B)$. We define $N_0 := \ker([n]_{B_0})$ and $N := N_{0,K}$. Then $N = \ker([n]_B)$ since $[n]_{B_0,K} = [n]_B$. The projection $p : B \to B/N$ induces an isogeny $f : A \to B/N$ such that $f \circ \varphi = p$ (see [**Con06**, 3.8]). We denote by $\ker(f)^\circ$ the connected component of the unit element of $\ker(f)$ and by $f_i : A \to A' := A/\ker(f)^\circ$ the canonical projection. Then f_i is a purely inseparable isogeny (see [**EvdGM19**, 5.6]) and f factors through A', i.e. there exists an isogeny $f_s : A' \to B/N$ such that $f_s \circ f_i = f$ (see [**Con06**, 3.7]). We now have constructed isogenies

$$B \xrightarrow{\varphi} A \xrightarrow{f_i} A' \xrightarrow{f_s} B/N$$

such that $f_s \circ f_i \circ \varphi = p$. The restriction $f_i \circ \varphi | N^\circ \to \ker(f_s)$ is trivial because N° is connected and $\ker(f_s) \cong \ker(f)/\ker(f)^\circ$ is étale (see [**Con06**, 3.10]). It follows that $\ker(f_i \circ \varphi) \supset N^\circ$ and thus $f_i \circ \varphi$ factors through an isogeny $g : B/N^\circ \to A'$ by [**Con06**, 3.7]. The kernel of g is a closed subgroup scheme of the finite étale group scheme N/N°, hence $\ker(g)$ is étale and $g : B/N^\circ \to A'$ is a *separable* isogeny. But $B/N^\circ = (B_0/(N_0)^\circ)_K$ and thus Lemma 2.1 implies that there exists an abelian variety A_0/K_0 such that $A' \cong A_{0,K}$. □

LEMMA 2.3. *Let $\varphi \colon D \dashrightarrow C$ be a dominant rational map of K-curves. If D is K/K_0-isotrivial, then C is K/K_0-isotrivial.*

PROOF. Since K and K_0 are algebraically closed it suffices to consider the case that K/K_0 is transcendental. Without loss of generality we can assume that C and D are smooth projective and that $\varphi : D \to C$ is a non-constant morphism. Let D_0 be a smooth projective K_0-curve with an isomorphism $\psi : D_{0,K} \to D$.

If $g(C) = 0$, then C is isomorphic to $\mathbb{P}^1_K$, hence C is K/K_0-isotrivial.

If $g(C) = 1$, then C is an elliptic curve. By Albanese functoriality there is a surjective homomorphism of abelian varieties $(J_{D_0})_K \cong J_D \to C$, and therefore Lemma 2.2 implies that there exists an elliptic curve C_0/K_0 such that C is K-isogenous to $C_{0,K}$. Lemma 2.1 then shows that C is K/K_0-isotrivial.

We may thus assume that $g(C) \geq 2$. Let $K_0 \subseteq F \subseteq K$ be an intermediate field such that F/K_0 is finitely generated and transcendental, and C, D, φ and ψ descend to F, i.e. there exist smooth projective F-curves C_1 and D_1 with $C_{1,K} = C$ and $D_{1,K} = D$, a morphism $\varphi_1 : D_1 \to C_1$ such that $\varphi_{1,K}$ corresponds to φ, and an isomorphism $D_{0,F} \cong D_1$. Since K_0 is algebraically closed, $|D_1(F)| = |D_0(F)| \geq$

$|D_0(K_0)| = \infty$. Since $\varphi_1 : D_1(F) \to C_1(F)$ has finite fibres, this implies that $|C_1(F)| = \infty$. Thus, since C_1 is of genus at least 2, the Grauert-Manin theorem (see [**Sam66**, pp. 107 and 118]) implies that C_1 is $\tilde{F}/K_0$-isotrivial, and hence C is K/K_0-isotrivial. □

LEMMA 2.4. *Let K_1 be an algebraically closed intermediate field of K/K_0. Let C be a K_1-curve. If C is K/K_0-isotrivial, then C is K_1/K_0-isotrivial.*

PROOF. We can assume that C is smooth and projective. As C is K/K_0-isotrivial, there exists a K_0-curve C_0 and a K-isomorphism $C_K \to C_{0,K}$. Therefore there exists a finitely generated extension F/K_1 inside K and an F-isomorphism $f : C_F \to C_{0,F}$. Choose a K_1-variety S with function field F. By [**Gro66**, 8.8.2, 8.10.5] there exists a non-empty open subscheme S' of S such that f extends to an S'-isomorphism $C \times_{K_1} S' \to C_{0,K_1} \times_{K_1} S'$. As K_1 is algebraically closed there exists $s \in S'(K_1)$, and specializing at s gives a K_1-isomorphism $C \to C_{0,K_1}$. Hence C is K_1/K_0-isotrivial. □

Let A/K be an abelian variety. A subvariety $X \subseteq A$ is K/K_0-**special** if there exist

- an abelian subvariety A' of A,
- an abelian variety B_0/K_0,
- a K_0-subvariety $X_0 \subseteq B_0$,
- a surjective K-homomorphism $h\colon A' \to B_{0,K}$,
- and an element $g \in A(K)$

such that $X = g + h^{-1}(X_{0,K})$. We write t_g for the translation by g on A.

The situation is as follows:

$$\begin{array}{ccccc} A & \hookleftarrow & A' & \xrightarrow{h} & B_{0,K} \\ \uparrow & & \uparrow & & \uparrow \\ X & \xleftarrow{t_g} & h^{-1}(X_{0,K}) & \xrightarrow{h} & X_{0,K} \end{array}$$

PROPOSITION 2.5. *Let A/K be an abelian variety and $C \subseteq A$ a K-subcurve. Then the following statements are equivalent.*

(1) *C is a K/K_0-special subvariety of A.*
(2) *C is of absolute genus 1 or K/K_0-isotrivial.*

PROOF. PROOF OF (1) $\Rightarrow$ (2).

If C is special, then there exist an abelian subvariety $A' \subseteq A$, an abelian variety B_0/K_0, a K_0-subvariety $C_0 \subseteq B_0$, a surjective K-homomorphism $h\colon A' \to B_{0,K}$, and an element $g \in A(K)$ such that $C = g + h^{-1}(C_{0,K})$. Without loss of generality assume that $0_{B_0} \in C_0$, $0_A \in C$, and $g = 0_A$. Then $C \subseteq A'$, so assume without loss of generality that $A = A'$.

If $\dim(C_0) = 0$, then $C = \ker(h)$ is an elliptic curve, hence of absolute genus 1.

If $\dim(C_0) \geq 1$, then $\dim(C_0) = \dim(C) = 1$, so h has finite fibers over $C_{0,K}$, and hence all fibers of h are finite. Consequently, h is an isogeny. By [**EvdGM19**, 5.12] there exists an isogeny $\hat{h}\colon B_{0,K} \to A$ such that $h \circ \hat{h} = [n]_{B_{0,K}}$ for some $n \in \mathbb{N}$. Let C_0' be an irreducible component of $[n]_{B_0}^{-1}(C_0)$ such that $[n]_{B_0}(C_0') = C_0$. Now we have a non-constant morphism

$$C'_{0,K} \xrightarrow{\hat{h}} C = h^{-1}(C_{0,K}) \xrightarrow{h} C_{0,K}$$

and Lemma 2.3 implies that C is K/K_0-isotrivial.

PROOF OF (2) $\Rightarrow$ (1).

If C is of absolute genus 1, then the normalization $\lambda : \bar{C} \to C$ of C is an elliptic curve. By [**Mil86a**, Cor. 2.2] there exists $g \in A(K)$ and a homomorphism $\lambda' : \bar{C} \to A$ of abelian varieties such that $\lambda = t_g \circ \lambda'$. It follows that $A' := \mathrm{im}(\lambda')$ is an abelian subvariety of A and $C = g + A'$. If B_0 is the zero abelian variety over K_0 and $h : A' \to B_{0,K}$ the zero homomorphism, then $A' = h^{-1}(0_{B_{0,K}})$ and thus $C = g + h^{-1}(0_{B_{0,K}})$ is special.

From now on assume that C is isotrivial. We have to prove that C is special. There exists a smooth projective K_0-curve C_0 and a finite birational morphism $f : C_{0,K} \to C$. Let $P_0 \in C_0(K_0)$ be a point and let $\alpha_0 : C_0 \to J_{C_0}$ be the corresponding Albanese map (sending P_0 to 0). It is a closed immersion because the genus of C_0 must be at least 1 by [**Mil86a**, 3.8]. By applying a translation on A we can assume without loss of generality that $f(P_0) = 0_A$. It follows from the universal mapping property [**Mil86b**, 1.1] of the Jacobian as Picard variety that $J_{C_0,K}$ is the Jacobian of $C_{0,K}$. By Albanese functoriality (see [**Mil86b**, 6.1]), the morphism f induces a homomorphism $f' \colon J_{C_0,K} \to A$ with $f' \circ \alpha_{0,K} = f$. Thus C lies on the abelian subvariety $A' := f'(J_{C_0,K})$, and $C = f'(\alpha_{0,K}(C_{0,K}))$. Lemma 2.2 gives an abelian variety B_0/K_0 and a purely inseparable isogeny $h \colon A' \to B_{0,K}$. Now consider the morphisms

$$C_{0,K} \xrightarrow{\alpha_{0,K}} J_{C_0,K} \xrightarrow{f'} A' \xrightarrow{h} B_{0,K}.$$

By [**Mil86a**, 20.4(b)] the homomorphism $h \circ f'$ is defined over K_0, that is, there exists a homomorphism $\beta_0 : J_{C_0} \to B_0$ such that $h \circ f' = \beta_{0,K}$. Now $C_0' := \beta_0(\alpha_0(C_0))$ is a subvariety of B_0 such that $C_{0,K}' = h(f'(\alpha_{0,K}(C_{0,K})))$. As h is a purely inseparable isogeny, the topological space underlying each fibre is a singleton (see [**Gro60**, 3.5.11]). Hence $h : A' \to B_{0,K}$ is bijective on the underlying topological spaces of A' and $B_{0,K}$. It follows that

$$h^{-1}(C_{0,K}') = (C_{0,K}' \times_{B_{0,K}} A')_{\mathrm{red}} = f'(\alpha_{0,K}(C_{0,K})) = C.$$

Thus C is special. □

3. Rössler's theorem on inseparable points on curves

In this section we recall and slightly generalize a recent result of Rössler about inseparable points on curves, which strengthens the aforementioned Grauert-Manin theorem.

THEOREM 3.1. *Let F/F_0 be a finitely generated extension of fields of positive characteristic, and let C be an F-curve of absolute genus $g \geq 2$. If $C(F_{\mathrm{ins}})$ is infinite, then C is $\tilde{F}/\tilde{F}_0$-isotrivial.*

In the following proof we deduce Theorem 3.1 from Rössler's [**Rös21**, Thm. 1.1], where this theorem is proven in the special case where F_0 is algebraically closed, $\mathrm{trdeg}(F/F_0) = 1$ and C is smooth and projective.

PROOF. There exists a regular projective curve over F_{ins} that is birational to $C_{F_{\mathrm{ins}}}$. This curve is automatically geometrically regular because F_{ins} is perfect (cf. [**Gro65**, 6.7.7]), hence smooth over F_{ins} by [**Gro67**, 17.5.2]. It follows that there exists a finite extension E/F inside F_{ins} and a smooth projective E-curve $\bar{C}$ that is birational to C_E. Thus $\bar{C}(E_{\mathrm{ins}})$ is infinite and it suffice to prove that $\bar{C}$ is

$\tilde{F}/\tilde{F}_0$-isotrivial. We can thus assume for the rest of the proof that C is smooth and projective.

Let $\mathcal{F}$ be the set of all algebraically closed intermediate fields F' of $\tilde{F}/\tilde{F}_0$ with $\mathrm{trdeg}(\tilde{F}/F') = 1$. Then $\tilde{F}_0 = \bigcap_{F' \in \mathcal{F}} F'$: Indeed, every $x \in \tilde{F} \setminus \tilde{F}_0$ is an element of a transcendence base $(x, x_2, \cdots, x_n)$ of $\tilde{F}/\tilde{F}_0$ and x is not contained in the algebraic closure F' of $F_0(x_2, \cdots, x_n)$.

For every $F' \in \mathcal{F}$, $C((F'F)_{\mathrm{ins}}) \supseteq C(F_{\mathrm{ins}})$ is infinite, so [**Rös21**, Thm. 1.1] gives that that C is $\tilde{F}/F'$-isotrivial. Therefore, C is in fact $\tilde{F}/\bigcap_{F' \in \mathcal{F}} F'$-isotrivial: This can be seen by looking at the point $c \in M_g(\tilde{F})$ corresponding to C on the coarse moduli space M_g for smooth proper curves of genus g, see [**MFK92**, 5.6, 7.14], or explicitly by [**Koi72**, Corollary 3.2.2]. □

REMARK 3.2. We note that this theorem could also be deduced from the older [**Kim97**] instead of [**Rös21**]: Let F be a function field of one variable over a field F_0 of positive characteristic p. Consider an affine plane curve

$$C = \mathrm{Spec}(F[X,Y]/(f(X,Y)))$$

where $f(X,Y) \in F[X,Y]$ is absolutely irreducible. We denote by $f^{(n)}(X,Y)$ the polynomial obtained from $f(X,Y)$ by raising the coefficients of $f(X,Y)$ to the p^n-th power and by

$$C^{(n)} := \mathrm{Spec}(F[X,Y]/(f^{(n)}(X,Y)))$$

the corresponding curve. For $n \geq 1$ there is the Frobenius injection

$$\mathrm{Fr}\colon C^{(n-1)}(F) \to C^{(n)}(F),\ (x,y) \mapsto (x^p, y^p)$$

and we define

$$C^{(n)}(F)_{\mathrm{new}} := C^{(n)}(F) \setminus \mathrm{Fr}(C^{(n-1)}(F)).$$

Finally, $C^{(0)}(F)_{\mathrm{new}} := C(F)$. Note that the Frobenius gives a bijection between $C^{(n)}(F)$ and $C(F^{p^{-n}})$, and hence between $C^{(n)}(F)_{\mathrm{new}}$ and $C(F^{p^{-n}}) \setminus C(F^{p^{-n+1}})$. Then [**Kim97**, Theorem 1] claims that if the absolute genus of C is at least 2 and $C^{(n)}(F)_{\mathrm{new}} \neq \emptyset$ for infinitely many n, then C is $\tilde{F}/\tilde{\mathbb{F}}_p$-isotrivial.

This, however, is incorrect, as the following example shows: Let F_0 be any non-perfect transcendental ample field (e.g. $F_0 = \mathbb{F}_p(t)_{\mathrm{sep}}$), F any function field of one variable over F_0 (e.g. $F = F_0(X)$), and C any affine plane F_0-curve of absolute genus at least 2 which is not $\tilde{F}_0/\tilde{\mathbb{F}}_p$-isotrivial and has a smooth F_0-rational point (see e.g. Lemma 5.1 below for the existence of such a curve). Then [**Kim97**, Theorem 1] would imply that there exists an n such that $C(F_{\mathrm{ins}}) = C(F^{p^{-n}})$, hence $C(F_{0,\mathrm{ins}}) = C(F_{0,\mathrm{ins}}) \cap C(F^{p^{-n}}) = C(F_0^{p^{-n}})$, contradicting for example [**Feh10**, Corollary 5].

4. The Mordell-Lang conjecture

The main ingredient in the proof of Theorem 1.1 is a partial result towards the general Mordell-Lang conjecture for function fields.

CONJECTURE 4.1 (General Mordell-Lang conjecture for function fields). *Let K/K_0 be an extension of algebraically closed fields, A/K an abelian variety, $X \subseteq A$ a subvariety, and $\Lambda \subseteq A(K)$ a subgroup of finite rank such that $\Lambda \cap X(K)$ is Zariski-dense in X. Then X is K/K_0-special.*

A slightly weaker result was proven by Hrushovski in [**Hru96**], where the condition that Λ has finite rank is replaced by the stronger condition that $\Lambda \otimes \mathbb{Z}_{(p)}$ is a finitely generated $\mathbb{Z}_{(p)}$-module, where p is the characteristic of K.

In [**GM06**], Ghioca and Moosa use work of Scanlon and the theory of generic difference fields to reduce the general Mordell-Lang conjecture to the special case where $\Lambda \subseteq A(F_{\mathrm{ins}})$ for some finitely generated field F. This reduction also works if on both sides one adds the condition that X is a curve. The following result is a slight generalization of [**GM06**, Corollary 2.6]:

PROPOSITION 4.2. *Let K be an algebraically closed field of positive characteristic p, and let $K_0 = \tilde{\mathbb{F}}_p$. Let A/K be an abelian variety, C a subcurve of A, and $\Lambda \subseteq A(K)$ a subgroup of finite rank. If $\Lambda \cap C(K)$ is infinite, then the absolute genus of C is 1 or C is K/K_0-isotrivial, i.e. C is K/K_0-special.*

PROOF. First of all, A and C are defined over a finitely generated subfield F of K. By the reduction [**GM06**, Theorem 2.2], we may assume that $\Lambda \subseteq A(F_{\mathrm{ins}})$. This implies that $C(F_{\mathrm{ins}})$ is infinite.

Since C is a subcurve of an abelian variety, the absolute genus of C is at least 1, [**Mil86a**, 3.8]. If the absolute genus of C is 1, then C is K/K_0-special by Proposition 2.5. If the absolute genus of C is at least 2, then C is K/K_0-isotrivial by Theorem 3.1, and hence K/K_0-special by Proposition 2.5. □

This proves Theorem 1.2.

REMARK 4.3. Proposition 4.2 is stated in [**GM06**, Corollary 2.6] in the special case $A = J_C$. The proof there builds on [**Kim97**, Theorem 1], cf. Remark 3.2. Note however, that the proof of Proposition 4.2 does not rely on Kim's result but rather on that of Rössler.

5. Proof of Theorem 1.1

The idea of the proof of Theorem 1.1 is to construct a curve on A that is non-special and has a smooth F-rational point. To achieve this, we 'lift' a non-isotrivial curve from $\mathbb{A}^2$ to A.

LEMMA 5.1. *If F/F_0 is a transcendental field extension, then there exists a smooth affine plane F-curve C of absolute genus at least 2 with $C(F) \neq \emptyset$ such that C is not $\tilde{F}/\tilde{F}_0$-isotrivial.*

PROOF. Let $t \in F \setminus \tilde{F}_0$. First assume that $\mathrm{char}(F) \neq 2$. Let

$$E : y^2 = f(x) = (x-1)(x+1)(x-t^2)$$

be the elliptic curve over F with j-invariant

$$j(E) = \frac{(16t^4+48)^3}{64t^8 - 128t^4 + 64}.$$

Since $j(E) \in F \setminus \tilde{F}_0$, it follows that E is not $\tilde{F}/\tilde{F}_0$-isotrivial. The map $(x, y) \mapsto (x^2, y)$ defines a separable morphism from the curve

$$C : y^2 = f(x^2)$$

to E which is ramified over $(0, t) \in E(F)$. Therefore, C is a smooth F-curve of absolute genus at least 2 by the Riemann-Hurwitz formula, and $C(F) \neq \emptyset$. By Lemma 2.3, C is not $\tilde{F}/\tilde{F}_0$-isotrivial.

If $\mathrm{char}(F) = 2$, we can proceed similarly and let $E : y^2 + xy = f(x) = x^3 + tx + 1$ be the elliptic curve with j-invariant $j(E) = (t^2+1)^{-1}$. The map $(x, y) \mapsto (x^3, y)$ defines a separable morphism from the curve $C : y^2 + x^3 y = f(x^3)$ to E that is ramified over $(0, 1) \in E(F)$. $\square$

LEMMA 5.2. *Let F be a field. Let X be an F-variety of dimension n and Y an F-variety of dimension $m < n$. Let $x \in X(F)$ be a smooth point of X and let $y \in Y(F)$ a smooth point of Y. Then there exist an F-subvariety $V \subseteq X$ of dimension* $\dim(V) = m$, *such that x is a smooth point of V, and a dominant F-rational map $V \dashrightarrow Y$ mapping x to y.*

PROOF. There exists a smooth open neighbourhood Y' of y in Y and an étale morphism $f : Y' \to \mathbb{A}^m$, see [**Gro67**, 17.11.4]. By [**Mil80**, I.3.14] this morphism f is locally standard étale. Hence, after replacing Y' by a smaller open neighbourhood of y if necessary, there exists an affine open subscheme $U = \mathrm{Spec}(R) \subseteq \mathbb{A}^m$ such that $f(Y') \subseteq U$ and $f|Y' \to U$ is standard étale, i.e. such that there exist a polynomial $Q \in R[T]$ and a U-isomorphism of Y' with an open subscheme of $\mathrm{Spec}(R[T]/(Q))$. In particular there exists an immersion

$$i : Y' \to \mathrm{Spec}(R[T]) = U \times \mathbb{A}^1 \to \mathbb{A}^{m+1} \to \mathbb{A}^n.$$

There exists a smooth open neighbourhood X' of x in X and an étale morphism $g : X' \to \mathbb{A}^n$, again by [**Gro67**, 17.11.4], and we can assume that $g(x) = i(y)$. We form the fibre product $Z = X' \times_{\mathbb{A}^n, g, i} Y' = g^{-1}(Y')$ and consider the diagram

$$\begin{array}{ccc} Z & \xrightarrow{\ j\ } & X' \subset X \\ \downarrow{\scriptstyle p} & & \downarrow{\scriptstyle g} \\ Y \supset Y' & \xrightarrow{\ i\ } & \mathbb{A}^n. \end{array}$$

The projection p is étale, Y' is smooth and $\dim(Y') = m$. Hence Z is smooth of dimension m. Furthermore j is an immersion. Hence we can identify Z with a closed subscheme of an open subscheme X'' of X'. We have $x \in Z(F)$. Let Z_0 be the connected component of Z passing through x and let V be the closure of Z_0 in X (with the reduced induced subscheme structure). Then Z_0 is smooth and connected and $Z_0(F) \neq \emptyset$, hence it is geometrically integral, so also V is geometrically integral, see [**Gro65**, 4.6.3]. Hence V is a *subvariety* of X with $\dim(V) = m$. Finally the étale morphism $p|Z_0 \to Y'$ induces a dominant rational map $V \dashrightarrow Y$ mapping x to y. $\square$

THEOREM 5.3. *Let F_1 be a field which is not algebraic over a finite field, and let A/F_1 be a non-zero abelian variety. Then there exists a smooth F_1-curve C_A with $C_A(F_1) \neq \emptyset$ and the following property: If F is an extension of F_1 with $|C_A(F)| = \infty$, then the rank of $A(F)$ is infinite.*

PROOF. Replace A by $A \times A$ to assume without loss of generality that A is of dimension at least 2. Take any smooth F_1-curve C of absolute genus at least 2 with $C(F_1) \neq \emptyset$. By Lemma 5.2 there exist an F_1-curve $D \subseteq A$ with a dominant F_1-rational map $\varphi\colon D \dashrightarrow C$ such that D has a smooth F_1-rational point. By the Riemann-Hurwitz formula, D is of absolute genus at least 2.

If F_1 is of characteristic zero and F is an extension of F_1 with $|D(F)| = \infty$, then the Mordell-Lang conjecture in characteristic zero implies that the rank of $A(F)$ is infinite, cf. [**FP10**, Theorem 2.6(a)].

If F_1 is of characteristic $p > 0$, then $K_1 = \tilde{F}_1$ is transcendental over $K_0 = \tilde{\mathbb{F}}_p$ and so we can choose the curve C to be not K_1/K_0-isotrivial, see Lemma 5.1, and then also D is not K_1/K_0-isotrivial by Lemma 2.3. If F is an extension of F_1 with $|D(F)| = \infty$, let $K = \tilde{F}$ and $\Lambda = A(F)$, and suppose that the rank of Λ is finite. Since $\Lambda \cap D(K) = D(F)$ is infinite and the absolute genus of D is at least 2, Proposition 4.2 implies that D is K/K_0-isotrivial, and so Lemma 2.4 implies that D is K_1/K_0-isotrivial, a contradiction. Thus the rank of $A(F)$ must be infinite.

Therefore, in both cases D contains a smooth open subcurve C_A that satisfies the requirements. □

Theorem 1.1 is an immediate consequence of Theorem 5.3 for $F_1 = F$, since if F is ample, then $|C_A(F)| = \infty$.

6. Corollaries and applications

6.1. Uncountable rank. It turns out that in the situation of Theorem 1.1, not only is the rank of $A(F)$ infinite, but it is as large as possible.

COROLLARY 6.1. *Let F be an ample field that is not algebraic over a finite field and let A/F be a non-zero abelian variety. Then the rank of $A(F)$ equals the cardinality of F.*

PROOF. If F is countable, then the claim follows from Theorem 1.1. If F is uncountable, then $|A(F)| = |F|$ by [**Har09**, Prop. 3.3]. Since $A_{\mathrm{tor}}(\tilde{F})$ is countable, this implies that $A(F) \otimes \mathbb{Q}$ cannot have a basis of less than $|F|$ elements. □

6.2. Large extensions of Hilbertian fields. In [**Pet06**, Theorem 1.4], the second author generalizes [**FJ74**, Theorem 9.1] from finitely generated fields to arbitrary Hilbertian fields. This generalization also follows easily from our results. In the following, **almost all** is to be understood in the sense of Haar measure, and for an e-tuple $\boldsymbol{\sigma} \in \mathrm{Gal}(F)^e$, where $\mathrm{Gal}(F) = \mathrm{Aut}(F_{\mathrm{sep}}/F)$ is the absolute Galois group of F, $F_{\mathrm{sep}}(\boldsymbol{\sigma})$ is the fixed field in F_{sep} of the group generated by $\sigma_1, \ldots, \sigma_e$.

COROLLARY 6.2. *If F is a Hilbertian field, A/F a non-zero abelian variety, and e a non-negative integer, then $A(F_{\mathrm{sep}}(\boldsymbol{\sigma}))$ has infinite rank for almost all $\boldsymbol{\sigma} \in \mathrm{Gal}(F)^e$.*

PROOF. Let C_A be the F-curve associated to A by Theorem 5.3. Since F is Hilbertian, there exists a linearly disjoint sequence $F_1, F_2, \ldots$ of extensions of F of fixed degree d such that $C_A(F_i) \setminus C_A(F) \neq \emptyset$ for all i, cf. [**FJ08**, Proof of Theorem 18.6.1, Part A]. By Borel-Cantelli, the set of $\boldsymbol{\sigma} \in \mathrm{Gal}(F)^e$ for which $F_{\mathrm{sep}}(\boldsymbol{\sigma})$ contains infinitely many of the F_i has measure one, cf. [**FJ08**, 18.5.3(b)]. For those $\boldsymbol{\sigma}$ we have that $|C_A(F_{\mathrm{sep}}(\boldsymbol{\sigma}))| = \infty$, and hence that the rank of $A(F_{\mathrm{sep}}(\boldsymbol{\sigma}))$ is infinite by Theorem 5.3. □

We do not know if almost all $F_{\mathrm{sep}}(\boldsymbol{\sigma})$ are ample if F is not countable, therefore we had to apply Theorem 5.3, and not Theorem 1.1 directly.

6.3. Fields of totally S-adic numbers. We give a simplified proof and a generalization of [**FP10**, Theorem 1.3]. Recall that a **local prime** $\mathfrak{p}$ of a field F is an equivalence class of absolute values on F such that the completion $\hat{F}_{\mathfrak{p}}$ is a local field, i.e. a finite extension of $\mathbb{R}$, $\mathbb{Q}_p$ or $\mathbb{F}_p((t))$. If S is a finite set of local primes of F, the field $F_{\mathrm{tot},S}$ of **totally S-adic numbers** over F is defined as the maximal Galois extension of F contained in all completions $\hat{F}_{\mathfrak{p}}$, $\mathfrak{p} \in S$. For $\boldsymbol{\sigma} \in \mathrm{Gal}(F)^e$, let $F_{\mathrm{sep}}[\boldsymbol{\sigma}]$ be the maximal Galois extension of F in the fixed field $F_{\mathrm{sep}}(\boldsymbol{\sigma})$.

COROLLARY 6.3. *Let F be a countable Hilbertian field, S a finite set of local primes of F, and e a non-negative integer. If A/F is a non-zero abelian variety, then $A(F_{\mathrm{tot},S} \cap F_{\mathrm{sep}}[\boldsymbol{\sigma}])$ has infinite rank for almost all $\boldsymbol{\sigma} \in \mathrm{Gal}(F)^e$.*

PROOF. Geyer and Jarden prove in [**GJ02**] that the field $F_{\mathrm{tot},S} \cap F_{\mathrm{sep}}[\boldsymbol{\sigma}]$ is ample for almost all $\boldsymbol{\sigma} \in \mathrm{Gal}(F)^e$. Hence, Theorem 1.1 implies the claim. □

This also generalizes [**FJ74**, Theorem 2.1], [**GJ06**, Theorem B], and a special case of [**Pet06**, Theorem A.1].

6.4. Larsen's conjecture. Our last application of Theorem 1.1 is a conditional result. Among the numerous conjectures about which fields are ample, the following conjecture (see e.g. [**JK10**, p. 496], [**BSF12**, §4.2]) is particularly interesting for the questions considered in this work.

CONJECTURE 6.4 (Koenigsmann). *If F is an infinite field with finitely generated absolute Galois group, then F is ample.*

One of the consequences of Theorem 1.1 is that a proof of Conjecture 6.4 would immediately prove the following conjecture of Larsen, see [**Lar03**, Question 1], [**Im06**], [**IL08**], [**IL21**].

CONJECTURE 6.5 (Larsen). *Let F be an infinite finitely generated field, A/F a non-zero abelian variety, and e a non-negative integer. Then $A(F_{\mathrm{sep}}(\boldsymbol{\sigma}))$ has infinite rank for* all $\boldsymbol{\sigma} \in \mathrm{Gal}(F)^e$.

COROLLARY 6.6. *Conjecture* 6.4 *implies Conjecture* 6.5.

PROOF. The absolute Galois group of $F_{\mathrm{sep}}(\boldsymbol{\sigma})$ is generated by $\boldsymbol{\sigma}$, and hence is finitely generated. Therefore, Theorem 1.1 implies the claim of our corollary. □

Acknowledgments

We would like to thank Moshe Jarden for his constant encouragement regarding this paper. We thank the anonymous referee for a careful reading of the manuscript and for comments that helped to improve our exposition.

References

[BBP14] Franck Benoist, Elisabeth Bouscaren, and Anand Pillay, *Semiabelian varieties over separably closed fields, maximal divisible subgroups, and exact sequences*, J. Inst. Math. Jussieu **15** (2016), no. 1, 29–69, DOI 10.1017/S147474801400022X. MR3427593

[BSF12] Lior Bary-Soroker and Arno Fehm, *Open problems in the theory of ample fields* (English, with English and French summaries), Geometric and differential Galois theories, Sémin. Congr., vol. 27, Soc. Math. France, Paris, 2013, pp. 1–11. MR3203546

[Con06] Brian Conrad, *Chow's K/k-image and K/k-trace, and the Lang-Néron theorem*, Enseign. Math. (2) **52** (2006), no. 1-2, 37–108. MR2255529

[EvdGM19] Bas Edixhoven, Gerard van der Geer, and Ben Moonen, *Abelian Varieties - Preliminary version of the first chapters*. 2019. Manuscript.

[Feh10] Arno Fehm, *Subfields of ample fields. Rational maps and definability*, J. Algebra **323** (2010), no. 6, 1738–1744, DOI 10.1016/j.jalgebra.2009.11.037. MR2588135

[FJ74] Gerhard Frey and Moshe Jarden, *Approximation theory and the rank of abelian varieties over large algebraic fields*, Proc. London Math. Soc. (3) **28** (1974), 112–128, DOI 10.1112/plms/s3-28.1.112. MR337997

[FJ08] Michael D. Fried and Moshe Jarden, *Field arithmetic*, 3rd ed., Ergebnisse der Mathematik und ihrer Grenzgebiete. 3. Folge. A Series of Modern Surveys in Mathematics [Results in Mathematics and Related Areas. 3rd Series. A Series of Modern Surveys in Mathematics], vol. 11, Springer-Verlag, Berlin, 2008. Revised by Jarden. MR2445111

[FP10] Arno Fehm and Sebastian Petersen, *On the rank of abelian varieties over ample fields*, Int. J. Number Theory **6** (2010), no. 3, 579–586, DOI 10.1142/S1793042110003071. MR2652895

[FP11] Arno Fehm and Elad Paran, *Galois theory over rings of arithmetic power series*, Adv. Math. **226** (2011), no. 5, 4183–4197, DOI 10.1016/j.aim.2010.11.010. MR2770445

[Gro60] A. Grothendieck, *Éléments de géométrie algébrique. I. Le langage des schémas* (French), Inst. Hautes Études Sci. Publ. Math. **4** (1960), 228. MR217083

[Gro65] A. Grothendieck, *Éléments de géométrie algébrique. IV. Étude locale des schémas et des morphismes de schémas. II* (French), Inst. Hautes Études Sci. Publ. Math. **24** (1965), 231. MR199181

[Gro66] A. Grothendieck, *Éléments de géométrie algébrique. IV. Étude locale des schémas et des morphismes de schémas. III*, Inst. Hautes Études Sci. Publ. Math. **28** (1966), 255. MR217086

[Gro67] A. Grothendieck, *Éléments de géométrie algébrique. IV. Étude locale des schémas et des morphismes de schémas IV* (French), Inst. Hautes Études Sci. Publ. Math. **32** (1967), 361. MR238860

[GJ02] Wulf-Dieter Geyer and Moshe Jarden, *PSC Galois extensions of Hilbertian fields*, Math. Nachr. **236** (2002), 119–160, DOI 10.1002/1522-2616(200203)236:1⟨119::AID-MANA119⟩3.3.CO;2-L. MR1888560

[GJ06] Wulf-Dieter Geyer and Moshe Jarden, *The rank of abelian varieties over large algebraic fields*, Arch. Math. (Basel) **86** (2006), no. 3, 211–216, DOI 10.1007/s00013-005-1492-x. MR2215309

[GM06] Dragos Ghioca and Rahim Moosa, *Division points on subvarieties of isotrivial semi-abelian varieties*, Int. Math. Res. Not., posted on 2006, Art. ID 65437, 23, DOI 10.1155/IMRN/2006/65437. MR2264715

[Har09] David Harbater, *On function fields with free absolute Galois groups*, J. Reine Angew. Math. **632** (2009), 85–103, DOI 10.1515/CRELLE.2009.051. MR2544144

[Hru96] Ehud Hrushovski, *The Mordell-Lang conjecture for function fields*, J. Amer. Math. Soc. **9** (1996), no. 3, 667–690, DOI 10.1090/S0894-0347-96-00202-0. MR1333294

[IL08] Bo-Hae Im and Michael Larsen, *Abelian varieties over cyclic fields*, Amer. J. Math. **130** (2008), no. 5, 1195–1210, DOI 10.1353/ajm.0.0020. MR2450206

[IL13] Bo-Hae Im and Michael Larsen, *Some applications of the Hales-Jewett theorem to field arithmetic*, Israel J. Math. **198** (2013), no. 1, 35–47, DOI 10.1007/s11856-013-0009-8. MR3096628

[IL21] Bo-Hae Im and Michael Larsen, Abelian varieties and finitely generated galois groups. This volume, 2021.

[Im06] Bo-Hae Im, *Mordell-Weil groups and the rank of elliptic-curves over large fields*, Canad. J. Math. **58** (2006), no. 4, 796–819, DOI 10.4153/CJM-2006-032-4. MR2245274

[Jar03] Moshe Jarden, *On ample fields*, Arch. Math. (Basel) **80** (2003), no. 5, 475–477, DOI 10.1007/s00013-003-4610-7. MR1995626

[JK10] Markus Junker and Jochen Koenigsmann, *Schlanke Körper (slim fields)*, J. Symbolic Logic **75** (2010), no. 2, 481–500, DOI 10.2178/jsl/1268917491. MR2648152

[Kim97] Minhyong Kim, *Purely inseparable points on curves of higher genus*, Math. Res. Lett. **4** (1997), no. 5, 663–666, DOI 10.4310/MRL.1997.v4.n5.a4. MR1484697

[Koi72] Shoji Koizumi, *The fields of moduli for polarized abelian varieties and for curves*, Nagoya Math. J. **48** (1972), 37–55. MR352095

[Lar03] Michael Larsen, *Rank of elliptic curves over almost separably closed fields*, Bull. London Math. Soc. **35** (2003), no. 6, 817–820, DOI 10.1112/S0024609303002431. MR2000029

[MFK92] D. Mumford, J. Fogarty, and F. Kirwan, *Geometric invariant theory*, 3rd ed., Ergebnisse der Mathematik und ihrer Grenzgebiete (2) [Results in Mathematics and Related Areas (2)], vol. 34, Springer-Verlag, Berlin, 1994. MR1304906

[Mil80] James S. Milne, *Étale cohomology*, Princeton Mathematical Series, vol. 33, Princeton University Press, Princeton, N.J., 1980. MR559531

[Mil86a] J. S. Milne, *Abelian varieties*, Arithmetic geometry (Storrs, Conn., 1984), Springer, New York, 1986, pp. 103–150. MR861974

[Mil86b] J. S. Milne, *Jacobian varieties*, Arithmetic geometry (Storrs, Conn., 1984), Springer, New York, 1986, pp. 167–212. MR861976

[Pet06] Sebastian Petersen, *On a question of Frey and Jarden about the rank of abelian varieties*, J. Number Theory **120** (2006), no. 2, 287–302, DOI 10.1016/j.jnt.2005.12.006. MR2257548

[Pop96] Florian Pop, *Embedding problems over large fields*, Ann. of Math. (2) **144** (1996), no. 1, 1–34, DOI 10.2307/2118581. MR1405941

[Pop10] Florian Pop, *Henselian implies large*, Ann. of Math. (2) **172** (2010), no. 3, 2183–2195, DOI 10.4007/annals.2010.172.2183. MR2726108

[Pop14] Florian Pop, *Little survey on large fields—old & new*, Valuation theory in interaction, EMS Ser. Congr. Rep., Eur. Math. Soc., Zürich, 2014, pp. 432–463. MR3329044

[Ros73] Michael Rosen, *S-units and S-class group in algebraic function fields*, J. Algebra **26** (1973), 98–108, DOI 10.1016/0021-8693(73)90036-7. MR327777

[Rös21] Damian Rössler, Perfect points on curves. This volume, 2021.

[RW02] Michael Rosen and Siman Wong, *The rank of abelian varieties over infinite Galois extensions*, J. Number Theory **92** (2002), no. 1, 182–196, DOI 10.1006/jnth.2001.2692. MR1880593

[Sam66] Pierre Samuel, La conjecture de mordell pour les corps de fonctions. *Seminaire N. Bourbaki*, 287:149–167, 1966.

Institut für Algebra, Fakultät Mathematik, Technische Universität Dresden, 01062 Dresden, Germany

Email address: arno.fehm@tu-dresden.de

Universität Kassel, Fachbereich 10, Wilhelmshöher Allee 73, 34121 Kassel, Germany

Email address: petersen@mathematik.uni-kassel.de

Contemporary Mathematics
Volume **767**, 2021
https://doi.org/10.1090/conm/767/15395

The group of automorphisms of the Heisenberg curve

Jannis A. Antoniadis and Aristides Kontogeorgis

ABSTRACT. The Heisenberg curve is defined to be the curve corresponding to an extension of the projective line by the Heisenberg group modulo n, ramified above three points. This curve is related to the Fermat curve and its group of automorphisms is studied. Also we give an explicit equation for the curve C_3.

1. Introduction

Probably the most famous curve in number theory is the Fermat curve given by affine equation

$$F_n : x^n + y^n = 1.$$

This curve can be seen as ramified Galois cover of the projective line with Galois group $\mathbb{Z}/n\mathbb{Z} \times \mathbb{Z}/n\mathbb{Z}$ with action given by $\sigma_{a,b} : (x, y) \mapsto (\zeta^a x, \zeta^b y)$ where $(a, b) \in \mathbb{Z}/n\mathbb{Z} \times \mathbb{Z}/n\mathbb{Z}$. The ramified cover

$$\pi : F_n \to \mathbb{P}^1$$

has three ramified points and the cover

$$F_n^0 := F_n - \pi^{-1}(\{0, 1, \infty\}) \xrightarrow{\pi} \left(\mathbb{P}^1 - \{0, 1, \infty\}\right),$$

is a Galois topological cover. We can see the hyperbolic space $\mathbb{H}$ as the universal covering space of $\mathbb{P}^1 - \{0, 1, \infty\}$. The Galois group of the above cover is isomorphic to the free group F_2 in two generators, and a suitable realization of this group in our setting is the group Δ which is the subgroup of $\mathrm{SL}(2, \mathbb{Z}) \subseteq \mathrm{PSL}(2, \mathbb{R})$ generated by the elements $a = \begin{pmatrix} 1 & 2 \\ 0 & 1 \end{pmatrix}$, $b = \begin{pmatrix} 1 & 0 \\ 2 & 1 \end{pmatrix}$ and $\pi(\mathbb{P}^1 - \{0, 1, \infty\}, x_0) \cong \Delta$. Related to the group Δ is the modular group

$$\Gamma(2) = \left\{ \gamma = \begin{pmatrix} a & b \\ c & d \end{pmatrix} \in \mathrm{SL}(2, \mathbb{Z}) : \gamma \equiv 1_2 \bmod 2 \right\},$$

which is isomorphic to $\{\pm I\}\Delta$ while $\Gamma(2)\backslash\mathbb{H} \cong \left(\mathbb{P}^1 - \{0, 1, \infty\}\right)$. The groups Δ and $\Gamma(2)$ act in exactly the same way on the hyperbolic plane $\mathbb{H}$.

REMARK 1. Covers of the projective line minus three points are very important in number theory because of the Belyi theorem [**3**],[**4**] that asserts that all algebraic curves defined over $\bar{\mathbb{Q}}$ fall into this category. It seems that the idea of studying algebraic curves as "modular curves" goes back to S. Lang and to D. Rohrlich [**22**].

2020 *Mathematics Subject Classification.* Primary 14H37.
Key words and phrases. Automorphisms, curves, differentials, numerical semigroups.

For every finitely generated group G generated by two elements there is a homomorphism

$$\Gamma(2) \to G.$$

Notice that $\Gamma(2)^{\mathrm{ab}} \cong \mathbb{Z} \times \mathbb{Z}$ so using the projection

$$\psi : \Delta \to \Delta^{\mathrm{ab}} \to \mathbb{Z}/n\mathbb{Z} \times \mathbb{Z}/n\mathbb{Z}$$

we can write the open Fermat curve F_n^0 as the quotient

$$F_n^0 = \ker \psi \backslash \mathbb{H}.$$

In the theory of modular curves [**20**], the hyperbolic space $\mathbb{H}$ is extented to $\bar{\mathbb{H}} = \mathbb{H} \cup \mathbb{P}^1(\mathbb{Q})$, where $\mathbb{P}^1(\mathbb{Q})$ is the set of cusps so that subgroups Γ of $\mathrm{SL}(2,\mathbb{Z})$ give rise to compact quotients. The orbits of $\mathbb{P}^1(\mathbb{Q})$ under the action of Γ are the cusps of the curve $\Gamma\backslash\mathbb{H}$. In this setting the cusps of the Fermat curve F_n are the points $F_n - F_n^0$.

Aim of this article is to initialize the study of the curve C_n, which is defined in the following way: Consider the Heisenberg group modulo n:

$$H_n = \left\{ \begin{pmatrix} 1 & x & z \\ 0 & 1 & y \\ 0 & 0 & 1 \end{pmatrix} : x, y, z \in \mathbb{Z}/n\mathbb{Z} \right\}.$$

It is a finite group generated by the elements

$$a_H = \begin{pmatrix} 1 & 1 & 0 \\ 0 & 1 & 0 \\ 0 & 0 & 1 \end{pmatrix} \text{ and } b_H = \begin{pmatrix} 1 & 0 & 0 \\ 0 & 1 & 1 \\ 0 & 0 & 1 \end{pmatrix}. \tag{1}$$

Notice that

$$[a_H, b_H] = \begin{pmatrix} 1 & 0 & 1 \\ 0 & 1 & 0 \\ 0 & 0 & 1 \end{pmatrix} \tag{2}$$

and

$$H_n^{\mathrm{ab}} \cong \mathbb{Z}/n\mathbb{Z} \times \mathbb{Z}/n\mathbb{Z}.$$

We have the short exact sequence

$$1 \to \mathbb{Z}/n\mathbb{Z} \cong Z_n \to H_n \longrightarrow H_n^{\mathrm{ab}} \to 1, \tag{3}$$

where

$$Z_n := \left\{ \begin{pmatrix} 1 & 0 & z \\ 0 & 1 & 0 \\ 0 & 0 & 1 \end{pmatrix} : z \in \mathbb{Z}/n\mathbb{Z} \right\}.$$

Observe that there is an epimorphism: $\phi : \Gamma(2) \to H_n$ sending each generator of $\Gamma(2)$ to the elements $a_H, b_H \in H_n$. This way an open curve C_n^0 is defined as $\ker \phi \backslash \mathbb{H}$ that can be compactified to a compact Riemann surface C_n.

We have the following diagram:

(4)

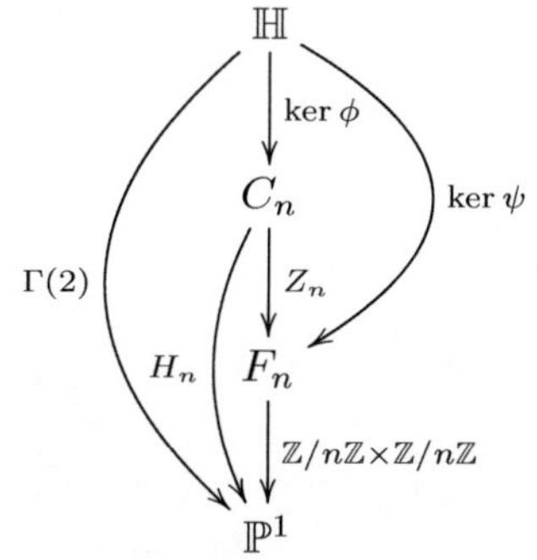

DEFINITION 2. Let X be a curve that comes as a compactification by adding some cusps of the open curve $\Gamma\backslash\mathbb{H}$ where Γ is a subgroup of finite index of $\mathrm{SL}(2,\mathbb{Z})$. The group of *modular automorphisms* is the group

$$\mathrm{Aut}^{\mathrm{m}}(X) = N_{\mathrm{SL}(2,\mathbb{R})}(\Gamma)/\Gamma,$$

where $\mathrm{SL}(2,\mathbb{R})$ is the group of automorphisms of $\mathbb{H}$ and $N_{\mathrm{SL}(2,\mathbb{R})}(\Gamma)$ is the normalizer of Γ in $\mathrm{SL}(2,\mathbb{R})$.

REMARK 3. An automorphism σ of a complete curve X which comes out from an open curve $\Gamma\backslash\mathbb{H}$ by adding the set of cusps $\mathbb{P}^1(\mathbb{Q})$ is modular if and only if σ sends cusps to cusps and non-cusps to non-cusps.

REMARK 4. Deciding if there are extra non-modular automorphism is a difficult classical question for the case of modular and Shimura curves, see [**1**], [**13**], [**12**], [**8**], [**16**], [**21**], [**17**] for some related results.

2. The triangle group approach

A Fuchsian group Γ is a finitely generated discrete subgroup of $\mathrm{PSL}(2,\mathbb{R})$. It is known that a Fuchsian group has a set of $2g$ hyperbolic generators $\{a_1, b_1, \ldots, a_g, b_g\}$, a set of elliptic generators $x_1, \ldots, x_r$ and parabolic generators $p_1, \ldots, p_s$ and some hyperbolic boundary elements $h_1, \ldots, h_t$, see [**25**]. The relations are given by

$$x_1^{m_1} = x_2^{m_2} = \cdots = x_r^{m_r} = 1$$

$$\prod_{i=1}^{g}[a_i, b_i]\prod_{j=1}^{r} x_j \prod_{k=1}^{s} p_k \prod_{i=1}^{t} h_t = 1.$$

The signature of Γ is

$$(g; m_1, \ldots, m_r; s; t),$$

where $m_1, \ldots, m_r$ are natural numbers ≥ 2 and are called the periods of Γ.

Triangle groups: A triangle group $\Delta(\ell, m, n)$ is a group with signature $[0; \ell, m, n]$. We also thing parabolic elements as elliptic elements of infinite period and in this point of view , the group $\Gamma(2)$ can also be considered as the triangle group $\Gamma(\infty, \infty, \infty)$.

The Fermat curve can be uniformized in terms of triangle groups. This is a quite different uniformization than the uniformization given in (4). Namely we have the following diagramm of curves and groups

(5)

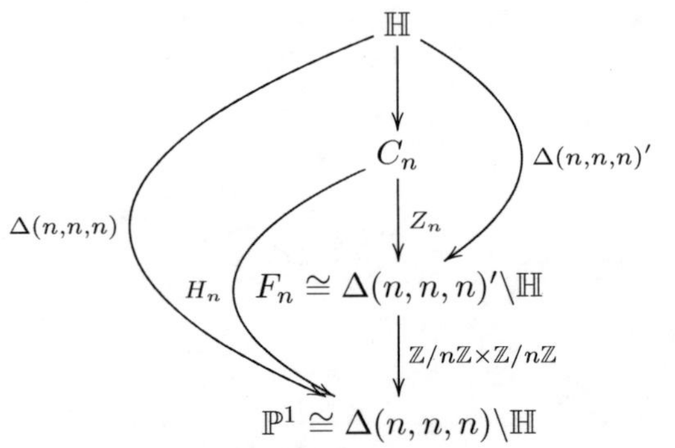

$\Delta(n,n,n)'$ is the commutator of the triangle group $\Delta(n,n,n)$. For $n > 3$ it is known that $\Delta(n,n,n)'$ is the universal covering group of the Fermat curve see [**31**].

We will show in lemma 11 that if $(n, 2) = 1$, then $C_n \to F_n$ is unramified. In this case, if $D(n)$ denotes the universal covering group of the Heisenberg curve then $D(n)$ is a normal subgroup of $\Delta(n,n,n)'$ and $\Delta(n,n,n)'/D(n) \cong \mathbb{Z}/n\mathbb{Z}$.

The presentation of a Riemann surface in terms of a co-compact triangle group has several advantages. Concerning automorphism groups, the advantage is that if Π is the funtamental group of the curve, which is a normal torsion free subgroup of the triangle group $\Delta(a,b,c)$, then the group $N_{\mathrm{PSL}(2,\mathbb{R})}(\Pi)/\Pi$ is the whole automorphism group not only the group of modular automorphisms, see [**11**]. The computation of the automorphism group is then simplified, as we will see for the case of Heisenberg and Fermat curves, since $N_{\mathrm{PSL}(2,\mathbb{R})}(\Pi)$ is known to be also a triangle curve which contains $\Delta(\ell, m, n)$, and these groups are fully classified [**7**], [**25**], [**11**, table 2]. This approach has the disadvantage that does not provide explicitly the automorphisms acting on the curve.

REMARK 5. The automorphism group of the Fermat curve can be computed by using the classification of the triangle groups which contain $\Delta(n,n,n)$ as a normal subgroup [**7**], [**25**], [**11**, table 2]. Indeed, the only such group is $\Delta(2,3,2n)$ and this computation provides and alternative method for proving:

$$\mathrm{Aut}(F_n) = \Delta(2,3,2n)/\Delta(n,n,n) = (\mathbb{Z}/n\mathbb{Z} \times \mathbb{Z}/n\mathbb{Z}) \rtimes S_3.$$

Notice that the triangle curve $\Delta(2,3,2n)$ is compatible with the ramification diagram given in figure 1.

REMARK 6. In addition to the above proof, the authors are aware of the following different methods for computing the automorphisms group of the Fermat curve: there are proofs using the Riemann-Hurwitz formula [**28**], [**18**] and proofs using the embedding of the Fermat curve in $\mathbb{P}^2$ and projective duality [**23**], [**15**].

3. Restriction and lifting of automorphisms in covers

In this section we will study the following:

QUESTION 7. Assume that $X \to Y$ is a Galois cover of curves. How are the groups $\mathrm{Aut}(X)$ and $\mathrm{Aut}(Y)$ related? When can an automorphism in $\mathrm{Aut}(Y)$ lift to an automorphism of $\mathrm{Aut}(X)$?

Assume that $X^0 = \Gamma_X \backslash \mathbb{H}, Y^0 = \Gamma_Y \backslash \mathbb{H}$ are either open curves corresponding to certain subgroups Γ_X, Γ_Y of $\mathrm{SL}(2,\mathbb{Z})$ or complete curves uniformized by cocompact groups Γ_X, Γ_Y. In the first case the complete curves X, Y are obtained by adding the cusps and in the second case $X^0 = X$ and $Y^0 = Y$.

PROPOSITION 8. *Let* $\{1\} < \Gamma_X \lhd \Gamma_Y$ *and consider the sequence of Galois covers*

$$\mathbb{H} \to \Gamma_X \backslash \mathbb{H} \to \Gamma_Y \backslash \mathbb{H}$$

and $G = \mathrm{Gal}(X/Y)$. *If* Γ_X, Γ_Y *are both normal subgroups of* $\mathrm{SL}(2,\mathbb{Z})$ *then a modular automorphism* σ *of* Y *lifts to* $|G|$ *automorphisms of* X, *if and only if* $\sigma \Gamma_X \sigma^{-1} \subset \Gamma_X$. *A modular automorphism* τ *of* X *restricts to a modular automorphism of* Y *if and only if* $\tau G \tau^{-1} \subset G$, *i.e. if and only if* $\tau^{-1}\Gamma_Y \tau \subset \Gamma_Y$.

Similarly if Γ_X, Γ_Y *are cocompact subgroups of* $\mathrm{SL}(2,\mathbb{R})$ *uniformizing the compact curves* X, Y, *then an automorphism* σ *of* Y *lifts to* G *automorphisms of* X, *if and only if* $\sigma \Gamma_X \sigma^{-1} \subset \Gamma_X$. *An automorphism* τ *of* X *restricts to a modular automorphism of* Y *if and only if* $\tau G \tau^{-1} \subset G$, *i.e. if and only if* $\tau^{-1}\Gamma_Y \tau \subset \Gamma_Y$.

PROOF. We will prove the first case, where Γ_X, Γ_Y are subgroups of $\mathrm{SL}(2,\mathbb{Z})$ uniformizing the open curves X^0, Y^0. The second case has a similar proof.

By definition, a modular automorphism of Y, is represented by element in the normalizer $N_{\mathrm{SL}(2,\mathbb{R})}\Gamma_Y$. This element has to normalize Γ_X as well in order to extend to an automorphism of Γ_X.

For restricting an automorphism from X to Y. Set $G = \mathrm{Gal}(X/Y) = \Gamma_Y/\Gamma_X$ and let N be the subgroup of $\mathrm{Aut}^0(X)$ which restrict to automorphisms of Y. We have the tower of fields shown on the right. The group N has to normalize G so that N/G is group acting on Y. Since G is a subgroup of the modular automorphism group, there is a conjugation action of every automorphism of X on G. In particular a modular automorphism $\tau \cdot \Gamma_X$ of X acts by conjugation on an element $\gamma \cdot \Gamma_X \in G$:

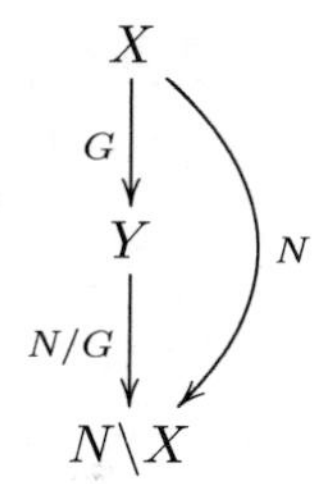

$$(\tau \cdot \Gamma_X)(\gamma \cdot \Gamma_X)(\tau \cdot \Gamma_X)^{-1} = \tau\gamma\tau^{-1} \cdot \Gamma_X,$$

and the later element is in G if an only if $\tau\gamma\tau^{-1} \in \Gamma_Y$. $\square$

4. The Fermat curve

In this section we will collect some known results about the Fermat curve and its automorphism group.

We will use the coorespondence of functions fields of one variable to curves and of points to places, see [**27**]. In particular we will use that coverings of curves correspond to algebraic extensions of their function fields.

LEMMA 9. *For* $n \geq 4$ *the automorphism group of the Fermat curve is isomorphic to the semidirect product* $(\mathbb{Z}/n\mathbb{Z} \times \mathbb{Z}/n\mathbb{Z}) \rtimes S_3$. *In the cover* $F_n \to F_n/\mathrm{Aut}(F_n) \cong \mathbb{P}^1$, *three points are ramified with ramification indices* $2n, 3, 3$ *respectively.*

PROOF. For the automorphism group of the Fermat curve see [**28**],[**18**]. In characteristic zero the automorphism group can be also studied as in remark 5.

The curves F_2 and F_3 are rational and elliptic respectively and so they have infinite automorphism group. The Fermat curve can be seen as a Kummer cover, i.e.

the function field $\mathbb{C}(F_n) = \mathbb{C}(x)[\sqrt[n]{x^n - 1}]$ is a Kummer extension of the rational function field $\mathbb{C}(x)$ and the ramification places in this Kummer extension correspond to the irreducible polynomials $x - \zeta^i$, $i = 0, \ldots, n-1$, where ζ is a primitive n-th root of unity. We have the following picture of function fields and ramification of places.

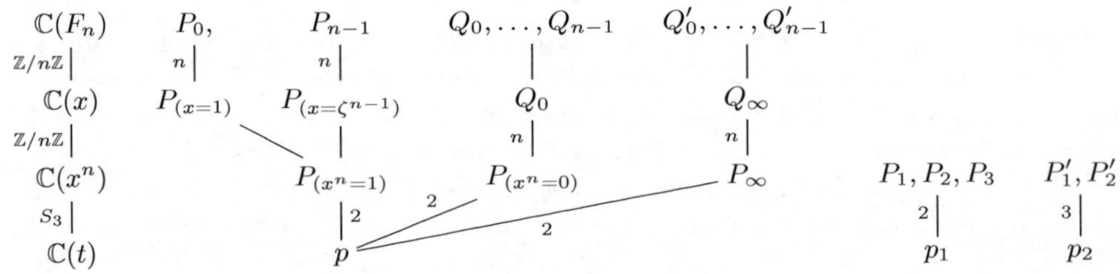

FIGURE 1. Ramification diagram for the Fermat Curve

□

Let us now consider the cover $F_n \to F_n^{\mathbb{Z}/n\mathbb{Z}\times\mathbb{Z}/n\mathbb{Z}} \cong \mathbb{P}^1$. Consider the function field $\mathbb{C}(x^n) = \mathbb{C}(F_n)^{\mathbb{Z}/n\mathbb{Z}\times\mathbb{Z}/n\mathbb{Z}}$. In the extension $\mathbb{C}(F_n)/\mathbb{C}(x^n)$ three places are ramified $P_{(x^n=0)}, P_{(x^n=1)}, P_\infty$. We will describe now the places of $\mathbb{C}(F_n)$ which restrict to the three points ramified above. For this we need the projective form of the Fermat curve given by

$$X^n + Y^n = Z^n.$$

The n places $P_0, \ldots, P_{n-1}$ which restrict to $P_{(x^n=1)}$ correspond to points with projective coordinates $P_k = (\zeta^k : 0 : 1)$ for $k = 0, \ldots, n-1$. We have also the n places $Q_0, \ldots, Q_{n-1}$ which restrict to $P_{x^n=0}$ which correspond to points with projective coordinates $Q_k = (0 : \zeta^k : 1)$ and the n places $Q'_0, \ldots, Q'_{n-1}$ that restrict to P_∞ and correspond to points with projective coordinates $Q'_k = (\epsilon\zeta^k : 1 : 0)$, $k = 0, \ldots, n-1$, $\epsilon^2 = \zeta$. An element $\sigma_{a,b} \in \mathbb{Z}/n\mathbb{Z} \times \mathbb{Z}/n\mathbb{Z}$ is acting on coordinates $(X : Y : Z)$ by the following rule:

$$\sigma_{a,b} : (X, Y, Z) \mapsto (\zeta^a X, \zeta^b Y, Z).$$

We will compute the stabilizers of the places $P_0, \ldots, P_{n-1}$, $Q_0, \ldots, Q_{n-1}$, $Q'_0, \ldots,$ Q'_{n-1} ramified in $\mathbb{C}(F_n)/\mathbb{C}(x^n)$:

LEMMA 10. *The points $(\zeta^k : 0 : 1)$ for $k = 0, \ldots, n-1$ are fixed by the cyclic group of order n generated by $\sigma_{0,1}$. The points $(0 : \zeta^k : 1)$, $k = 0, \ldots, n-1$ are fixed by the cyclic group of order n generated by $\sigma_{1,0}$. Finaly the points $(\zeta^k : 1 : 0)$ for $k = 0, \ldots, n-1$ are fixed by the cyclic group of order n generated by $\sigma_{1,1}$.*

PROOF. By computation. □

5. Automorphism group of the Heisenberg curve

LEMMA 11. *If $(n, 2) = 1$, then the cover $C_n \to F_n$ is unramified.*

PROOF. Notice first that

$$\begin{pmatrix} 1 & x & z \\ 0 & 1 & y \\ 0 & 0 & 1 \end{pmatrix}^{\nu} = \begin{pmatrix} 1 & \nu x & \nu z + \frac{\nu(\nu-1)}{2} xy \\ 0 & 1 & \nu y \\ 0 & 0 & 1 \end{pmatrix}. \tag{6}$$

If $(n, 2) = 1$ then by eq. (6) every element in H_n has order at most n. Notice that the top right corner element is $\nu z + \frac{\nu(\nu-1)}{2} xy$ and $2 \mid (n-1)$.

Also the only points that can ramify in $C_n \to F_n$ are the points of the Fermat curve, which lie above $\{0, 1, \infty\}$ since outside this set the cover is unramified. But if such a point P of the Heisenberg curve was ramified in $C_n \to F_n$, then its stabilizer $H_n(P)$ should be a cyclic group of order greater that n and no such group exist. □

Assume now that $(n, 2) = 2$. Consider the map $\pi : H_n \to H_n^{\mathrm{ab}} \cong \mathbb{Z}/n\mathbb{Z} \times \mathbb{Z}/n\mathbb{Z}$ defined in equation (3). Let $\sigma_{a,b} \in H_n^{\mathrm{ab}}$ be the automorphism corresponding to the pair $(a, b) \in \mathbb{Z}/n\mathbb{Z} \times \mathbb{Z}/n\mathbb{Z}$ with $a, b \in \mathbb{Z}/n\mathbb{Z}$.

Select an element $\alpha_{a,b}$ such that $\pi(\alpha) = \sigma_{a,b}$. Such an element has the following matrix form for some $z \in \mathbb{Z}/n\mathbb{Z}$:

$$\alpha_{a,b} = \begin{pmatrix} 1 & a & z \\ 0 & 1 & b \\ 0 & 0 & 1 \end{pmatrix}.$$

By eq. (6) if $ab = 0$ and $a = 1$ or $b = 1$ then the order $\mathrm{ord}(\alpha_{a,b})$ is at most n. So $\mathrm{ord}(\alpha_{1,0}) = n$ and $\mathrm{ord}(\alpha_{0,1}) = n$. Using again (6) we see that $\mathrm{ord}(\alpha_{1,1}) = 2n$.

This fact, combined to the computation of the stabilizers given in lemma 10 gives the following:

LEMMA 12. *If $(n, 2) = 2$ in the cover $C_n \to F_n$ only the points $(\zeta^k : 1 : 0)$ above P_∞ can ramify with ramification index at most 2.*

In order to understand the even n case we will treat first the $n = 2$ case.

LEMMA 13. *The Heisenberg curve C_2 is rational and in $C_2 \to F_2$ only two points of F_2 are branched in the cover $C_2 \to F_2$, namely $(1 : 1 : 0)$ and $(-1 : 1 : 0)$.*

PROOF. The case $n = 2$ is special since the Fermat curve F_2 is rational. In this case the Heisenberg group H_2 has order 8 and is isomorphic to the dihedral group D_4 generated by the elements

$$a_1 = \begin{pmatrix} 1 & 1 & 1 \\ 0 & 1 & 1 \\ 0 & 0 & 1 \end{pmatrix} \bmod 2 \text{ and } a_2 = \begin{pmatrix} 1 & 1 & 0 \\ 0 & 1 & 0 \\ 0 & 0 & 1 \end{pmatrix} \bmod 2,$$

where the order of a_1 is 4 and the order of a_2 is 2. From the classification of finite subgroups of the projective line [**29**], we have the following: The curve F_2 has the group $D_2 = \mathbb{Z}/2\mathbb{Z} \times \mathbb{Z}/2\mathbb{Z}$ inside its automorphism group, while in the cover $F_2 \to F_2/D_2$ three points $\{0, 1, \infty\}$ are ramified with ramification indices 2. At least one point should ramify in the cover $C_2 \to F_2$ since F_2 is simply connected, this should be a cusp and the only cusps that are permitted by lemma 12 are $(1 : 1 : 0)$, $(-1 : 1 : 0)$ and both should ramifify. The genus of C_2 is zero by Riemann-Hurwitz formula. The ramification indices in the intermediate extensions are shown in the following table:

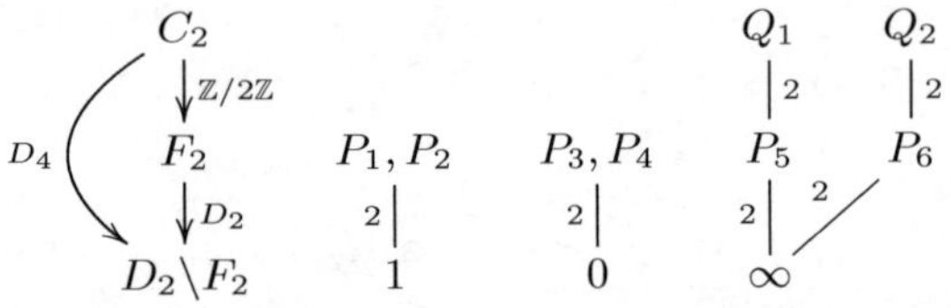

Notice that the ramification type of D_4 acting on the rational function field is $(2, 2, 4)$, which is in accordance to the classification in [**29**]. □

LEMMA 14. *If $2 \mid n$ then the cusps of C_n of the form $(\zeta^k : 1 : 0)$ are ramified in the cover $C_n \to F_n$ with ramification index equal to two.*

PROOF. Observe now that the elements of order $2n$

$$\sigma_j := \begin{pmatrix} 1 & 1 & j \\ 0 & 1 & 1 \\ 0 & 0 & 1 \end{pmatrix}$$

all restrict to $\sigma_{1,1} \in \mathrm{Aut}(F_n)$. Consider the points $Q'_1, \ldots, Q'_n$ of F_n that are above ∞ and for a fixed point Q'_{ν_0} consider the set of elements $\bar{Q}_{\nu_0,j}$ $j = 1, \ldots, t$ extending Q'_{ν_0} for $i = 1, \ldots, n$. Select now a point $\bar{Q}_{\nu_0,\mu_0}$ among them. If σ_0 does not fix $\bar{Q}_{\nu_0,\mu_0}$ the it moves it to the point $\bar{Q}_{\nu_0,j}$. But then there is an element $\tau \in \mathrm{Gal}(C_n, F_n) = Z_n$ moving $\bar{Q}_{\nu_0,\mu_0}$. Therefore its fixed by a matrix σ_j for an appropriate j. We compute

$$\sigma_j^n = \begin{pmatrix} 1 & 1 & j \\ 0 & 1 & 1 \\ 0 & 0 & 1 \end{pmatrix}^n = \begin{pmatrix} 1 & 0 & -n/2 \\ 0 & 1 & 0 \\ 0 & 0 & 1 \end{pmatrix} \in Z_n.$$

This means that in the cover $C_n \to (\mathbb{Z}/n\mathbb{Z} \times \mathbb{Z}/n\mathbb{Z})\backslash F_n$ the point $\bar{Q}_{\nu_0,\mu_0}$ is ramified with ramification index $2n$. □

LEMMA 15. *The genus g_{C_n} of the curve C_n equals:*

$$g_{C_n} = \begin{cases} \frac{n^2(n-3)}{2} + 1 & \text{if } (n,2) = 1 \\ \frac{n^2(n-3)}{2} + \frac{n^2}{4} + 1 & \text{if } 2 \mid n \end{cases}$$

PROOF. If $(n, 2) = 1$, then the cover $C_n \to F_n$ is unramified with Galois group $Z_n = \mathbb{Z}/n\mathbb{Z}$ and Riemann-Hurwitz formula implies that

$$2g_{C_n} - 2 = n(2g_{F_n} - 2).$$

We know that $g_{F_n} = \frac{(n-1)(n-2)}{2}$ and this gives the result in this case.

If $2 \mid n$ then in the cover $C_n \to F_n$ n cusps of the Fermat curve are ramified with ramification index 2. These cusps have $\frac{n^2}{2}$ points in total above them, so Riemann-Hurwitz in this case gives

$$2g_{C_n} - 2 = n(2g_{F_n} - 2) + \frac{n^2}{2}, \tag{7}$$

and the desired result follows. □

We have the following diagram:

$$
\begin{array}{cccc}
C_n & P_{1,j},\ldots P_{n,j} & Q_{1,j},\ldots Q_{n,j} & Q'_{1,j},\ldots Q'_{n/e,j} \\
\downarrow & | & | & |\, {\scriptstyle e=1 \text{ or } 2} \\
F_n & P_j & Q_j & Q'_j \\
\downarrow & | & | & | \\
\mathbb{P}^1 \cong \frac{F_n}{\mathbb{Z}/n\mathbb{Z}\times\mathbb{Z}/n\mathbb{Z}} & 0 & 1 & \infty
\end{array}
$$

where $j = 0, \ldots, n-1$ and e is 1 or 2 according to the value of n mod2.

6. Modular automorphisms

6.1. The Fermat curve. The open Fermat curve is the curve $F_n^0 := \ker \psi \backslash \mathbb{H}$, where

$$\ker\psi := \langle a^n, b^n, [a,b] \rangle \subset \langle a, b \rangle = \Gamma(2).$$

Notice that the above group differs from the universal covering group $\Delta(n,n,n)'$ given in section 2 which correspond to the closed Fermat curve.

Every automorphism of the Fermat curve is modular. Indeed, the generators $\sigma_{1,0}, \sigma_{0,1}$ of the $\mathbb{Z}/N\mathbb{Z}\times\mathbb{Z}/N\mathbb{Z}$ part of the automorphism group, are coming from the deck transformations $a, b \in \pi^1(\mathbb{P}^1\backslash\{0,1,\infty\}) = \Delta \subset \mathrm{SL}(2,\mathbb{Z}) \subset \mathrm{SL}(2,\mathbb{R})$. On the other hand it is known [**32**, exer. 3 p. 32] that $S_3 = \mathrm{SL}(2,\mathbb{Z})/\Gamma(2)$ and the action is given by lifts of elements of S_3 to $\mathrm{SL}(2,\mathbb{Z})$. The fact that all automorphisms are indeed modular comes from the fact that $\mathrm{SL}(2,\mathbb{Z})$ leaves both the set $\mathbb{H}$ and the cusps $\mathbb{P}^1$ invariant.

Let $\mathbf{F}_n$ denote the free group in n generators. Notice that the group S_3 acts by conjugation on Δ so it can be seen as a subgroup of the group of outer automorphisms of Δ. It is known [**5**, exam. 1 p. 117], [**6**, th. 3.1.7 p. 125] that the epimorphism

$$\mathbf{F}_2 \to \mathbf{F}_2^{\mathrm{ab}} \cong \mathbb{Z}\times\mathbb{Z}$$

induces an isomorphism of

$$\mathrm{Out}(\mathbf{F}_2)/\mathrm{In}(\mathbf{F}_2) = \mathrm{Aut}(\mathbf{F}_2) \to \mathrm{GL}(2,\mathbb{Z}).$$

The group S_3 is generated by the folowing automorphisms of the free group $\mathbf{F}_2 = \langle a, b\rangle$:

$$i_1 : a \leftrightarrow b \qquad i_2 : \begin{array}{l} a \mapsto b^{-1}a^{-1} \\ b \mapsto b \end{array} \tag{8}$$

The above generators i_1, i_2 of S_3 keep the group $\ker\psi$ invariant.

On the other hand, an arbitrary element of S_3 reduces to an action on the Fermat curve F_n by permutation of the variables X, Y, Z in the projective model of the curve. The automorphism intechanging X, Y coressponds to the involution interchanging a, b. Let us now consider the automorphism $\bar{\tau}$ interchanging X, Z in the projective model the Fermat curve.

Denote by $\sigma_{i,j}$ the automorphism of the Fermat curve sending

$$\sigma_{i,j} : (X : Y : Z) \mapsto (\zeta^i X : \zeta^j Y : Z).$$

By computation we have

$$\bar{\tau}\sigma_{i,j}\bar{\tau}^{-1} : (X : Y : Z) \mapsto (X : \zeta^j Y : \zeta^i Z) = (\zeta^{-i} X : \zeta^{-i+j} Y : Z).$$

Therefore, the conjugation action of $\bar{\tau}$ on $\mathrm{GL}(2, \mathbb{Z}/n\mathbb{Z}) = \mathrm{Aut}(\mathbb{Z}/n\mathbb{Z} \times \mathbb{Z}/n\mathbb{Z})$ is given by the matrix $\begin{pmatrix} -1 & -1 \\ 0 & 1 \end{pmatrix}$.

6.2. The Heisenberg curve. We will now describe the automorphisms of the Fermat curves F_n that can be lifted to automorphisms of C_n. Every representative $\sigma \in \mathrm{Aut}(\mathbb{H})$ of an element $\bar{\sigma} \in \mathrm{Aut}(F_n) = N_{\mathrm{SL}(2,\mathbb{R})}(\ker \psi)/\ker \psi$ should keep the group $\ker \psi$ invariant when acting by conjugation. The subgroup of the automorphism group of F_n should also keep the group

$$\ker(\phi) = \langle a^n, b^n, [a,b]^n \rangle$$

invariant, in order to extend to an automorphism of C_n.

The elements a, b generating $\mathbb{Z}/n\mathbb{Z} \times \mathbb{Z}/n\mathbb{Z}$ modulo $\ker\psi$ keep both groups $\ker \phi, \ker \psi$ invariant, so $\mathbb{Z}/n\mathbb{Z} \times \mathbb{Z}/n\mathbb{Z} < \mathrm{Aut}(F_n)$ is lifted to a subgroup of automorphisms of C_n.

Let $\sigma \in \Gamma$ be a representative of an element $\bar{\sigma} \in S_3 \subset \mathrm{Aut}(F_n)$. This element is lifted to an automorphism of the group C_n if and only if the conjugation action of σ keeps the defining group $\ker \phi$ invariant.

Let i_1, i_2 be the involutions generating S_3 as defined in eq. (8). Checking that the involution i_1 keeps $\ker \phi$ invariant is trivial.

We will now check the involution i_2: It is clear that $(b^n)^{i_2} = b^{i_2}$ and $[a,b]^{i_2} = [b^{-1}, a^{-1}]$ so $([a,b]^n)^{i_2} = [b^{-1}, a^{-1}]^n$.

Let a_H, b_H be the generators of the Heisenberg group, seen as elements in $F_2 = \langle a_H, b_H \rangle / \ker \phi$ as given in eq. (1). In order to check whether the generator a^n of $\ker \phi$ are sent to $\ker \phi$ under the action of i_2 it is enough to prove that its image modulo $\ker \phi$ is the zero element in the Heisenberg group. We compute

$$a_H^{i_2} = b_H^{-1} a_H^{-1} \text{ and } b_H^{i_2} = b_H.$$

Therefore

$$a_H^{i_2} = \begin{pmatrix} 1 & 0 & 0 \\ 0 & 1 & -1 \\ 0 & 0 & 1 \end{pmatrix} \begin{pmatrix} 1 & -1 & 0 \\ 0 & 1 & 0 \\ 0 & 0 & 1 \end{pmatrix} = \begin{pmatrix} 1 & -1 & 0 \\ 0 & 1 & -1 \\ 0 & 0 & 1 \end{pmatrix}$$

so by eq. (6) we have

$$\left(a_H^{i_2}\right)^n = \begin{cases} \begin{pmatrix} 1 & 0 & -\frac{n}{2} \\ 0 & 1 & 0 \\ 0 & 0 & 1 \end{pmatrix} & \text{if } n \equiv 0 \bmod 2 \\ \begin{pmatrix} 1 & 0 & 0 \\ 0 & 1 & 0 \\ 0 & 0 & 1 \end{pmatrix} & \text{if } (n,2) = 1 \end{cases}$$

and this gives the identity matrix if and only if $(n,2) = 1$. Therefore in the case $2 \mid n$ the element $(a^n)^{i_2}$ does not belong to the group $\ker \phi$.

LEMMA 16. *The modular automorphism group for the Heisenberg curve is given by an extension*

$$1 \to \mathbb{Z}/n\mathbb{Z} \to \mathrm{Aut}^{\mathrm{m}}(C_n) \to G_n \to 1$$

where G_n is the group

$$G_n = \begin{cases} \mathrm{Aut}(F_n) & \text{if } (n,2)=1 \\ (\mathbb{Z}/n\mathbb{Z} \times \mathbb{Z}/n\mathbb{Z}) \rtimes \mathbb{Z}/2\mathbb{Z} & \text{otherwise} \end{cases}$$

PROOF. The Heisenberg group is already in the automorphism group, fitting in the short exact sequence

$$\begin{array}{ccccccccc} 1 & \longrightarrow & \mathbb{Z}/n\mathbb{Z} & \longrightarrow & \mathrm{Aut}^{\mathrm{m}}(C_n) & \longrightarrow & G_n & \longrightarrow & 1 \\ & & \| & & \uparrow & & \uparrow & & \\ 1 & \longrightarrow & \mathbb{Z}/n\mathbb{Z} & \longrightarrow & H_n & \longrightarrow & \mathbb{Z}/n\mathbb{Z} \times \mathbb{Z}/n\mathbb{Z} & \longrightarrow & 1 \end{array}$$

We have computed the part of $\mathrm{Aut}(F_n)$ which lifts to automorphisms of C_n according to the value of n modulo 2.

We will now prove that every modular automorphism of C_n restricts to an automorphism of F_n. Indeed, using proposition 8 we have to show that every automorphism σ that fixes by conjugation elements of $\langle a^n, b^n, [a,b]^n\rangle$ fixes elements of $\langle a^n, b^n, [a,b]\rangle$ as well.

Assume that σ fixes the group $\langle a^n, b^n, [a,b]^n\rangle$. If $(a^n)^\sigma$ is a word in $a^n, b^n, [a,b]^n$ then it is obviously a word in $a^n, b^n, [a,b]$. For the commutator we will use the following result due to Nielsen [**19**, th. 3.9]

$$[a,b]^\sigma = T[a,b]^{\pm 1}T^{-1},$$

where T is a word in a, b. So the invariance of the commutator $[a,b]$ under the action of the outer automorphism σ follows since $\langle a^n, b^n, [a,b]\rangle$ is a normal subgroup of $\langle a, b\rangle$. □

LEMMA 17. *For $2 \mid n$ three points are ramified in $C_n \mapsto \mathrm{Aut}^{\mathrm{m}}(C_n)$, with ramification indices $4n, n, 2$.*

PROOF. In the cover $C_n \to F_n \to (\mathbb{Z}/n\mathbb{Z} \times \mathbb{Z}/n\mathbb{Z})\backslash F_n \cong \mathbb{P}^1$ we have ramification on the points $P_{x=0}, P_{x=1}$ and P_∞, with ramification indices $n, n, 2n$.

Consider the group S_3 acting on $(\mathbb{Z}/n\mathbb{Z} \times \mathbb{Z}/n\mathbb{Z})\backslash F_n$ and generated by the automorphisms $x \mapsto f(x)$, where

$$f(x) \in \left\{ x, 1-x, \frac{1}{x}, \frac{1}{1-x}, \frac{x}{x-1}, \frac{x-1}{x} \right\}.$$

The involution $i_1 : a \leftrightarrow b$ in terms of generators of the free group, corresponds to the involution $x \mapsto 1-x$, which sends $1 \leftrightarrow 0$ and keeps $\infty, 1/2$ invariant. Therefore, the three ramified points have ramification indices $2, n, 4n$. □

THEOREM 18. *Every automorphism of C_n is modular, i.e., sends the cusps to the cusps. In particular the automorphism group of the curve C_n in this case equals:*

$$\mathrm{Aut}(C_n) = N_{\mathrm{SL}(2,\mathbb{R})}(\ker\phi)/\ker\phi.$$

PROOF. If $(n,2)=1$ then the group $D(n)$ corresponding to the Heisenberg curve in the triangle uniformization given in eq. (5) is a normal subgroup of $\Delta(n,n,n)$. The normalizer $N_{\mathrm{PSL}(2,\mathbb{R})}(D(n))$ contains the trigonal curve $\Delta(n,n,n)$ and since every group containing a triangle group with finite index is also triangle [**2**, th. 10.6.5 p.279] the normalizer is also a triangle group. By the computation of the modular automorphism group and the classification given in [**11**, table 2] we have that $N_{\mathrm{PSL}(2,\mathbb{R})}(D(n)) = D(2,n,2n)$, which gives us

$$\mathrm{Aut}(C_n) = \frac{N_{\mathrm{PSL}(2,\mathbb{R})}(D(n))}{D(n)} = \mathrm{Aut}^{\mathrm{m}}(C_n).$$

For the $2 \mid n$ case we will employ the Riemann-Hurwitz formula. Let G be the automorphism group, $|G| = 2n^3m$, where $m = [G : \mathrm{Aut}^{\mathrm{m}}(C_n)]$. Let $Y = C_n^G$. Since $\mathbb{C}(Y) \subset C_n^{H_n}$ we have that $g_Y = 0$. By eq. (7) $2g_{C_n} - 2 = n^2(n-3) + n^2/2$ and Riemann-Hurwitz theorem [**9**, ex. IV.2.5] gives us that

$$n^2(n-3) + n^2/2 = n^3 2m(-2 + \sum_{i=1}^{r}(1 - 1/e_i))$$

where $e_i \geq 2$ are the ramification indices of the r-points of $\mathbb{P}^1$ ramified in extension $C_n \to G\backslash C_n = \mathbb{P}^1$. Set $\Omega_n = -2 + \sum_{i=1}^{r}(1 - 1/e_i)$. Then the Riemann-Hurwitz formula can be written as

$$n - 5/2 = 2mn\Omega_n.$$

Observe that $\Omega_n \geq 0$ and $m \geq 1$. So for $n > 3$ we must have $\Omega_n > 0$. If $\Omega_n \geq 1/4$, then it is obvious that $m = 1$. Indeed, we should have

$$1 \leq m = \frac{2n-5}{4n\Omega_n} \leq 2 - 5/n < 2. \tag{9}$$

In the above formula we have that for $n \leq 4$ the term $2 - 5/n < 1$, which is not compatible with the $1 \leq m$ inequality. This means that for $n = 4$ the inequality $\Omega_n \geq 1/4$ is not possible so $\Omega_n < 1/4$. For Ω_4 the ramification index index for one point is at least $16 = 4n$.

$$\Omega_4 = -2 + \left(1 - \frac{1}{16e}\right) + \sum_{i=2}^{r}\left(1 - \frac{1}{e_i}\right) \geq -\frac{17}{16} + \frac{r-1}{2}.$$

The above value for Ω_4 is negative for $r \leq 3$ and $1/4$ for $r = 4$. So we need less than 3 ramification points, which we assume that have ramification indices $16e, \kappa, \lambda$. In this case we have

$$\Omega_4 = 1 - \frac{1}{16e} - \frac{1}{\ell} - \frac{1}{\kappa}.$$

Since this value has to be smaller than $1/4$ the values for e, λ, κ can not take very big values. If both $\kappa, \ell \geq 3$ then

$$\Omega_4 \geq 1 - \frac{1}{16} - \frac{2}{3} > \frac{1}{4}$$

which is impossible. So $\kappa = 2$ and $\lambda = 3$ is the only possible case. For this case $\Omega_4 \geq 5/48$ and eq. (9) implies that $m \leq 9/5 < 2$.

Now we consider the $n > 4$ case and we will show that $\Omega_n \geq 1/4$. If three or more points, other than the one with $e_1 = 4ne$, are ramified in the cover $C_n \to Y$, then

$$\Omega_n = -2 + 1 - \frac{1}{4ne} + \sum_{i=2}^{r}\left(1 - \frac{1}{e_i}\right) \geq \frac{1}{2} - \frac{1}{4ne} \geq \frac{1}{4}.$$

Consider now that exactly three points are ramified in $C_n \to Y$. If two of them are ramified with ramification index 2 then $\Omega_n = -1/4n$ and this is not allowed since $\Omega_n \geq 0$.

Assume now that we have exactly 3 ramification points with ramification $(4ne, \kappa, \ell)$. In this case

$$\Omega_n \geq -2 + \left(1 - \frac{1}{4n}\right) + \left(1 - \frac{1}{\ell}\right) + \left(1 - \frac{1}{\kappa}\right) = 1 - \frac{1}{4n} - \frac{1}{\ell} - \frac{1}{k}.$$

If $n > 4$ then $n \geq 6$ (n is even) so

$$\Omega_n \geq \frac{23}{24} - \frac{1}{\ell} - \frac{1}{k}.$$

It is clear that the above quantity is bigger than $1/4$ if ℓ and κ are big enough. For instance if $\ell, \kappa \geq 3$ then $23/24 - 2/3 = 7/24$ and $\Omega_n \geq 1/4$. We have to check the case $\kappa = 2$ and in this case $\Omega_n = 23/24 - 1/2 - 1/\ell = 11/24 - 1/\ell$ so the inequality $\Omega_n \geq 1/4$ holds provided $\ell \geq 5$.

The cases $\ell = 3, 4$ give the corresponding bounds $B_\ell = (2n-5)\Omega_\ell/(4n)$,

$$B_3 = \frac{3(2n-5)}{n(3n+2)}, \qquad B_4 = \frac{2n-5}{n^2+n},$$

All the above values are < 1 and can not bound the quantity m, hence they can't occur. □

7. The curve C_3

The Fermat curve $F_3 : x^3 + y^3 = 1$ is elliptic and it has the projective cannonical Weierstrass form $zy^2 = x^3 - 432z^3$, see [**14**, p. 50-52] and [**10**, ex. 3 p.32]. Its torsion 3-points are the flexes which can be computed as the zeros of the Hessian determinant:

$$\mathrm{Hess}(y^2z - x^3 + 432z^3) = \det \begin{pmatrix} -6x & 0 & 0 \\ 0 & 2z & 2y \\ 0 & 2y & 2592y \end{pmatrix} = 24(y^2 - 1296z^2)x.$$

- If $x = 0$, then $zy^2 = -432z^3$ which gives the solutions
$$(0 : 1 : 0), (0 : 12\sqrt{-3} : 1), (0 : -12\sqrt{-3} : 1).$$
- If $x \neq 0$, then $y^2 = 1296z^2$, which gives the solutions $(1 : 0 : 0)$, which does not satisfy the equation of the elliptic curve. We also have the solution $y = \pm\sqrt{1296}z$, which we plug into the equation of the elliptic curve to obtain:
$$z^3 1296 = x^3 - 432z^3,$$
so for $z = 1$ we obtain $x^3 = 1728$ so $x = 12\zeta_3^i$, and $\zeta_3 = (-1 + \sqrt{-3})/2$ is a primitive third root of unity. We therefore have 6-more 3-torsion points namely
$$(12\zeta_3^i : \pm 36 : 1).$$

The curve C_3 is by Riemann-Hurwitz formula also an elliptic curve and the covering map

$$C_3 \to F_3$$

is an isogeny. For each point of order 3 computed above Vélu method [**30**] can be applied and using sage [**26**] we compute the following table:

point of order 3	Equation of isogenus curve	j-invariant
$(0 : \pm 12\sqrt{-3} : 1)$	$y^2 = x^3 + 11664$	0
$(-6(1+\sqrt{-3}) : \pm 36 : 1)$	$y^2 = x^3 + 2160(1-\sqrt{-3})x - 109296$	-12288000
$(6(1-\sqrt{-3}) : \pm 36 : 1)$	$y^2 = x^3 + 2160(1+\sqrt{-3})x - 109296$	-12288000
$(12 : \pm 36 : 1)$	$y^2 = x^3 - 4320x - 109296$	-12288000

The three last curves have the same j-invariant and are isomorphic (they are quadratic twists of each other). The first one has j-invariant zero and therefore the automorphism group $\mathrm{Aut}^0(E)$ of E, consisted of automorphisms $E \to E$ which fix the identity, is a cyclic group of order 6, see [**24**, ch. III. par. 10.1]. This is compatible with the structure of the Heisenberg curve C_3, since the neutral element of C_3 is fixed by a group of order 6.

On the other hand the other 3 isomorphic curves of the above table have j-invariant $\neq 0, 1728$ so $\mathrm{Aut}^0(E)$ is a cyclic group of order 2. Therefore the equation of C_3 is given by

$$C_3 : y^2 = x^3 + 2^4 \cdot 3^6.$$

Acknowledgment

The authors would like to thank Professor Dinakar Ramakrishnan for proposing the study of Heisenberg curves to them.

References

[1] Matthew Baker and Yuji Hasegawa, *Automorphisms of $X_0^*(p)$*, J. Number Theory **100** (2003), no. 1, 72–87, DOI 10.1016/S0022-314X(02)00120-8. MR1971247

[2] Alan F. Beardon, *The geometry of discrete groups*, Graduate Texts in Mathematics, vol. 91, Springer-Verlag, New York, 1995. Corrected reprint of the 1983 original. MR1393195

[3] G. V. Belyĭ, *Galois extensions of a maximal cyclotomic field* (Russian), Izv. Akad. Nauk SSSR Ser. Mat. **43** (1979), no. 2, 267–276, 479. MR534593

[4] G. V. Belyĭ, *A new proof of the three-point theorem* (Russian, with Russian summary), Mat. Sb. **193** (2002), no. 3, 21–24, DOI 10.1070/SM2002v193n03ABEH000633; English transl., Sb. Math. **193** (2002), no. 3-4, 329–332. MR1913596

[5] Joan S. Birman, *Braids, links, and mapping class groups*, Princeton University Press, Princeton, N.J.; University of Tokyo Press, Tokyo, 1974. Annals of Mathematics Studies, No. 82. MR0375281

[6] Oleg Bogopolski, *Introduction to group theory*, EMS Textbooks in Mathematics, European Mathematical Society (EMS), Zürich, 2008. Translated, revised and expanded from the 2002 Russian original. MR2396717

[7] Leon Greenberg, *Maximal Fuchsian groups*, Bull. Amer. Math. Soc. **69** (1963), 569–573, DOI 10.1090/S0002-9904-1963-11001-0. MR148620

[8] Michael Corin Harrison, *A new automorphism of $X_0(108)$*, (2011).

[9] Robin Hartshorne, *Algebraic geometry*, Springer-Verlag, New York-Heidelberg, 1977. Graduate Texts in Mathematics, No. 52. MR0463157

[10] Dale Husemöller, *Elliptic curves*, 2nd ed., Graduate Texts in Mathematics, vol. 111, Springer-Verlag, New York, 2004. With appendices by Otto Forster, Ruth Lawrence and Stefan Theisen. MR2024529

[11] Sadok Kallel and Denis Sjerve, *On the group of automorphisms of cyclic covers of the Riemann sphere*, Math. Proc. Cambridge Philos. Soc. **138** (2005), no. 2, 267–287, DOI 10.1017/S0305004104008096. MR2132169

[12] S. Kamienny, *On the automorphism groups of some modular curves*, Math. Proc. Cambridge Philos. Soc. **134** (2003), no. 1, 61–64, DOI 10.1017/S0305004102006023. MR1937792

[13] M. A. Kenku and Fumiyuki Momose, *Automorphism groups of the modular curves $X_0(N)$*, Compositio Math. **65** (1988), no. 1, 51–80. MR930147

[14] Anthony W. Knapp, *Elliptic curves*, Mathematical Notes, vol. 40, Princeton University Press, Princeton, NJ, 1992. MR1193029

[15] Aristides Kontogeorgis, *Automorphisms of Fermat-like varieties*, Manuscripta Math. **107** (2002), no. 2, 187–205, DOI 10.1007/s002290100233. MR1894739

[16] Aristides Kontogeorgis and Victor Rotger, *On the non-existence of exceptional automorphisms on Shimura curves*, Bull. Lond. Math. Soc. **40** (2008), no. 3, 363–374, DOI 10.1112/blms/bdn012. MR2418792

[17] Aristides Kontogeorgis and Yifan Yang, *Automorphisms of hyperelliptic modular curves $X_0(N)$ in positive characteristic*, LMS J. Comput. Math. **13** (2010), 144–163, DOI 10.1112/S1461157009000023. MR2638986

[18] Heinrich-Wolfgang Leopoldt, *Über die Automorphismengruppe des Fermatkörpers* (German, with English summary), J. Number Theory **56** (1996), no. 2, 256–282, DOI 10.1006/jnth.1996.0017. MR1373551

[19] Wilhelm Magnus, Abraham Karrass, and Donald Solitar, *Combinatorial group theory*, 2nd ed., Dover Publications, Inc., Mineola, NY, 2004. Presentations of groups in terms of generators and relations. MR2109550

[20] James S. Milne, *Modular functions and modular forms (v1.31)*, 2017, Available at `www.jmilne.org/math/`, p. 134.

[21] Santiago Molina and Victor Rotger, *Automorphisms and reduction of Heegner points on Shimura curves at Cerednik-Drinfeld primes*, Proc. Amer. Math. Soc. **142** (2014), no. 2, 381–390, DOI 10.1090/S0002-9939-2013-11772-7. MR3133980

[22] David Ephraim Rohrlich, *Modular functions and the Fermat curves*, ProQuest LLC, Ann Arbor, MI, 1976. Thesis (Ph.D.)–Yale University. MR2626317

[23] Tetsuji Shioda, *Arithmetic and geometry of Fermat curves*, Algebraic Geometry Seminar (Singapore, 1987), World Sci. Publishing, Singapore, 1988, pp. 95–102. MR966448

[24] Joseph H. Silverman, *The arithmetic of elliptic curves*, 2nd ed., Graduate Texts in Mathematics, vol. 106, Springer, Dordrecht, 2009. MR2514094

[25] David Singerman, *Finitely maximal Fuchsian groups*, J. London Math. Soc. (2) **6** (1972), 29–38, DOI 10.1112/jlms/s2-6.1.29. MR322165

[26] W. A. Stein et al., *Sage Mathematics Software (Version 8.9)*, The Sage Development Team, 2019, `http://www.sagemath.org`.

[27] Henning Stichtenoth, *Algebraic function fields and codes*, 2nd ed., Graduate Texts in Mathematics, vol. 254, Springer-Verlag, Berlin, 2009. MR2464941

[28] Pavlos Tzermias, *The group of automorphisms of the Fermat curve*, J. Number Theory **53** (1995), no. 1, 173–178, DOI 10.1006/jnth.1995.1085. MR1344839

[29] Robert C. Valentini and Manohar L. Madan, *A hauptsatz of L. E. Dickson and Artin-Schreier extensions*, J. Reine Angew. Math. **318** (1980), 156–177. MR579390

[30] Jacques Vélu, *Isogénies entre courbes elliptiques* (French), C. R. Acad. Sci. Paris Sér. A-B **273** (1971), A238–A241. MR294345

[31] Jürgen Wolfart, *The "obvious" part of Belyi's theorem and Riemann surfaces with many automorphisms*, Geometric Galois actions, 1, London Math. Soc. Lecture Note Ser., vol. 242, Cambridge Univ. Press, Cambridge, 1997, pp. 97–112. MR1483112

[32] Masaaki Yoshida, *Hypergeometric functions, my love*, Aspects of Mathematics, E32, Friedr. Vieweg & Sohn, Braunschweig, 1997. Modular interpretations of configuration spaces. MR1453580

Department of Mathematcis, University of Crete, Herakleion Crete Greece 71409
Email address: `antoniad@uoc.gr`

Department of Mathematics, University of Athens, Panepistimioupolis, 15784 Athens, Greece
Email address: `kontogar@math.uoa.gr`

Contemporary Mathematics
Volume **767**, 2021
https://doi.org/10.1090/conm/767/15396

Tate-Shafarevich groups of constant elliptic curves and isogeny volcanoes

Brendan Creutz and José Felipe Voloch

To Gerhard Frey on his 75th birthday

Abstract. We describe the structure of Tate-Shafarevich groups of a constant elliptic curves over function fields by exploiting the volcano structure of isogeny graphs of elliptic curves over finite fields.

1. Introduction

Let k be a finite field and E/k an elliptic curve. Let F/k be another curve and let $K = k(F)$ its function field. The base change E/K is usually referred to as a constant elliptic curve over the global function field K. Already since the 60's, from the work of Tate completed by Milne [**Mil68**], it has been known that the full Birch and Swinnerton-Dyer conjecture holds for such constant elliptic curves and, in particular, that the Tate-Shafarevich group $Ш(E/K)$ is finite. But not much information on this group beyond what the Birch and Swinnerton-Dyer conjecture gave was obtained at the time. Later, $Ш(E/K)$ was computed for supersingular E and various fields K by Elkies, Shioda and others ([**Shi91**] and references therein) motivated by the fact that the Mordell-Weil group $E(K)$ with its canonical height pairing gave interesting examples of lattice packings. In those examples, $Ш(E/K)$ is a p-group, where p is the characteristic of k.

Parallel to these developments, motivated by applications to computational number theory and cryptography, there have been a number of papers studying the ℓ-isogeny graphs on elliptic curves ([**Sut13**] and references therein). These graphs are defined by taking as vertices elliptic curves over a field k and edges representing isogenies of degree ℓ between them. A fundamental insight of Kohel (Theorem 2.3 below) was extended and reformulated by Fouquet and Morain [**FM02**] to the statement that a connected component of the ℓ-isogeny graph for a prime ℓ over a finite field k consisting of ordinary curves has a specific structure referred to as ℓ-volcano graph (Definition 2.1 below).

The purpose of this paper is to bring these two trends together and use the structure of the isogeny graphs to obtain an explicit description of the Tate-Shafarevich

2020 *Mathematics Subject Classification.* Primary 11G05.

Key words and phrases. Elliptic curves, function fields, Tate-Shafarevich group, isogeny graphs.

The authors were supported by the Marsden Fund Council administered by the Royal Society of New Zealand.

group Ш(E/K) in many instances and, indeed, go further and compute all terms of the descent exact sequence for ℓ-isogenies, when F is also an elliptic curve.

2. Preliminaries

2.1. Isogeny graphs.

DEFINITION 2.1. A graph with vertex set V is called an ℓ-volcano graph if its vertex set can be partitioned as $V = V_0 \cup V_1 \cdots \cup V_m$, where m is called the height of the volcano, V_0 the base and V_m the crater or top. In addition, the induced graph on V_m is a connected regular graph of degree at most 2 (the edges on this subgraph are called horizontal), the degree of all vertices not on V_0 is $\ell + 1$, the degree of the vertices in V_0 is 1, for each vertex on $V_i, i < m$, there is a unique edge from it to a vertex in V_{i+1} (these are called upward edges) and for each vertex on $V_i, i > 0$, the other edges go to vertices in V_{i-1} (these are called downward edges). The vertices in V_i are said to have height (or level) i.

REMARK 2.2. Note that, if we remove the edges of V connecting the vertices of V_m, the connected components of the resulting graph are trees with roots at the the vertices of V_m. We call those the trees of the volcano.

We will assume throughout that ℓ is prime and not the characteristic of k. (For the case $\ell = p$ see Section 5). As stated above, a connected component of the ℓ-isogeny graph of ordinary curves over the finite field k has the structure of a ℓ-volcano graph (with a minor modification if the component contains a curve with j-invariant 0 or 1728). Moreover, denoting by $h_\ell(E)$ the height (or level) of E in its connected component of the ℓ-isogeny graph, viewed as a volcano, we have $h_\ell(E) = v_\ell([\mathrm{End}(E) : \mathbb{Z}[\pi]])$, where π is the Frobenius endomorphism of E. We refer to an isogeny of degree a power of ℓ as upward, downward or horizontal if the corresponding path on the ℓ-isogeny graph is a sequence of edges with the same property.

THEOREM 2.3. [**Koh96**, Propositions 21 and 22] *Let E and E' be isogenous ordinary elliptic curves over k. Then $h_\ell(E) = h_\ell(E')$ if and only if there is an isogeny $E \to E'$ of degree prime to ℓ.*

2.2. Descent. Briefly, let us recall the descent mechanism and the definition of the Selmer and Tate-Shafarevich groups. For details, see [**Sil09**, Chapter X].

Given an isogeny $\phi : E \to E_1$ defined over a field K, we have the exact sequence

$$0 \to E[\phi] \to E \to E_1 \to 0$$

which gives, by taking cohomology, the Kummer sequence,

$$0 \to E_1(K)/\phi(E(K)) \to \mathrm{H}^1(K, E[\phi]) \to \mathrm{H}^1(K, E)[\phi] \to 0.$$

If K is a global field, the corresponding exact sequences for each completion K_v of K can also be considered. By comparing the corresponding sequences, one obtains the descent sequence

$$0 \to E_1(K)/\phi(E(K)) \to \mathrm{Sel}^\phi(E_1/K) \to Ш(E/K)[\phi] \to 0 \tag{2.1}$$

where $\mathrm{Sel}^\phi(E_1/K)$ is the Selmer group consisting of the elements of $\mathrm{H}^1(K, E[\phi])$ whose image in $\prod_v \mathrm{H}^1(K_v, E[\phi])$ lies in the image of $\prod_v E_1(K_v)$ (where the direct product is taken over all places v of K) and

$$Ш(E/K) := \ker\left(\mathrm{H}^1(K,E) \to \bigoplus_v \mathrm{H}^1(K_v,E)\right)$$

is the Tate-Shafarevich group.

2.3. Constant elliptic curves. If E is an elliptic curve over a finite field k and F is another curve defined over k with function field $K = k(F)$, then a basic fact, central to our approach, is the identification $E(K) = \mathrm{Mor}(F,E)$. If, in addition F is also an elliptic curve with identity 0_F then, evaluating a map $\alpha \in E(K) = \mathrm{Mor}(F,E)$ at 0_F gives a retraction of the inclusion $E(k) \subset E(K)$. This gives a split exact sequence,

$$0 \to E(k) \to E(K) \to \mathrm{Hom}(F,E) \to 0\,. \tag{2.2}$$

The following lemma is central to our determination of the first term in (2.1).

LEMMA 2.4. *Suppose $\phi : E \to E_1$ is a an isogeny of ℓ-primary degree. If ϕ is downward, then* $\mathrm{Hom}(E,E_1) = \phi\,\mathrm{End}(E)$. *If ϕ has prime degree, then* $\mathrm{Hom}(E,E_1) = \phi\,\mathrm{End}(E)$ *if and only if ϕ is downward or $E \simeq E_1$ and ϕ is self dual up to an automorphism.*

PROOF. First suppose ϕ is downward and $\alpha \in \mathrm{Hom}(E,E_1)$. We can factor α as $\alpha = \alpha_\ell \circ \alpha'$ with $\alpha' : E \to E'$ an isogeny of degree prime to ℓ and $\alpha_\ell : E' \to E_1$ an isogeny of ℓ-primary degree. By Theorem 2.3, $h_\ell(E) = h_\ell(E')$ which is greater than $h_\ell(E_1)$ by assumption. From the structure of the ℓ-isogeny graph (see Remark 2.2) the final edge in any path from E' to E_1 in the ℓ-isogeny graph must be ϕ. Therefore $\alpha_\ell \in \phi\,\mathrm{Hom}(E',E_1)$ and so $\alpha \in \phi\,\mathrm{End}(E)$.

For the remainder of the proof we suppose ϕ has degree ℓ. If ϕ is not downward, then it is either upward or horizontal. We consider each case in turn.

First suppose ϕ is upward and let $L = \mathrm{End}(E_1) \otimes \mathbb{Q}$. Then ϕ determines an injective ring homomorphism $\iota : \mathrm{End}(E) \to L$ by $\gamma \mapsto \phi\gamma\phi^* \otimes \ell^{-1}$, where ϕ^* is the dual isogeny of ϕ. This induces an inclusion $\mathrm{End}(E) \subset \mathrm{End}(E_1) \subset L$ with $[\mathrm{End}(E_1) : \mathrm{End}(E)] = \ell$ since ϕ is upward. Applying the case already proven to the downward isogeny ϕ^* we get $\mathrm{Hom}(E_1,E) = \phi^*\,\mathrm{End}(E_1)$. Dualizing gives $\mathrm{Hom}(E,E_1) = \mathrm{End}(E_1)\phi$. If we also had $\mathrm{Hom}(E,E_1) = \phi\,\mathrm{End}(E)$, then multiplying by ϕ^* on the right would give $\phi\,\mathrm{End}(E)\phi^* = \mathrm{End}(E_1)\ell$. But, this implies $\iota(\mathrm{End}(E)) = \mathrm{End}(E_1)$, which is a contradiction.

Now consider the case when ϕ is horizontal. This can only happen when $\mathrm{End}(E) = \mathcal{O}_L$ is the maximal order in $L = \mathrm{End}(E) \otimes \mathbb{Q}$ and $\ell = \mathfrak{p}\mathfrak{q}$ is not inert. Moreover, $\phi = \phi_\mathfrak{p}$ or $\phi = \phi_\mathfrak{q}$, where for an ideal $\mathfrak{a} \subset \mathrm{End}(E)$, the notation here means that $\ker(\phi_\mathfrak{a}) = \{P \in E \,:\, \alpha(P) = 0, \text{ for all } \alpha \in \mathfrak{a}\}$. Relabeling if necessary we can suppose $\phi = \phi_\mathfrak{p}$. If $\mathfrak{p}$ is principal, then $E_1 \simeq E$ and $\phi^* \in \phi\,\mathrm{End}(E)$ is equivalent to $\phi^* \in \phi\,\mathrm{Aut}(E)$. Now suppose these ideals are not principal. Then every endomorphism of E of degree divisible by ℓ is itself divisible by $[\ell]$. Let m be the order of $\mathfrak{p}$ in the class group of $\mathcal{O}_L$. Then $\phi_{\mathfrak{q}^{m-1}} \in \mathrm{Hom}(E,E_1)$ corresponds to the path along the crater rim opposite from $\phi = \phi_\mathfrak{p}$. We claim that $\phi_{\mathfrak{q}^{m-1}} \notin \phi\,\mathrm{End}(E)$. Indeed, if $\phi_{\mathfrak{q}^{m-1}} = \phi_\mathfrak{p}\gamma$ with $\gamma \in \mathrm{End}(E)$, then being an endomorphism of ℓ-primary degree $\gamma = [\ell^r]$ for some r. But this implies $\mathfrak{q}^m = \mathfrak{q}^r\mathfrak{p}^{r+1}$, which contradicts unique factorization of ideals in $\mathcal{O}_L$. $\square$

3. Calculation of the Selmer group

Suppose $\phi : E \to E_1$ is an isogeny of degree m over the finite field k with m relatively prime to the characteristic of k. Let F/k be an elliptic curve and set $K = k(F)$. Set $G_k = \mathrm{Gal}(\overline{k}/k)$.

PROPOSITION 3.1. *The ϕ-Selmer group of E_1 is the unramified subgroup,*

$$\mathrm{Sel}^{\phi}(E_1/K) = \ker\left(\mathrm{H}^1(K, E[\phi]) \to \prod_v \mathrm{H}^1(\overline{k}K_v, E[\phi])\right).$$

PROOF. It is enough to show that for any place v of K the image of $E_1(K_v)/\phi(E(K_v))$ in $\mathrm{H}^1(K_v, E[\phi])$ under the map in (2.1) coincides with the kernel of the restriction map $\mathrm{H}^1(K_v, E[\phi]) \to \mathrm{H}^1(\overline{k}K_v, E[\phi])$. Suppose k_v is the residue field of K_v. The inflation-restriction sequence yields a commutative diagram with exact row

$$\begin{array}{ccccccc} & & E_1(k_v)/\phi(E(k_v)) & \longrightarrow & E_1(K_v)/\phi(E(K_v)) & & \\ & & \downarrow & & \downarrow & & \\ 0 & \longrightarrow & \mathrm{H}^1(k_v, E[\phi]) & \longrightarrow & \mathrm{H}^1(K_v, E[\phi]) & \longrightarrow & \mathrm{H}^1(\overline{k}K_v, E[\phi]). \end{array}$$

We claim that the arrows eminating from the top left are both isomorphisms, from which the result easily follows.

The top horizontal map is induced by the inclusion $E_1(k_v) \subset E_1(K_v)$. It is an isomorphism since the kernel of the reduction map $E_1(K_v) \to E_1(k_v)$ is ℓ-divisible for every ℓ not divisible by $p = \mathrm{char}(k)$ [**Sil09**, Propositions IV.2.3 and VI.2.2] and $\deg(\phi)$ is prime to p. That the vertical arrow is an isomorphism follows from exactness of (2.1) and the fact that $\mathrm{H}^1(k, E) = 0$ since every smooth genus 1 curve over the finite field k_v has a rational point. $\square$

Define $r_\phi : \mathrm{Hom}(F, E_1) \to \mathrm{Hom}(F[m], E[\phi])$ by $r_\phi(\alpha) = \phi^*\alpha|_{F[m]}$, where ϕ^* is the dual isogeny of ϕ and the notation means the restriction of $\phi^*\alpha$ to $F[m]$ and $\mathrm{Hom}(F[m], E[\phi])$ denotes the group of morphisms of G_k-modules.

LEMMA 3.2. $r_\phi(\alpha) = (\alpha^*|_{E_1[\phi^*]})^D$*, the Cartier dual of the morphism of Galois modules obtained by restricting α^* to kernel of ϕ^*.*

PROOF. The Weil pairing gives isomorphisms

$$F[m] \ni x \mapsto e_m(x, \bullet) \in \mathrm{Hom}(F[m], \mu_m) = F[m]^D$$

and

$$E[\phi] \ni x \mapsto e_\phi(x, \bullet) \in E_1[\phi^*]^D.$$

The Cartier dual of α^* restricted to $F[m]$ is the map $F[m]^D \ni e_m(x, \bullet) \mapsto e_m(x, \alpha^*(\bullet))$. For $y \in E_1[\phi^*]$ functoriality of the Weil pairing gives that

$$e_m^F(x, \alpha^*(y)) = e_m^{E_1}(\alpha(x), y) = e_\phi(\phi^*\alpha(x), y).$$

Therefore $(\alpha^*|_{E'[\phi^*]})^D$ is the map sending x to $\phi^*\alpha(x) = r_\phi(\alpha)(x)$. $\square$

LEMMA 3.3. $r_\phi(\alpha) = 0$ *if and only if* $\alpha \in \phi\,\mathrm{Hom}(F, E)$.

PROOF. If $\alpha = \phi\alpha'$, then $r_\phi(\alpha) = \phi^*\phi\alpha' = m\alpha' = 0$ when restricted to $F[m]$. Conversely, suppose $\phi^*\alpha = 0$ on $F[m]$. Then the isogeny $\phi^*\alpha$ must factor through multiplication by m. This means there is some $\beta \in \mathrm{Hom}(F, E_1)$ such that $\phi^*\alpha = m\beta = \phi^*\phi\beta$. Therefore the map $\alpha - \phi\beta$ takes values in $E_1[\phi^*]$. But a difference of isogenies is either surjective or the zero map. Hence $\alpha = \phi\beta$. □

REMARK 3.4. In the case that $\phi = [n]$ is multiplication by n on E, $r_\phi(\alpha) = n\alpha \in \mathrm{Hom}(F[n^2], E[n])$. This can be identified with $\alpha|_{F[n]} \in \mathrm{Hom}(F[n], E[n])$ as follows. Every element of the Galois module $\mathrm{Hom}(F[n^2], E[n])$ is divisible by n. The map sending $\beta = n\beta' \in \mathrm{Hom}(F[n^2], E[n])$ to $\beta'|_{F[n]} \in \mathrm{Hom}(F[n], E[n])$ is an isomorphism sending $r_n(\alpha)$ to $\alpha|_{F[n]}$.

THEOREM 3.5. *Let $\phi : E \to E_1$ be an isogeny of degree m with m not divisible by the characteristic of k. There is an exact and commutative diagram*

$$\begin{array}{ccccccccc}
0 & \longrightarrow & E_1(k)/\phi(E(k)) & \longrightarrow & E_1(k)/\phi(E(k)) & & & & \\
& & \downarrow & & \downarrow{\scriptstyle \partial_\phi} & & & & \\
0 & \longrightarrow & E_1(K)/\phi(E(K)) & \xrightarrow{\partial_\phi} & \mathrm{Sel}^\phi(E_1/K) & \longrightarrow & Ш(E/K)[\phi] & \longrightarrow & 0 \\
& & \downarrow & & \downarrow & & \| & & \\
0 & \longrightarrow & \frac{\mathrm{Hom}(F,E_1)}{\phi\,\mathrm{Hom}(F,E)} & \xrightarrow{r_\phi} & \mathrm{Hom}(F[m], E[\phi]) & \longrightarrow & Ш(E/K)[\phi] & \longrightarrow & 0\,.
\end{array}$$

PROOF. Let $G = \mathrm{Gal}(K^{\mathrm{sep}}/\bar{k}K)$. Since $E[\phi]^G = E[\phi]$, the inflation-restriction sequence gives

$$0 \to \mathrm{H}^1(k, E[\phi]) \to \mathrm{H}^1(K, E[\phi]) \to \mathrm{Hom}(G, E[\phi]) \to \mathrm{H}^2(k, E[\phi]) = 0\,,$$

where the final term is 0 because k, being a finite field, has cohomological dimension 1. Since every smooth genus 1 curve over the finite field k has a rational point, we have that $\mathrm{H}^1(k, E) = 0$. So the connecting homomorphism induces an isomorphism $\partial_\phi : E_1(k)/\phi(E(k)) \simeq \mathrm{H}^1(k, E[\phi])$. By Proposition 3.1 the ϕ-Selmer group and its image in $\mathrm{Hom}(G, E[\phi])$ are the unramified subgroups. The maximal unramfied extension of $\bar{k}K$ of exponent m corresponds to multiplication by m on F, so every unramified (continuous) homomorphism $G \to E[\phi]$ factors through $F[m]$. This gives an isomorphism $\mathrm{Hom}(G, E[\phi])^{\mathrm{unr}} \to \mathrm{Hom}(F[m], E[\phi])$.

It remains to show that the square involving r_ϕ commutes. For this suppose $\alpha \in \mathrm{Mor}(F, E_1) = E_1(K)$ is a morphism. Adjusting by $E_1(k)$ if necessary we may assume $\alpha \in \mathrm{Hom}(F, E_1)$ is an isogeny. The fibered product $F' = F \times_{E_1} E$ along α and ϕ corresponds to an extension L/K for which there is a point $\beta \in E(L)$ such that $\phi\beta = \alpha$. The cocycle $\eta : \mathrm{Gal}(\overline{K}/K) \to E[\phi]$ given by $\sigma \mapsto {}^\sigma(\beta) - \beta$ represents $\partial_\phi(\alpha)$. The action of σ is translation by some $P_\sigma \in \ker(\psi)$ so that $({}^\sigma\beta)(x) = \beta(x + P_\sigma)$. Then ${}^\sigma\beta - \beta = \beta(P_\sigma)$. We have a commutative diagram

$$\begin{array}{ccccc}
F & \xrightarrow{\psi^*} & F' & \xrightarrow{\psi} & F \\
\downarrow{\scriptstyle \alpha} & & \downarrow{\scriptstyle \beta} & & \downarrow{\scriptstyle \alpha} \\
E_1 & \xrightarrow{\phi^*} & E & \xrightarrow{\phi} & E_1
\end{array}$$

where the rows are multiplication by $m = \deg(\phi)$. The cocycle η factors through $F[m]$ and from the commutativity of the diagram we see that the map $\sigma \mapsto \beta(P_\sigma)$ is given by $r_\phi(\alpha) : F[m] \to E[\phi]$. □

When ϕ is multiplication by n the theorem together with Remark 3.4 gives the following.

COROLLARY 3.6. *Suppose E, F are elliptic curves over the finite field k and n is an integer not divisible by the characteristic of k. With $K = k(F)$, there is an exact sequence*

$$0 \to \frac{\mathrm{Hom}(F, E)}{n\,\mathrm{Hom}(F, E)} \to \mathrm{Hom}(F[n], E[n]) \to Ш(E/K)[n] \to 0\,.$$

If E and F are isogenous ordinary elliptic curves, then $Ш(E/K)[n]$ has rank at most 2 over $\mathbb{Z}/n\mathbb{Z}$. If E and F are isogenous supersingular elliptic curves, then $Ш(E/K)[n] = 0$.

This says that the nontrivial elements of $Ш(E/K)[n]$ arise from Galois module morphisms $F[n] \to E[n]$ that do not come from an isogeny $F \to E$. From this one sees quite directly how Tate's isogeny theorem is equivalent to the finiteness of $Ш(E/K)[n^\infty]$. A similar result, formulated in terms of the Brauer group of the product of two elliptic curves over a number field occurs in [**SZ12**, Proposition 3.3].

COROLLARY 3.7. *Suppose $\phi : E \to E_1$ is an isogeny with kernel generated by a k-rational point of order n. Then $\mathrm{Sel}^\phi(E_1/K)$ sits in a split exact sequence*

$$0 \to k^\times/k^{\times n} \to \mathrm{Sel}^\phi(E_1/K) \to F[n](k) \to 0\,.$$

If $E[n] \subset E(k)$, then $\mathrm{Sel}^n(E/K)$ and sits in a split exact sequence

$$0 \to k^\times/k^{\times n} \oplus k^\times/k^{\times n} \to \mathrm{Sel}^n(E/K) \to F[n](k) \oplus F[n](k) \to 0\,.$$

THEOREM 3.8. *Suppose $\phi : E \to E_1$ is an isogeny of ordinary elliptic curves over k whose kernel is generated by a k-rational point of order ℓ. Let F be any elliptic curve isogenous to E and set $K = k(F)$. Then the following are equivalent*

(1) *$h_\ell(E_1) < h_\ell(E) \le h_\ell(F)$;*
(2) *$\mathrm{Hom}(F, E_1) = \phi\,\mathrm{Hom}(F, E)$;*
(3) *$E_1(K)/\phi(E(K)) = E_1(k)/\phi(E(k))$;*
(4) *$Ш(E/K)[\phi] = Ш(E/K)[\ell]$ has rank 2.*

Denote by σ the triple

$$(\mathrm{rk}\, E_1(K)/\phi(E(K)), \mathrm{rk}\, \mathrm{Sel}^\phi(E/K), \mathrm{rk}\, Ш(E/K)[\phi])$$

where rk is the rank over $\mathbb{Z}/\ell\mathbb{Z}$. When the above equivalent conditions hold $\sigma = (1, 3, 2)$. Otherwise $\sigma = (2, 2, 0)$ if $h_\ell(F) = 0$, $\sigma = (3, 3, 0)$ if $h_\ell(E) < h_\ell(E_1) \le h_\ell(F)$ and $\sigma = (2, 3, 1)$, otherwise.

PROOF. The equivalence of (2) and (3) follows immediately from the split exact sequences (2.2) for E and E_1.

Suppose $h_\ell(E_1) < h_\ell(E) \le h_\ell(F)$. Let $\alpha \in \mathrm{Hom}(F, E_1)$ be any isogeny. We can factor α as $\alpha_\ell \circ \alpha'$ with $\alpha' : F \to F'$ of degree prime to ℓ and $\alpha_\ell : F' \to E_1$ of ℓ-primary degree. By Theorem 2.3 $h_\ell(F) = h_\ell(F')$. Thus α_ℓ is a path on the ℓ-volcano eventually consisting of downward edges on one the trees of the volcanoe (see Remark 2.2) and thus must factor through ϕ. Since α_ℓ has ℓ-primary degree this implies that α_ℓ, and hence also α, lies in $\phi\,\mathrm{Hom}(F, E)$.

Let us now consider the cases $h_\ell(E) < h_\ell(E_1), h_\ell(F)$ or $h_\ell(E) > h_\ell(E_1), h_\ell(F)$. In either case there is an isogeny $\alpha : F \to E_1$ with $v_\ell(\deg(\alpha)) = |h_\ell(F) - h_\ell(E_1)|$ which never reaches the level of E. This isogeny can be constructed as follows. Take F' on the tree containing E_1 at the level of F which, in addition, is a child of E_1 if $h(E_1) < h(F)$. Then by Theorem 2.3 there is an isogeny $F \to F'$ of degree prime to ℓ. There is also a path on the tree from F' to E_1 not going through E which defines an isogeny and the isogeny claimed earlier is the composition of those two. This isogeny cannot possibly lie in $\phi\,\mathrm{Hom}(F, E)$, since any $\gamma \in \mathrm{Hom}(F, E)$ has $v_\ell(\deg(\gamma)) \geq |h_\ell(F) - h_\ell(E)| > |h_\ell(F) - h_\ell(E_1)|$. Thus (2) does not hold.

It remains to consider the case when $h_\ell(E_1) \geq h_\ell(E)$. Lemma 2.4 handles this in the case $F = E$ or F is in the branch of the isogeny graph below E. Otherwise, it is clear that there is a path on the isogeny graph from F to E_1 not passing through E, showing that (2),(3) do not hold and, from this and the descent sequence (2.1), it follows that (4) does not hold either.

To complete the proof, let $\sigma = (a, b, c)$ and note that $a+c = b$ and that $b = 2$ if $h_\ell(F) = 0$ and $b = 3$, otherwise. Also, $a \geq 1$ because of the contribution of torsion. In order to have $a = 1$ we must have that all isogenies $F \to E_1$ factor through E. This can only be the case when $h_\ell(F) \geq h_\ell(E) > h_\ell(E_1)$. From this we can already conclude that $\sigma = (2, 2, 0)$ if $h_\ell(F) = 0$. We also get that $\sigma = (1, 3, 2)$ if $h_\ell(F) \geq h_\ell(E) > h_\ell(E_1)$.

We claim that $a = 3$ precisely when $h_\ell(F) \geq h_\ell(E_1) > h_\ell(E)$. Just as in the previous argument, an isogeny $F \to E$ must factor through E_1 via ϕ^* and so an isogeny in $\phi(E(K))$ must factor through $\phi \circ \phi^* = [\ell]$ and $E_1(K)/\phi(E(K)) = E_1(K)/\ell(E_1(K))$ which has rank 3. Hence $\sigma = (3, 3, 0)$.

If $h_\ell(E_1) = h_\ell(E)$, then $a = 2$, so $\sigma = (2, 3, 1)$ unless $h_\ell(F) = 0$. If $0 < h_\ell(F) \leq h_\ell(E) < h_\ell(E_1)$, then $\sigma = (2, 3, 1)$ as we can find a path from E_1 to some E' at the level of F not passing through E and we can then use an isogeny $F \to E'$ of degree prime to ℓ. Similarly, for $0 < h_\ell(F) \leq h_\ell(E_1) < h_\ell(E)$, $\sigma = (2, 3, 1)$ also. These are all the possibilities since $|h_\ell(E) - h_\ell(E_1)| \leq 1$. □

REMARK 3.9. If there is an isogeny $\phi : E \to E_1$ of degree ℓ whose kernel is not generated by a k-rational point, then E has height 0 in the volcano and one verifies that $\sigma = (1, 1, 0)$.

4. Calculation of Ш

Recall that π represents the Frobenius endomorphism of an elliptic curve.

THEOREM 4.1. *Suppose that E, F are isogenous ordinary elliptic curves over k and set $\mathcal{O} = \mathrm{End}(E) \cap \mathrm{End}(F)$, the intersection taking place in $\mathbb{Q}(\pi) = \mathrm{End}(E) \otimes \mathbb{Q}$. There is an isomorphism of groups $Ш(E/k(F)) \simeq \mathcal{O}/\mathbb{Z}[\pi] \times \mathcal{O}/\mathbb{Z}[\pi]$. In particular, for any prime ℓ, $Ш(E/k(F))[\ell^\infty]$ is isomorphic to the product of two cyclic groups of order $\ell^{\min\{h_\ell(E), h_\ell(F)\}}$.*

REMARK 4.2. This generalizes the observation of Milne [**Mil68**, p 102] that $\#Ш(E/k(E)) = [\mathrm{End}(E) : \mathbb{Z}[\pi]]^2$.

REMARK 4.3. It would be interesting to find a meaningful interpretation of the isomorphism of the theorem and, in particular, identify a natural symplectic pairing on $\mathcal{O}/\mathbb{Z}[\pi] \times \mathcal{O}/\mathbb{Z}[\pi]$ which corresponds to the Cassels-Tate pairing. Motivated by results of [**RS17**], Ariel Shnidman suggested the following. Let $\mathfrak{f} = [\mathcal{O} : \mathbb{Z}[\pi]]$ so that $\mathcal{O}/\mathbb{Z}[\pi]\times\mathcal{O}/\mathbb{Z}[\pi]\simeq\mathcal{O}/\mathfrak{f}\mathbb{Z}$. The latter has an alternating nondegenerate pairing given by $(\alpha,\beta)=\frac{1}{2d\mathfrak{f}}\operatorname{Tr}_{\mathbb{Q}(\sqrt{d})/\mathbb{Q}}(\sqrt{d}\alpha\overline{\beta})$, where $\mathbb{Q}(\pi)=\mathbb{Q}(\sqrt{d})$.

PROOF. Let ℓ be a prime. Since $\operatorname{Hom}(F,E)/\ell$ has dimension 2 over $\mathbb{F}_\ell$, Corollary 3.6 shows that $Ш(E/k(F))[\ell]$ has dimension 0 or 2. Therefore $Ш(E/k(F))[\ell^\infty]$ is a product of two cyclic groups. Since $\mathcal{O}/\mathbb{Z}[\pi]$ is cyclic and $v_\ell([\mathcal{O}:\mathbb{Z}[\pi]])=\min\{h_\ell(E),h_\ell(F)\}$ we see that the theorem follows from the claim that, for each ℓ, $v_\ell(\#Ш(E/k(F))[\ell^\infty])=2\min\{h_\ell(E),h_\ell(F)\}$. This holds trivially for the p-part as $Ш(E/k(F))[p]=0$ and the p-isogeny graph has just one level (see Section 5).

Note that $Ш(E/k(F))\simeq Ш(F/k(E))$, since both are isomorphic to the Brauer group of the surface $E\times F$ [**Tat95**, Theorem 3.1]. By swapping the roles of E and F if necessary we may therefore assume that $h_\ell(E)\le h_\ell(F)$. Let E' be any elliptic curve lying below F in the ℓ-isogeny graph (i.e., such that there is a downward isogeny $F\to E'$, possibly the identity) with $h_\ell(E')=h_\ell(E)$. Since $h_\ell(E)=h_\ell(E')$, Theorem 2.3 implies that there is an isogeny $\alpha:E'\to E$ of degree prime to ℓ. This isogeny induces an isomorphism $Ш(E'/k(F))[\ell^\infty]\simeq Ш(E/k(F))[\ell^\infty]$. So, replacing E by E' if needed, we are reduced to the case that E lies below F in the ℓ-isogeny graph.

Suppose the characteristic polynomial of Frobenius on F/k factors as $x^2+tx+q=(x-a)(x-b)$. By [**Mil68**, Theorem 3] we have

$$\#Ш(E/k(F))\cdot R_{E/k(F)}=q\left(1-\frac{a}{b}\right)\left(1-\frac{b}{a}\right), \tag{4.1}$$

where $R_{E/k(F)}=|\det\operatorname{Tr}(\alpha_i\beta_j)|$, for any bases $\{\alpha_i\}$ and $\{\beta_i\}$ of the free $\mathbb{Z}$-modules $\operatorname{Hom}(F,E)$ and $\operatorname{Hom}(E,F)$. By Lemma 2.4 we have $\operatorname{Hom}(F,E)=\phi\operatorname{End}(F)$. It follows that $R_{E/k(F)}=\deg(\phi)^2R_{F/k(F)}$. Comparing (4.1) for $F/k(F)$ and $E/k(F)$ shows that $\#Ш(F/k(F))=\deg(\phi)^2\#Ш(E/k(F))=\ell^{2(h_\ell(F)-h_\ell(E))}\#Ш(E/k(F))$. It thus suffices to verify the claim when $F=E$.

Let $\mathfrak{f},\mathfrak{g}\in\mathbb{Z}$ be the conductors of the orders $\mathbb{Z}[\pi]$ and $\operatorname{End}(F)$ in the quadratic imaginary field $L=\operatorname{End}(F)\otimes\mathbb{Q}$, so that $\mathbb{Z}[\pi]=\mathbb{Z}[\mathfrak{f}D]$, $\operatorname{End}(F)=\mathbb{Z}[\mathfrak{g}D]$ and $\mathcal{O}_L=\mathbb{Z}[D]$, where $D=\frac{d_L+\sqrt{d_L}}{2}$, d_L is the fundamental discriminant. In particular $\operatorname{End}(F)/\mathbb{Z}[\pi]\simeq\mathbb{Z}/n\mathbb{Z}$ where $n=\mathfrak{f}/\mathfrak{g}$. Then

$$R_{F/k(F)}=\left|\det\begin{bmatrix}\operatorname{Tr}([1]) & \operatorname{Tr}(\mathfrak{g}D)\\ \operatorname{Tr}(\mathfrak{g}D) & \operatorname{Tr}(\mathfrak{g}^2D^2)\end{bmatrix}\right|.$$

For any integer h we have

$$\left|\det\begin{bmatrix}\operatorname{Tr}([1]) & \operatorname{Tr}(hD)\\ \operatorname{Tr}(hD) & \operatorname{Tr}(h^2D^2)\end{bmatrix}\right|=\left|\begin{bmatrix}2 & hd_L\\ hd_L & h^2(d_L^2+d_L)/2)\end{bmatrix}\right|=h^2d_L$$

As $\pi=\mathfrak{f}D$ we have

$$\mathfrak{f}^2d_L=\left|\det\begin{bmatrix}\operatorname{Tr}([1]) & \operatorname{Tr}(\pi)\\ \operatorname{Tr}(\pi) & \operatorname{Tr}(\pi^2)\end{bmatrix}\right|=|2(a^2+b^2)-(a+b)^2|=(a-b)(b-a)=q\left(1-\frac{b}{a}\right)\left(1-\frac{a}{b}\right)$$

Comparing with (4.1) we see that $\#Ш(F/k(F))=[\operatorname{End}(F):\mathbb{Z}[\pi]]^2$, as required. □

COROLLARY 4.4. *For an arbitrary integer $n\ge 0$, E has height n in the ℓ-isogeny graph if and only if $Ш(E/k(E))[\ell^\infty]\simeq\mathbb{Z}/\ell^n\mathbb{Z}\times\mathbb{Z}/\ell^n\mathbb{Z}$.*

COROLLARY 4.5. *(*[**MMS⁺08**, Theorem 5]*) For an arbitrary integer $n>0$, if E has full rational ℓ^n-torsion, then E has height at least n in the ℓ-isogeny graph.*

PROOF. If $E(k)$ contains the ℓ^n-torsion, then by Corollary 3.6 we have that $Ш(E/k(E))[\ell^n] \simeq \mathbb{Z}/\ell^n\mathbb{Z}\times\mathbb{Z}/\ell^n\mathbb{Z}$. So the result follows from the previous corollary. □

COROLLARY 4.6. *For an arbitrary integer $n > 0$, a downward isogeny of degree ℓ^n must annihilate* $Ш(E/k(E))[\ell^n]$.

PROOF. Suppose E is at level $i > 0$ in the ℓ-isogeny graph. Then

$$Ш(E/k(E))[\ell^\infty] \simeq \mathbb{Z}/\ell^i\mathbb{Z} \times \mathbb{Z}/\ell^i\mathbb{Z}.$$

By Theorem 3.8 any composition of i downward ℓ-isogenies must kill $Ш(E/k(E))[\ell^\infty]$, but each step can only kill ℓ-torsion. Hence each step must kill all of the ℓ-torsion. □

5. Concluding remarks

The questions addressed in this paper can be considered also for supersingular curves and/or in the case $\ell = p$ although (apart from the supersingular $\ell = p$ case) the situation is much simpler.

If E is ordinary and $\ell = p$, there are only two degree p isogenies from E, the Frobenius and Verschiebung. The component of the p-isogeny graph containing E is thus a cycle consisting of the conjugates of E. The Selmer group and $E_1(K)/\phi(E(K))$ are one dimensional if ϕ is Frobenius or if ϕ is Verschiebung and the p-torsion of E is not defined over K. When ϕ is Verschiebung and the p-torsion of E is defined over K, both these groups are two dimensional. The p-torsion of $Ш(E/K)$ is trivial in both cases. See [**Ulm91**, Proposition 3.3], but note that the Selmer group in the case $A = 1$ of the Verschiebung descent has dimension one more than reported there.

Now consider the case when E, F are isogenous supersingular curves over k. If $\mathbb{F}_{p^2} \subset k$, then $\mathrm{Hom}(F, E)$ has rank 4 and so Corollary 3.6 shows that $Ш(E/k(F))$ is a p-group. If $\mathbb{F}_{p^2} \not\subset k$, then the connected components of the ℓ-isogeny graph of E are volcanoes with at most two levels (for $\ell = 2$, [**DG16**]). Moreover, if there are two levels, then $\ell = 2$. Indeed, if E is supersingular and $E[\ell] \subset E(k)$, then $\ell|(q-1)$. If, in addition, $\mathbb{F}_{p^2} \not\subset k$ then $\#E(k) = q+1$ and we get that $\ell|(q+1)$, so $\ell = 2$. The analogues of Theorems 3.8 and 4.1 hold when $\mathbb{F}_{p^2} \not\subset k$, with $\mathrm{End}(E)$ an order of the quadratic field $\mathbb{Q}(\sqrt{-p})$.

Finally, we remark that, as noted in [**Mil68**], if E, F are non-isogenous elliptic curves over k, then it follows from the Birch-Swinnerton-Dyer formula that

$$\#Ш(E/k(F)) = (\#E(k) - \#F(k))^2.$$

Acknowledgments

The authors would like to thank Doug Ulmer for correspondence regarding [**Ulm91**] and helpful comments as well as Ariel Shnidman for a discussion concerning Remark 4.3.

References

[DG16] Christina Delfs and Steven D. Galbraith, *Computing isogenies between supersingular elliptic curves over $\mathbb{F}_p$*, Des. Codes Cryptogr. **78** (2016), no. 2, 425–440, DOI 10.1007/s10623-014-0010-1. MR3451433

[FM02] Mireille Fouquet and François Morain, *Isogeny volcanoes and the SEA algorithm*, Algorithmic number theory (Sydney, 2002), Lecture Notes in Comput. Sci., vol. 2369, Springer, Berlin, 2002, pp. 276–291, DOI 10.1007/3-540-45455-1_23. MR2041091

[Koh96] David Russell Kohel, *Endomorphism rings of elliptic curves over finite fields*, ProQuest LLC, Ann Arbor, MI, 1996. Thesis (Ph.D.)–University of California, Berkeley. MR2695524

[Mil68] J. S. Milne, *The Tate-Šafarevič group of a constant abelian variety*, Invent. Math. **6** (1968), 91–105, DOI 10.1007/BF01389836. MR244264

[MMS+08] J. Miret, R. Moreno, D. Sadornil, J. Tena, and M. Valls, *Computing the height of volcanoes of l-isogenies of elliptic curves over finite fields*, Appl. Math. Comput. **196** (2008), no. 1, 67–76, DOI 10.1016/j.amc.2007.05.037. MR2382590

[RS17] Julian Rosen and Ariel Shnidman, *Extensions of CM elliptic curves and orbit counting on the projective line*, Res. Number Theory **3** (2017), Paper No. 9, 13, DOI 10.1007/s40993-017-0073-y. MR3642282

[Sil09] Joseph H. Silverman, *The arithmetic of elliptic curves*, 2nd ed., Graduate Texts in Mathematics, vol. 106, Springer, Dordrecht, 2009. MR2514094

[Shi91] Tetsuji Shioda, *Theory of Mordell-Weil lattices*, Proceedings of the International Congress of Mathematicians, Vol. I, II (Kyoto, 1990), Math. Soc. Japan, Tokyo, 1991, pp. 473–489. MR1159235

[SZ12] Alexei N. Skorobogatov and Yuri G. Zarhin, *The Brauer group of Kummer surfaces and torsion of elliptic curves*, J. Reine Angew. Math. **666** (2012), 115–140, DOI 10.1515/CRELLE.2011.121. MR2920883

[Sut13] Andrew V. Sutherland, *Isogeny volcanoes*, ANTS X—Proceedings of the Tenth Algorithmic Number Theory Symposium, Open Book Ser., vol. 1, Math. Sci. Publ., Berkeley, CA, 2013, pp. 507–530, DOI 10.2140/obs.2013.1.507. MR3207429

[Tat95] John Tate, *On the conjectures of Birch and Swinnerton-Dyer and a geometric analog*, Séminaire Bourbaki, Vol. 9, Soc. Math. France, Paris, 1995, pp. Exp. No. 306, 415–440. MR1610977

[Ulm91] Douglas L. Ulmer, *p-descent in characteristic p*, Duke Math. J. **62** (1991), no. 2, 237–265, DOI 10.1215/S0012-7094-91-06210-1. MR1104524

School of Mathematics and Statistics, University of Canterbury, Private Bag 4800, Christchurch 8140, New Zealand

Email address: `brendan.creutz@canterbury.ac.nz`

URL: `http://www.math.canterbury.ac.nz/~b.creutz`

School of Mathematics and Statistics, University of Canterbury, Private Bag 4800, Christchurch 8140, New Zealand

Email address: `felipe.voloch@canterbury.ac.nz`

URL: `http://www.math.canterbury.ac.nz/~f.voloch`

Contemporary Mathematics
Volume **767**, 2021
https://doi.org/10.1090/conm/767/15397

Endomorphisms of ordinary superelliptic Jacobians

Yuri G. Zarhin

To Gerhard Frey

ABSTRACT. Let K be a field of prime characteristic p, $n \ge 5$ an integer, $f(x)$ an irreducible polynomial over K of degree n, whose Galois group is either the full symmetric group $\mathbf{S}_n$ or the alternating group $\mathbf{A}_n$. Let ℓ be an odd prime different from p, $\mathbb{Z}[\zeta_\ell]$ the ring of integers in the ℓth cyclotomic field, $C_{f,\ell} : y^\ell = f(x)$ the corresponding superelliptic curve and $J(C_{f,\ell})$ its Jacobian. We prove that the ring of all $\bar{K}$-endomorphisms of $J(C_{f,\ell})$ coincides with $\mathbb{Z}[\zeta_\ell]$ if $J(C_{f,\ell})$ is an *ordinary* abelian variety and $(\ell, n) \ne (5,5)$.

1. Introduction

Throughout this paper K is a field and $\bar{K}$ is its algebraic closure. We write $\mathrm{Gal}(K)$ for the absolute Galois group $\mathrm{Gal}(\bar{K}/K) := \mathrm{Aut}(\bar{K}/K)$ of K. If X is an abelian variety of positive dimension over $\bar{K}$, then $\mathrm{End}(X)$ stands for the ring of all its $\bar{K}$-endomorphisms and $\mathrm{End}^0(X)$ for the corresponding $\mathbb{Q}$-algebra $\mathrm{End}(X) \otimes \mathbb{Q}$. We write 1_X for the identity endomorphism of X. If X is defined over K then we write $\mathrm{End}_K(X)$ for the ring of all K-endomorphisms of X. We have

$$\mathbb{Z} = \mathbb{Z} \cdot 1_X \subset \mathrm{End}_K(X) \subset \mathrm{End}(X),$$

$$\mathbb{Q} = \mathbb{Q} \cdot 1_X \subset \mathrm{End}_K(X) \otimes \mathbb{Q} =: \mathrm{End}^0_K X \subset \mathrm{End}^0(X).$$

Let $\mathbb{C}$ be the field of complex numbers, q a positive integer, $\zeta_q \in \mathbb{C}$ a primitive qth root of unity, $\mathbb{Q}(\zeta_q) \subset \mathbb{C}$ the qth cyclotomic field, and $\mathbb{Z}[\zeta_q]$ the ring of integers in $\mathbb{Q}(\zeta_q)$.

The aim of this paper is to construct in prime characteristic p explicit examples of abelian varieties X with $\mathrm{End}(X) \cong \mathbb{Z}[\zeta_\ell]$ where ℓ is a prime different from p. In characteristic 0 such examples are provided by the Jacobians $J(C_{f,\ell})$ of superelliptic curves $C_{f,\ell} : y^\ell = f(x)$ where $f(x) \in K[x]$ is an irreducible polynomial with "large" Galois group [**18, 21, 22**]. Namely, if $\mathrm{char}(K) = 0$, $\deg(f) \ge 5$ and the Galois group $\mathrm{Gal}(f)$ of $f(x)$ over K is either the full symmetric group $\mathbf{S}_n$ or the alternating group $\mathbf{A}_n$, then $\mathrm{End}(J(C_{f,\ell}) \cong \mathbb{Z}[\zeta_\ell]$. In the present paper we prove (Theorem 2.5)

2020 *Mathematics Subject Classification.* Primary 14H40, 14K05, 11G30, 11G10.
Key words and phrases. Ordinary abelian varieties, superelliptic Jacobians, endomorphisms of abelian varieties.
The author was partially supported by Simons Foundation Collaboration grant #585711. Part of this work was done during the author's stay at the Institute Henri Poincaré (June - July 2019) and the Weizmann Institute of Science (December 2019 - January 2020), whose hospitality and support are gratefully acknowledged.

that this assertion about the endomorphism rings remains true in characteristic p under additional assumptions that ℓ is odd, $J(C_{f,\ell})$ is an *ordinary* abelian variety, and $(\ell, n) \neq (5,5)$. (The hyperelliptic case $\ell = 2$ was treated earlier in [**23**].) We provide explicit examples of such polynomials (Example 2.11). We also discuss the case of the Jacobians of superelliptic curves $y^q = f(x)$ where q is a power of ℓ (Theorem 2.9).

The paper is organized as follows. Section 2 contains the necessary background from the theory of finite permutation groups, their natural (permutational) linear representations in characteristic ℓ, superelliptic curves and their Jacobians.

In Section 3 we outline the plan of the proof of our main results (Theorems 2.5 and 2.9). This section also contains certain auxiliary results (Theorems 3.1 and 3.2) about the endomorphism rings of abelian varieties that do not have to be superelliptic Jacobians; these results may be of certain independent interest. In the same section we deduce Theorem 2.5 from Theorems 3.1 and 3.2. We prove Theorems 3.1 in Section 6. Theorem 3.2 is proven in Section 7. In order to prove Theorem 3.1, we need to use an information about endomorphism rings of abelian varieties and their torsion points that is discussed in Section 4. In order to deduce Theorem 3.2 from Theorem 3.1, we use properties of abelian schemes (Section 5), including canonical liftings of ordinary abelian varieties to characteristic 0 (Remark 5.5). Theorem 2.9 is proven in Section 8.

2. Definitions and statements

2.1. **Polynomials, Galois groups and Permutational Representations.** Let ℓ be a prime such that $\mathrm{char}(K) \neq \ell$. Let $f(x) \in K[x]$ be a separable polynomial of degree $n \geq 3$, $\mathfrak{R}_f \subset \bar{K}$ the (n-element) set of roots of f and $K(\mathfrak{R}_f) \subset \bar{K}$ the splitting field of f. We write $\mathrm{Gal}(f) = \mathrm{Gal}(f/K)$ for the Galois group $\mathrm{Gal}(K(\mathfrak{R}_f)/K)$ of f; it permutes the roots of f and may be viewed as a certain permutation group of $\mathfrak{R}_f$, i.e., as a subgroup of the group $\mathrm{Perm}(\mathfrak{R}_f) \cong \mathbf{S}_n$ of all permutations of $\mathfrak{R}_f$. We write $\mathrm{Alt}(\mathfrak{R}_f) \cong \mathbf{A}_n$ for the only subgroup of index 2 in $\mathrm{Perm}(\mathfrak{R}_f)$ that consists of all even permutations of $\mathfrak{R}_f$. Slightly abusing notation, we say that $\mathrm{Gal}(f)$ is the full symmetric group $\mathbf{S}_n$ (resp. the alternating group $\mathbf{A}_n$) if $\mathrm{Gal}(f) = \mathrm{Perm}(\mathfrak{R}_f)$ (resp. $\mathrm{Alt}(\mathfrak{R}_f)$).

Let us consider the standard faithful permutational representation of the group of permutation $\mathrm{Gal}(f) \subset \mathrm{Perm}(\mathfrak{R}_f)$ in the n-dimensional $\mathbb{F}_\ell$-vector space

$$\mathbb{F}_\ell^{\mathfrak{R}_f} = \{\phi : \mathfrak{R}_f \to \mathbb{F}_\ell\}$$

of $\mathbb{F}_\ell$-valued functions on $\mathfrak{R}_f$ and its $(n-1)$-dimensional subrepresesentation in the subspace

$$\left(\mathbb{F}_\ell^{\mathfrak{R}_f}\right)^0 := \{\phi : \mathfrak{R}_f \to \mathbb{F}_\ell \mid \sum_{\alpha \in \mathfrak{R}_f} \phi(\alpha) = 0\} \tag{1}$$

(see [**4**]). The natural surjection

$$\mathrm{Gal}(K) \twoheadrightarrow \mathrm{Gal}(K(\mathfrak{R}_f)/K) = \mathrm{Gal}(f)$$

provides $\mathbb{F}_\ell^{\mathfrak{R}_f}$ with the natural structure of $\mathrm{Gal}(K)$-module and $\left(\mathbb{F}_\ell^{\mathfrak{R}_f}\right)^0$ becomes its $\mathrm{Gal}(K)$-submodule.

2.2. **Superelliptic curves and Jacobians.** Let $C_{f,\ell}$ be a smooth projective model of the smooth affine K-curve $y^\ell = f(x)$; its genus g is equal to $(n-1)(\ell-1)/2$ if ℓ does *not* divide n and to $(n-2)(\ell-1)/2$ if ℓ divides n. The Jacobian $J(C_{f,\ell})$ of $C_{f,\ell}$ is a g-dimensional abelian variety that is defined over K.

Suppose that K contains a primitive ℓth root of unity, say ζ. The map $(x, y) \mapsto (x, \zeta y)$ gives rise to a non-trivial birational K-automorphism $\delta_\ell : C_{f,\ell} \to C_{f,\ell}$ of period ℓ. By Albanese functoriality, δ_ℓ induces an automorphism of $J(C_{f,\ell})$ which we still denote by δ_ℓ. It is known [**8**, p. 149], [**10**, p. 458]) (see also [**24, 26**]) that δ_ℓ satisfies the ℓth cyclotomic equation, i.e., ${\delta_\ell}^{\ell-1} + \cdots + \delta_\ell + 1 = 0$, in $\mathrm{End}(J(C_{f,\ell}))$. This gives rise to the ring embeddings

$$\mathbb{Z}[\zeta_\ell] \hookrightarrow \mathrm{End}_K(J(C_{f,\ell})) \subset \mathrm{End}(J(C_{f,\ell})), \ \zeta_\ell \mapsto \delta_\ell,$$

$$\mathbb{Q}(\zeta_\ell) \hookrightarrow \mathrm{End}^0_K(J(C_{f,\ell})) \subset \mathrm{End}^0(J(C_{f,\ell})), \ \zeta_\ell \mapsto \delta_\ell,$$

which send 1 to the identity automorphism of $J(C_{f,\ell})$. In particular, if $\mathbb{Q}[\delta_\ell]$ is the $\mathbb{Q}$-subalgebra of $\mathrm{End}^0(J(C_{f,\ell}))$ generated by δ_ℓ then the latter embedding establishes a canonical isomorphism of $\mathbb{Q}$-algebras

$$\mathbb{Q}(\zeta_\ell) \cong \mathbb{Q}[\delta_\ell]$$

that sends ζ_ℓ to δ_ℓ.

The set of fixed points of δ_ℓ

$$J(C_{f,\ell})^{\delta_\ell} := \{z \in J(C_{f,\ell})(\bar{K}) \mid \delta_\ell(z) = z\}$$

is a $\mathrm{Gal}(K)$-submodule of $J(C_{f,\ell})(\bar{K})$.

LEMMA 2.3 (See [**8, 10**]). *If ℓ does not divide n then the the $\mathrm{Gal}(K)$-module $J(C_{f,\ell})^{\delta_\ell}$ is isomorphic to $\left(\mathbb{F}_\ell^{\mathfrak{R}_f}\right)^0$ defined by Eq. (1).*

2.4. **Differentials of the first kind.** Let $\Omega^1(J(C_{f,\ell}))$ be the g-dimensional $\bar{K}$-vector space of (invariant) differentials of the first kind on $J(C_{f,\ell})$ and

$${\delta_\ell}^* : \Omega^1(J(C_{f,\ell})) \to \Omega^1(J(C_{f,\ell}))$$

the linear operator induced by functoriality by δ_ℓ. Clearly, $(\delta_\ell^*)^\ell$ is the identity map. Since $\mathrm{char}(\bar{K}) \ne \ell$, the linear operator ${\delta_\ell}^*$ is diagonalizable. It is known [**24**, Remarks 4.5, 4.6 and 4.7 on pp. 352–353] that the spectrum of δ_ℓ^* consists of (primitive) ℓth roots of unity ζ^{-i} where $i < \ell$ is a positive integer such that $[ni/\ell] > 0$ and the multiplicity of ζ^{-i} equals $[ni/\ell]$. In particular, 1 is *not* an eigenvalue of ${\delta_\ell}^*$.

The structure of the endomorphism algebra of superelliptic Jacobians was studied in [**15, 16, 20–22, 24, 26–29**]. In particular, the author proved in [**18, 22**] that if $\mathrm{char}(K) = 0$, $n \ge 5$, and $\mathrm{Gal}(f)$ is either $\mathbf{S}_n$ or $\mathbf{A}_n$ then $\mathrm{End}(J(C_{f,\ell})) = \mathbb{Z}[\zeta_\ell]$.

The aim of this paper is to extend these results to the case of prime characteristic under an additional assumption that $J(C_{f,\ell})$ is an *ordinary* abelian variety. Our main result is the following assertion.

THEOREM 2.5. *Suppose that K is a field of prime characteristic p. Let $n \ge 5$ be an integer and ℓ be an odd prime that does not coincide with p. Suppose that $\mathrm{Gal}(f)$ is either $\mathbf{S}_n$ or $\mathbf{A}_n$. Assume also that $J(C_{f,\ell})$ is an ordinary abelian variety.*

If $(\ell, n) \ne (5, 5)$ then

$$\mathrm{End}^0(J(C_{f,\ell})) = \mathbb{Q}[\delta_\ell] \cong \mathbb{Q}(\zeta_\ell) \ \ \textit{and} \ \ \mathrm{End}(J(C_{f,\ell})) = \mathbb{Z}[\delta_\ell] \cong \mathbb{Z}[\zeta_\ell].$$

2.6. Automorphisms and Endomorphisms of Superelliptic Jacobians. Now let us assume that $q = \ell^r$ is a power of a prime $\ell \neq \mathrm{char}(K)$, and consider the smooth projective model $C_{f,q}$ of the affine smooth curve $y^q = f(x)$. If K contains a primitive qth root of unity, say ζ, then $C_{f,q}$ admits a periodic K-automorphism

$$\delta_q : C_{f,q} \to C_{f,q}, \ (x,y) \mapsto (x, \zeta y). \tag{2}$$

It induces, by Albanese functoriality, the periodic automorphism of $J(C_{f,q})$, which we continue denote by

$$\delta_q \in \mathrm{Aut}_K(J(C_{f,q})).$$

It is known [**24**, Lemma 4.8 on p.354] that

$$\delta_q^q = 1_{J(C_{f,q})}, \ \sum_{i=0}^{\ell-1} \delta_q^{i\ell^{r-1}} = 0. \tag{3}$$

In addition, the $\mathbb{Q}$-subalgebra $\mathbb{Q}[\delta_q]$ of $\mathrm{End}_K^0(J(C_{f,q}))$ generated by δ_q, is canonically isomorphic to the direct sum $\oplus_{j=1}^r \mathbb{Q}(\zeta_{\ell^j})$ of cyclotomic fields.

The following assertion was proven by the author in [**24**, Th. 1.1 on p. 340].

THEOREM 2.7. *Let K be a subfield of $\mathbb{C}$. Suppose that $n \geq 5$ is an integer, ℓ is a prime and r is a positive integer. Let us assume that either $\ell \nmid n$ or $q := \ell^r$ divides n. Suppose that* $\mathrm{Gal}(f)$ *is either* $\mathbf{S}_n$ *or* $\mathbf{A}_n$. *If $(q,n) \neq (5,5)$ then the semisimple $\mathbb{Q}$-algebra* $\mathrm{End}^0(J(C_{f,q}))$ *coincides with* $\mathbb{Q}[\delta_q]$ *and is isomorphic to the direct sum $\oplus_{j=1}^r \mathbb{Q}(\zeta_{\ell^j})$ of cyclotomic fields. In particular,* $\mathrm{End}^0(J(C_{f,q}))$ *is commutative.*

REMARK 2.8. J. Xue [**15**] extended the result of Theorem 2.7 to the remaining case when $\ell \mid n$ but q does *not* divide n.

The second main result of this paper is an analogue of Theorem 2.7 in finite characteristic for ordinary Jacobians.

THEOREM 2.9. *Suppose that K is a field of prime characteristic p. Let $n \geq 5$ be an integer, ℓ is an odd prime that does not coincide with p, and r is a positive integer. Let us assume that either $\ell \nmid n$ or $q := \ell^r$ divides n. Suppose that* $\mathrm{Gal}(f)$ *is either* $\mathbf{S}_n$ *or* $\mathbf{A}_n$. *Assume also that the Jacobian $J(C_{f,q})$ of $C_{f,q}$ is an ordinary abelian variety.*

If $(q,n) \neq (5,5)$ then the semisimple $\mathbb{Q}$-algebra $\mathrm{End}^0(J(C_{f,\ell}))$ *coincides with* $\mathbb{Q}[\delta_q]$ *and is isomorphic to the direct sum $\oplus_{j=1}^r \mathbb{Q}(\zeta_{\ell^j})$ of cyclotomic fields.*

REMARKS 2.10.
- (i) Replacing K by its *perfectization* $\sqrt[p^\infty]{K}$, we may and will assume in the course of the proof of Theorems 2.5 and 2.9 that K is a *perfect* field.
- (ii) Replacing K by a suitable finite abelian extension, we may and will assume in the course of the proof of Theorems 2.5 and 2.9 that $\mathrm{Gal}(f) = \mathbf{A}_n$ and K contains a *primitive* ℓth (respectively qth) root of unity.
- (iii) Using an elementary substitution (see [**24**, Remark 4.3]), we may and will assume in the course of the proof of Theorems 2.5 and 2.9 that ℓ does *not* divide n.

EXAMPLE 2.11. Let us choose positive integers $q \geq 2$ and d, and assume that either $q \geq 7$ or $d \geq 3$. Pick an odd prime p that is congruent to 1 modulo $q(dq-1)$ and put $n = dq$; clearly, $n \geq 6$. Let $K = \bar{\mathbb{F}}_p(t)$ be the field of rational functions in one variable over an algebraic closure $\bar{\mathbb{F}}_p$ of the (finite) prime field $\mathbb{F}_p$ of characteristic p. Since $p > n$, the product $n(n-1)$ is *not* divisible by p. This implies that the polynomial $f(x) = x^n - x - t \in K[x]$ is a *Morse function* in a sense of [**12**, p. 39]; in particular, $\mathrm{Gal}(f) = \mathbf{S}_n$ [**12**, Th. 4.4.5]. Then the Jacobian J of the superelliptic K-curve $C : y^q = x^n - x - t$ is an *ordinary* abelian variety over K. In order to prove that, it suffices to check that the Jacobian J_0 of its specialization

$$C_0 : y^q = x^n - x = x^{dq} - x$$

(at $t = 0$) is an ordinary abelian variety over $\bar{\mathbb{F}}_p$. Dividing the equation of C_0 by x^n and using a substitution

$$u = \frac{y}{x^d}, \ v = \frac{1}{x},$$

we obtain that C_0 is birationally isomorphic to the curve

$$C' : u^q = 1 - v^{dq-1}.$$

Clearly, C' is covered by Fermat curve

$$w^m = 1 - z^m$$

with $m := q(dq-1)$. Since $p-1$ is divisible by $q(dq-1) = m$, the Jacobian of the Fermat curve is ordinary [**17**, Prop. 4.1 and Th. 4.2]. Since J_0 is isomorphic to a quotient of the Jacobian of the Fermat curve, it is also ordinary. This implies that J is ordinary. Now if $q = \ell$ is an odd prime then $n = d\ell \ge 2\ell \ge 6$ and the equality $\mathrm{End}(J(C_{f,\ell})) = \mathbb{Z}[\zeta_\ell]$ follows from Theorem 2.5. (Compare with [**25**] where the case of $\ell = 3, n = 9$ is discussed.)

Similarly, if $r \ge 2$ is an integer and $q = \ell^r$ is a power of an odd prime ℓ then it follows from Theorem 2.9 that

$$\mathrm{End}^0(J(C_{f,\ell})) \cong \oplus_{j=1}^r \mathbb{Q}(\zeta_{\ell^j}).$$

3. The plan of the proof

Our proof is based on *Serre-Tate canonical lifting of ordinary abelian varieties* [**3**] and the following assertions.

THEOREM 3.1. *Suppose that ℓ is a prime, r a positive integer, $n \ge 5$ a positive integer that is not divisible by ℓ, $q = \ell^r$, and $\mathbf{K}$ be a field of characteristic zero that contains a primitive qth root of unity, say ζ. Let $\mathbb{Y}$ be a positive-dimensional abelian variety over $\mathbf{K}$ that admits an endomorphism (actually an automorphism) $\delta_{\mathbf{K}} \in \mathrm{End}_{\mathbf{K}}(\mathbb{Y})$ satisfying the qth cyclotomic equation*

$$\Phi_q(\delta_{\mathbf{K}}) = \sum_{i=0}^{\ell-1} \delta_{\mathbf{K}}^{i\ell^{r-1}} = 0 \tag{4}$$

in $\mathrm{End}_{\mathbf{K}}(\mathbb{Y})$. Let us consider the $\mathbf{K}$-linear map

$$\delta_{\mathbf{K}}^* : \Omega^1(\mathbb{Y}) \to \Omega^1(\mathbb{Y})$$

induced by $\delta_{\mathbf{K}}$.

Suppose that $\mathbb{Y}$, $\delta_{\mathbf{K}}$ and $\delta_{\mathbf{K}}^*$ enjoy the following properties.

(1) *All eigenvalues of $\delta_{\mathbf{K}}^*$ are primitive qth roots of unity and their multiplicities are as follows. For each positive integer i with*

$$1 \le i < q, \ (i, \ell) = 1,$$

the multiplicity of ζ^{-i} *as an eigenvalue of* $\delta^*_{\mathbf{K}}$ *is* $[ni/q]$. *In particular,* ζ^{-i} *is an eigenvalue of* $\delta^*_{\mathbf{K}}$ *if and only if* $[ni/q] > 0$.

(2) *The* $\mathrm{Gal}(\mathbf{K})$*-submodule of fixed points of* $\delta_{\mathbf{K}}$

$$\mathbb{Y}^{\delta_{\mathbf{K}}} := \{y \in \mathbb{Y}(\bar{\mathbf{K}}) \mid \delta_{\mathbf{K}}(y) = y\} \subset \mathbb{Y}(\bar{\mathbf{K}})$$

is a $(n-1)$*-dimensional vector space over the prime finite field* $\mathbb{F}_\ell$, *the image of* $\mathrm{Gal}(\mathbf{K})$ *in* $\mathrm{Aut}_{\mathbb{F}_\ell}(\mathbb{Y}^{\delta_{\mathbf{K}}})$ *contains a subgroup isomorphic to* $\mathbf{A}_n$, *and the corresponding* $\mathbf{A}_n$*-module* $\mathbb{Y}^\delta$ *is isomorphic to the natural representation of* $\mathbf{A}_n$ *in*

$$(\mathbb{F}_\ell^n)^0 := \{(a_1, \dots, a_n) \in \mathbb{F}_\ell^n \mid \sum_{i=1}^n a_i = 0\}.$$

Then the following assertions are valid.

(a) $\delta^q_{\mathbf{K}} = 1_{\mathbb{Y}}$.

(b) *The ring homomorphism*

$$\mathrm{i}_{\mathbb{Y}} : \mathbb{Z}[\zeta_q] \to \mathrm{End}_{\mathbf{K}}(\mathbb{Y}) \subset \mathrm{End}(\mathbb{Y}) \tag{5}$$

that sends 1 *to* $1_{\mathbb{Y}}$ *and* ζ_q *to* $\delta_{\mathbf{K}}$ *is a ring embedding.*

(c) $\varphi(q) := [\mathbb{Q}(\zeta_q) : \mathbb{Q}]$ *divides* $2\dim(\mathbb{Y})$.

(d) *The spectrum of the linear operator* $\delta^*_{\mathbf{K}}$ *consists of more that* $\varphi(q)/2$ *distinct eigenvalues.*

(e)
$$\mathrm{i}_{\mathbb{Y}}(\mathbb{Z}[\zeta_q]) = \mathrm{End}_{\mathbf{K}}(\mathbb{Y}) = \mathrm{End}(\mathbb{Y}),$$

i.e., $\mathrm{End}(\mathbb{Y})$ *coincides with its own subring generated by* $\delta_{\mathbf{K}}$ *and the ring homomorphism* $i_{\mathbb{Y}}$ *is a ring isomorphism. In particular,* $\mathbb{Y}$ *is absolutely simple and* $\mathrm{End}^0(\mathbb{Y}) \cong \mathbb{Q}(\zeta_q)$.

We use Theorem 3.1 in order to prove the following analogue of Theorem 3.1 in finite characteristic.

THEOREM 3.2. *Suppose that* ℓ *is a prime,* r *a positive integer, and* $n \geq 5$ *is a positive integer that is not divisible by* ℓ. *Let us put* $q := \ell^r$. *Let* K *be a perfect field of prime characteristic* $p \neq \ell$ *that contains a primitive* q*th root of unity, say* $\bar{\zeta}$. *Let* Y *be a positive-dimensional ordinary abelian variety over* K *that admits an endomorphism (actually an automorphism)* $\delta \in \mathrm{End}_K(Y)$ *satisfying the* q*th cyclotomic equation*

$$\Phi_q(\delta) = \sum_{i=0}^{\ell-1} \delta^{i\ell^{r-1}} = 0 \tag{6}$$

in $\mathrm{End}_K(Y)$. *Let us consider the* K*-linear map*

$$\delta^* : \Omega^1(Y) \to \Omega^1(Y)$$

induced by δ.

Suppose that Y, δ *and* δ^* *enjoy the following properties.*

(1) *All eigenvalues of* δ^* *are primitive* q*th roots of unity and their multiplicities are as follows. If* i *is a positive integer with*

$$1 \leq i < q, \ (i, \ell) = 1,$$

then the multiplicity of $\bar{\zeta}^{-i}$ *as an eigenvalue of* δ^* *is* $[ni/q]$. *In particular,* $\bar{\zeta}^{-i}$ *is an eigenvalue of* δ^* *if and only if* $[ni/q] > 0$.

(2) *The* $\mathrm{Gal}(K)$*-submodule of fixed points of* δ

$$Y^{\delta} := \{y \in Y(\bar{\mathbf{K}}) \mid \delta(y) = y\} \subset Y(\bar{\mathbf{K}})$$

is a $(n-1)$*-dimensional vector space over the prime finite field* $\mathbb{F}_\ell$*, the image of* $\mathrm{Gal}(K)$ *in* $\mathrm{Aut}_{\mathbb{F}_\ell}(Y^{\delta})$ *contains a subgroup isomorphic to* $\mathbf{A}_n$*, and the corresponding* $\mathbf{A}_n$*-module* Y^{δ} *is isomorphic to the natural representation of* $\mathbf{A}_n$ *in*

$$(\mathbb{F}_\ell^n)^0 = \{(a_1, \dots, a_n) \in \mathbb{F}_\ell^n \mid \sum_{i=1}^{n} a_i = 0\}.$$

Then the following assertions are valid.

(a) $\delta_K^q = 1_Y$.

(b) *The ring homomorphism*

$$\mathrm{i}_Y : \mathbb{Z}[\zeta_q] \to \mathrm{End}_K(Y) \subset \mathrm{End}(Y) \tag{7}$$

that sends 1 *to* 1_Y *and* ζ_q *to* δ *is a ring embedding.*

(c) $\varphi(q) = [\mathbb{Q}(\zeta_q) : \mathbb{Q}]$ *divides* $2\dim(Y)$.

(d) *The spectrum of the linear operator* δ^* *consists of more that* $\varphi(q)/2$ *distinct eigenvalues.*

(e)

$$\mathrm{i}_Y(\mathbb{Z}[\zeta_q]) = \mathrm{End}_K(Y) = \mathrm{End}(Y),$$

i.e., $\mathrm{End}(Y)$ *coincides with its own subring generated by* δ *and the ring homomorphism* i_Y *is a ring isomorphism. In particular,* Y *is absolutely simple and* $\mathrm{End}^0(Y) \cong \mathbb{Q}(\zeta_q)$.

Proof of Theorem 2.5 (modulo Theorem 3.2). In light of Remarks 2.10, we may and will assume that K is a perfect field containing a primitive ℓth root of unity, $\mathrm{Gal}(f) = \mathbf{A}_n$ and ℓ does *not* divide n. Let us put

$$q = \ell,\ Y = J(C_{f,\ell}),\ \delta = \delta_\ell.$$

Let us choose an order on the n-element set $\mathfrak{R}_f$ that defines a bijection between $\mathfrak{R}_f$ and $\{1, 2, \dots n\}$. This gives us a group isomorphism $\mathrm{Gal}(f) \cong \mathbf{A}_n$ and an isomorphism of the corresponding $\mathbf{A}_n$-modules $\left(\mathbb{F}_\ell^{\mathfrak{R}_f}\right)^0$ and $(\mathbb{F}_\ell^n)^0$. Lemma 2.3 and the assertions in the beginning of Subsection 2.4 about eigenvalues of δ_ℓ^* imply that all the conditions of Theorem 3.2 are fulfilled. Now the desired result follows from Theorem 3.2. □

4. Endomorphism algebras of abelian varieties

4.1. **Endomorphisms and torsion points.** Let X be an abelian variety of positive dimension over an arbitrary field K. If n is a positive integer that is not divisible by $\mathrm{char}(K)$ then $X[n]$ stands for the kernel of multiplication by n in $X(\bar{K})$. It is well-known [**5**] that $X[n]$ is a free $\mathbb{Z}/n\mathbb{Z}$-module of rank $2\dim(X)$. In particular, if $n = \ell$ is a prime then $X[\ell]$ is a $2\dim(X)$-dimensional $\mathbb{F}_\ell$-vector space.

Since X is defined over K, $X[n]$ is a Galois submodule in $X(\bar{K})$ and all points of $X[n]$ are defined over a finite separable extension of K. We write $\rho_{n,X,K} : \mathrm{Gal}(K) \to \mathrm{Aut}_{\mathbb{Z}/n\mathbb{Z}}(X[n])$ for the corresponding homomorphism defining the structure of the Galois module on $X[n]$,

$$G_{n,X,K} \subset \mathrm{Aut}_{\mathbb{Z}/n\mathbb{Z}}(X[n])$$

for its image $\rho_{n,X,K}(\mathrm{Gal}(K))$ and $K(X[n])$ for the field of definition of all points of $X[n]$. Clearly, $K(X[n])$ is a finite Galois extension of K with Galois group $\mathrm{Gal}(K(X[n])/K) = G_{n,X,K}$. If $n = \ell$ then we get a natural faithful linear representation

$$G_{\ell,X,K} \subset \mathrm{Aut}_{\mathbb{F}_\ell}(X[\ell])$$

of $G_{\ell,X,K}$ in the $\mathbb{F}_\ell$-vector space $X[\ell]$.

Since X is defined over K, one may associate with every $u \in \mathrm{End}(X)$ and $\sigma \in \mathrm{Gal}(K)$ an endomorphism ${}^\sigma u \in \mathrm{End}(X)$ such that ${}^\sigma u(x) = \sigma u(\sigma^{-1}x)$ for all $x \in X(\bar{K})$. We get the group homomorphism

(8)
$$\kappa_X : \mathrm{Gal}(K) \to \mathrm{Aut}(\mathrm{End}(X)); \quad \kappa_X(\sigma)(u) = {}^\sigma u \quad \forall \sigma \in \mathrm{Gal}(K), u \in \mathrm{End}(X).$$

It is well-known that $\mathrm{End}_K(X)$ coincides with the subring of $\mathrm{Gal}(K)$-invariants in $\mathrm{End}(X)$, i.e., $\mathrm{End}_K(X) = \{u \in \mathrm{End}(X) \mid {}^\sigma u = u \quad \forall \sigma \in \mathrm{Gal}(K)\}$. It is also well-known that $\mathrm{End}(X)$ (viewed as a group with respect to addition) is a free commutative group of finite rank [**5**]. In addition, $\mathrm{End}_K(X)$ is its *pure* subgroup [**11**, Sect. 4, p. 501], i.e., the quotient $\mathrm{End}(X)/\mathrm{End}_K(X)$ is also a free commutative group of finite rank.

4.2. **Abelian varieties with multiplications.** Let E be a number field. Let (X, i) be a pair consisting of an abelian variety X of positive dimension over $\bar{K}$ and an embedding $i : E \hookrightarrow \mathrm{End}^0(X)$. We also assume that $1 \in E$ goes to 1_X.

REMARK 4.3. It is well known [**14**, Prop. 2 on p. 36]) that the degree $[E : \mathbb{Q}]$ divides $2\dim(X)$, i.e.

$$d_{X,E} := \frac{2\dim(X)}{[E : \mathbb{Q}]}$$

is a positive integer.

Let us denote by $\mathrm{End}^0(X, i)$ the centralizer of $i(E)$ in $\mathrm{End}^0(X)$. Clearly, $i(E)$ lies in the center of the finite-dimensional $\mathbb{Q}$-algebra $\mathrm{End}^0(X, i)$. It follows that $\mathrm{End}^0(X, i)$ carries a natural structure of finite-dimensional E-algebra.

Let $\mathcal{O}$ be the ring of integers in E. If $\mathfrak{a}$ is a non-zero ideal in $\mathcal{O}$ then the quotient $\mathcal{O}/\mathfrak{a}$ is a finite commutative ring. Let λ be a maximal ideal in $\mathcal{O}$. We write $k(\lambda)$ for the corresponding (finite) residue field $\mathcal{O}/\lambda$ and ℓ for $\mathrm{char}(k(\lambda))$. We have

$$\mathcal{O} \supset \lambda \supset \ell \cdot \mathcal{O}.$$

REMARK 4.4. (See [**27**, Remark 3.3].) Let us assume that λ is the only maximal ideal of $\mathcal{O}$ dividing ℓ, i.e., $\ell \cdot \mathcal{O} = \lambda^b$ where the positive integer b satisfies $[E : \mathbb{Q}] = b \cdot [k(\lambda) : \mathbb{F}_\ell]$.

(i) We have $\mathcal{O} \otimes \mathbb{Z}_\ell = \mathcal{O}_\lambda$ where $\mathcal{O}_\lambda$ is the completion of $\mathcal{O}$ with respect to λ-adic topology. The ring $\mathcal{O}_\lambda$ is a local principal ideal domain, its only maximal ideal is $\lambda\mathcal{O}_\lambda$, and

$$k(\lambda) = \mathcal{O}/\lambda = \mathcal{O}_\lambda/\lambda\mathcal{O}_\lambda, \ \ell \cdot \mathcal{O}_\lambda = (\lambda\mathcal{O}_\lambda)^b.$$

(ii) Let us choose an element $c \in \lambda$ that does not lie in λ^2. Then

$$\lambda = \ell \cdot \mathcal{O} + c \cdot \mathcal{O}.$$

This implies that for all positive integers $j \le b$

$$\lambda^j = \ell \cdot \mathcal{O} + c^j \cdot \mathcal{O}.$$

This implies that

$$(\lambda\mathcal{O}_\lambda)^j = c^j \cdot \mathcal{O}_\lambda.$$

In particular,

$$\ell \cdot \mathcal{O}_\lambda = c^b \cdot \mathcal{O}_\lambda.$$

It follows that

$$c^{-j}\ell \cdot \mathcal{O}_\lambda = c^{b-j} \cdot \mathcal{O}_\lambda.$$

(iii) Notice that $E_\lambda = E \otimes_{\mathbb{Q}} \mathbb{Q}_\ell = \mathcal{O} \otimes \mathbb{Q}_\ell = \mathcal{O}_\lambda \otimes_{\mathbb{Z}_\ell} \mathbb{Q}_\ell$ is the field coinciding with the completion of E with respect to λ-adic topology.

Suppose that X is defined over K and $i(\mathcal{O}) \subset \mathrm{End}_K(X)$. Then we may view elements of $\mathcal{O}$ as K-endomorphisms of X. We write $\mathrm{End}(X,i)$ for the centralizer of $i(\mathcal{O})$ in $\mathrm{End}(X)$ and $\mathrm{End}_K(X,i)$ for the centralizer of $i(\mathcal{O})$ in $\mathrm{End}_K(X)$. Clearly,

$$\mathrm{End}_K(X,i) = \mathrm{End}(X,i) \bigcap \mathrm{End}_K(X), \tag{9}$$

and $\mathrm{End}(X,i)$ and $\mathrm{End}_K(X,i)$ are pure subgroups of $\mathrm{End}(X)$ and $\mathrm{End}_K(X)$ respectively, i.e., the quotients $\mathrm{End}(X)/\mathrm{End}(X,i)$ and $\mathrm{End}_K(X)/\mathrm{End}_K(X,i)$ are torsion-free. Recall that $\mathrm{End}(X)/\mathrm{End}_K(X)$ is also torsion-free (see the very end of Subsection 4.1). It follows from (9) that the quotient $\mathrm{End}(X,i)/\mathrm{End}_K(X,i)$ embeds into $\mathrm{End}(X)/\mathrm{End}_K(X)$ and therefore is also torsion-free, i.e, $\mathrm{End}_K(X,i)$ is a pure subgroup in $\mathrm{End}(X,i)$.

We have

$$i(\mathcal{O}) \subset \mathrm{End}_K(X,i) \subset \mathrm{End}(X,i) \subset \mathrm{End}(X).$$

It is also clear that

$$\mathrm{End}(X,i) = \mathrm{End}^0(X,i) \bigcap \mathrm{End}(X),$$

$$\mathrm{End}^0(X,i) = \mathrm{End}(X,i) \otimes \mathbb{Q} \subset \mathrm{End}(X) \otimes \mathbb{Q} = \mathrm{End}^0(X).$$

Clearly, $\mathrm{End}(X,i)$ carries a natural structure of finitely generated torsion-free $\mathcal{O}$-module. Since $\mathcal{O}$ is a Dedekind ring, it follows that there exist non-zero ideals $\mathfrak{b}_1, \dots, \mathfrak{b}_t$ in $\mathcal{O}$ such that

$$\mathbb{Z} \cdot 1_X \subset \mathrm{End}(X,i) \cong \mathfrak{b}_1 \oplus \cdots \oplus \mathfrak{b}_t$$

(as $\mathcal{O}$-modules).

We have $\kappa_X(\sigma)(\mathrm{End}(X,i)) = \mathrm{End}(X,i)$ for all $\sigma \in \mathrm{Gal}(K)$, and

$$\mathrm{End}_K(X,i) = \{u \in \mathrm{End}(X,i) \mid {}^\sigma u = u \quad \forall \sigma \in \mathrm{Gal}(K)\},$$

where the group homomorphism $\kappa_X : \mathrm{Gal}(K) \to \mathrm{Aut}(\mathrm{End}(X))$ is defined in (8).

Let us assume that $\mathrm{char}(K)$ does not divide the order of $\mathcal{O}/\mathfrak{a}$ and put

$$X[\mathfrak{a}] := \{x \in X(\bar{K}) \mid i(e)x = 0 \quad \forall e \in \mathfrak{a}\}.$$

For example, assume that $\ell \neq \mathrm{char}(K)$. Then the order of $\mathcal{O}/\lambda = k(\lambda)$ is a power of ℓ and $X[\lambda] \subset X[\ell]$.

Clearly, $X[\mathfrak{a}]$ is a Galois submodule of $X(\bar{K})$. It is also clear that $X[\mathfrak{a}]$ carries a natural structure of $\mathcal{O}/\mathfrak{a}$-module. (It is known [**13**, Prop. 7.20] that this module is free.)

REMARK 4.5. Assume in addition that λ is the only maximal ideal of $\mathcal{O}$ dividing ℓ and pick

$$c \in \lambda \setminus \lambda^2 \subset \lambda \subset \mathcal{O}.$$

By Remark 4.4(ii), λ is generated by ℓ and c. Hence

$$X[\lambda] = \{x \in X_\ell \mid cx = 0\} \subset X[\ell]. \tag{10}$$

More generally, for all positive integers $j \le b$ the ideal λ^j is generated by ℓ and c^j, which implies that

$$X[\lambda^j] = \{x \in X_\ell \mid c^j x = 0\} \subset X[\ell]. \tag{11}$$

Obviously, every endomorphism from $\mathrm{End}(X, i)$ leaves invariant the subgroup $X[\mathfrak{a}] \subset X(\bar{K})$ and induces an endomorphism of the $\mathcal{O}/\mathfrak{a}$-module $X[\mathfrak{a}]$. This gives rise to a natural homomorphism

$$\mathrm{End}(X, i) \to \mathrm{End}_{\mathcal{O}/\mathfrak{a}}(X[\mathfrak{a}]),$$

whose kernel contains $\mathfrak{a}\cdot\mathrm{End}(X, i)$. Actually, the kernel coincides with $\mathfrak{a}\cdot\mathrm{End}(X, i)$, i. e., there is an embedding

$$\mathrm{End}(X, i) \otimes_{\mathcal{O}} \mathcal{O}/\mathfrak{a} \hookrightarrow \mathrm{End}_{\mathcal{O}/\mathfrak{a}}(X[\mathfrak{a}]). \tag{12}$$

See [**27**, pp. 699-700] for the proof.

Now we concentrate on the case of $\mathfrak{a} = \lambda$, assuming that

$$\ell \ne \mathrm{char}(K).$$

Then $X[\lambda]$ carries the natural structure of a $k(\lambda)$-vector space endowed with the structure of Galois module and (12) gives us the embedding

$$\mathrm{End}(X, i) \otimes_{\mathcal{O}} k(\lambda) \hookrightarrow \mathrm{End}_{k(\lambda)}(X[\lambda]). \tag{4}$$

Further we will identify $\mathrm{End}(X, i) \otimes_{\mathcal{O}} k(\lambda)$ with its image in $\mathrm{End}_{k(\lambda)}(X[\lambda])$. We write

$$\tilde{\rho}_{\lambda,X} : \mathrm{Gal}(K) \to \mathrm{Aut}_{k(\lambda)}(X[\lambda])$$

for the corresponding (continuous) homomorphism defining the Galois action on $X[\lambda]$. It is known [**13**, Prop. 7.20] (see also [**9**]) that

$$\dim_{k(\lambda)} X[\lambda] = \frac{2\dim(X)}{[E : \mathbb{Q}]} := d_{X,E}. \tag{13}$$

Let us put

$$\tilde{G}_{\lambda,X} = \tilde{G}_{\lambda,i,X} := \tilde{\rho}_{\lambda,X}(\mathrm{Gal}(K)) \subset \mathrm{Aut}_{k(\lambda)}(X[\lambda]).$$

Clearly, $\tilde{G}_{\lambda,X}$ coincides with the Galois group of the field extension $K(X[\lambda])/K$, where $K(X[\lambda])$ is the field of definition of all points in $X[\lambda]$.

It is also clear that the image of $\mathrm{End}_K(X, i) \otimes_{\mathcal{O}} k(\lambda)$ lies in the centralizer $\mathrm{End}_{\tilde{G}_{\lambda,i,X}}(X[\lambda])$ of $\tilde{G}_{\lambda,i,X}$ in $\mathrm{End}_{k(\lambda)}(X[\lambda])$.

LEMMA 4.6. *(See* [**27**, Lemma 3.8].*) Suppose that* $i(\mathcal{O}) \subset \mathrm{End}_K(X)$. *If* λ *is a maximal ideal in* $\mathcal{O}$ *containing a prime* ℓ *with* $\ell \ne \mathrm{char}(K)$, *and* $\mathrm{End}_{\tilde{G}_{\lambda,i,K}}(X[\lambda]) = k(\lambda)$ *then* $\mathrm{End}_K(X, i) = \mathcal{O}$.

5. Abelian schemes

Let $\mathbf{K}$ be a complete discrete valuation field, $O_{\mathbf{K}}$ its discrete valuation ring with maximal ideal $\mathfrak{m}$ and perfect residue field $\kappa := O_{\mathbf{K}}/\mathfrak{m}$. We write

$$O_{\mathbf{K}} \to O_{\mathbf{K}}/\mathfrak{m} = \kappa, \ a \mapsto \bar{a} = a + \mathfrak{m}$$

for the residue map. Suppose that q is a positive integer that is *not* divisible by $\mathrm{char}(\kappa)$ and such that κ contains a primitive qth root of unity. Then it follows from Hensel's Lemma that $O_{\mathbf{K}}$ contains a primitive qth root of unity, and the reduction map defines a canonical isomorphim

$$\mu_{q,\mathbf{K}} \cong \mu_{q,\kappa}, \ \gamma \mapsto \bar{\gamma}$$

between the order q cyclic groups of qth roots of unity

$$\mu_{q,\mathbf{K}} \subset O_{\mathbf{K}}^* \subset \mathbf{K}^*$$

and $\mu_{q,\kappa} \subset \kappa^*$.

Since $\mathbf{K}$ is complete, the absolute Galois group $\mathrm{Gal}(\mathbf{K}) = \mathrm{Aut}(\bar{\mathbf{K}}/\mathbf{K})$ coincides with the *decomposition group*, and the inertia subgroup $\mathcal{I}$ of $\mathrm{Gal}(\mathbf{K})$ is a closed normal subgroup of $\mathrm{Gal}(\mathbf{K})$. The natural extension of the reduction map to $\bar{\mathbf{K}}$ induces the surjective continuous group homomorphism [**11**, Sect. 1]

$$\mathrm{red}_{\mathrm{Gal}} : \mathrm{Gal}(\mathbf{K})/\mathcal{I} \to \mathrm{Gal}(\kappa), \tag{14}$$

which is actually an *isomorphism* of compact (profinite) groups.

We write S for $\mathrm{Spec}(O_{\mathbf{K}})$. It is well known that S consists of the *generic point* η that corresponds to $\{0\}$ and the closed point s that corresponds to $\mathfrak{m}$.

Let $f : \mathcal{X} \to S$ be an abelian scheme over S of positive relative dimension g. We write $X_{\mathbf{K}}$ for its generic fiber, which is a g-dimensional abelian variety over $\mathbf{K}$, and X_s for its closed fiber, which is a g-dimensional abelian variety over κ. By definition, $\mathcal{X}$ is a separated scheme; it is known [**2**, Remark 1.2 on p. 1] that $\mathcal{X}$ is a scheme of finite type over S and, therefore, is *noetherian*. In addition, $\mathcal{X}$ is a *Néron model* of $X_{\mathbf{K}}$ [**1**, Prop. 8 on p. 15]. In particular, the natural ring homomorphism

$$\mathrm{End}_S(\mathcal{X}) \to \mathrm{End}_{\mathbf{K}}(X_{\mathbf{K}}), \ \mathfrak{u} \mapsto \mathfrak{u}_{\mathbf{K}} \tag{15}$$

is a *isomorphism*. Notice that the natural ring homomorphism

$$\mathrm{End}_S(\mathcal{X}) \to \mathrm{End}_{\kappa}(X_s), \ \mathfrak{u} \mapsto \mathfrak{u}_s \tag{16}$$

is an *embedding*, thanks to Rigidity Lemma [**6**, Ch. 6, Sect. 1, Prop. 6.1 and Cor. 6.2].

As usual, we write $\mathcal{O}_{\mathcal{X}}$ for the structure sheaf of $\mathcal{X}$. Clearly, the ring $\Gamma(\mathcal{X}, \mathcal{O}_{\mathcal{X}})$ of global sections of $\mathcal{X}$ is a $O_{\mathbf{K}}$-algebra. Since f is proper, the $O_{\mathbf{K}}$-module $\Gamma(\mathcal{X}, \mathcal{O}_{\mathcal{X}})$ is finitely generated.

The following assertion is, actually, contained in [**11**, Theorem 1 on p. p. 493 and Lemma 2 on p. 495].

LEMMA 5.1. *Let n be a positive integer that is not divisible by* $\mathrm{char}(\kappa)$.

(i) *The action of the inertia subgroup $\mathcal{I}$ on the* $\mathrm{Gal}(K)$*-module $X_{\mathbf{K}}[n]$ is trivial and, therefore, one may view $X_{\mathbf{K}}[n]$ as the* $\mathrm{Gal}(\kappa)$*-module via* (14).

(ii) *There is a canonical isomorphism of* $\mathrm{Gal}(\kappa)$*-modules*

$$\mathrm{red}_n : X_{\mathbf{K}}[n] \cong X_s[n] \tag{17}$$

such that the diagram

$$\begin{array}{ccc} X_{\mathbf{K}}[n] & \xrightarrow{\mathfrak{u}_{\mathbf{K}}} & X_{\mathbf{K}}[n] \\ {\scriptstyle \mathrm{red}_n}\downarrow & & {\scriptstyle \mathrm{red}_n}\downarrow \\ X_s[n] & \xrightarrow{\mathfrak{u}_s} & X_s[n] \end{array} \tag{18}$$

is commutative for all $\mathfrak{u} \in \mathrm{End}_S(\mathcal{X})$.

(iii) *The* $\mathrm{Gal}(\kappa)$*-modules of fixed points*

$$X_{\mathbf{K}}[n]^{\mathfrak{u}_{\mathbf{K}}} = \{x \in X_{\mathbf{K}}[n] \mid \mathfrak{u}_{\mathbf{K}}(x) = x\}$$

and

$$X_s[n]^{\mathfrak{u}_s} = \{x \in X_s[n] \mid \mathfrak{u}_s(x) = x\}$$

are isomorphic for all $\mathfrak{u} \in \mathrm{End}_S(\mathcal{X})$.

PROOF. (i) follows from [**11**, Theorem 1 on p. p. 4 and Lemma 2 on p. 495]. This implies that $X_{\mathbf{K}}[n]^{\mathcal{I}} = X_{\mathbf{K}}[n]$.

The desired isomorphism of Galois modules

$$\mathrm{red}_n : X_{\mathbf{K}}[n]^{\mathcal{I}} = X_{\mathbf{K}}[n] \cong X_s[n]$$

is constructed in the proof of Lemma 2 on p. 495 of [**11**]. It follows readily from the construction that all the diagrams (18) are commutative. This proves (ii), which immediately implies (iii). □

The following assertion is probably well known but I failed to find a suitable reference.

LEMMA 5.2. *The scheme* $\mathcal{X}$ *is normal reduced irreducible and* $\Gamma(\mathcal{X}, \mathcal{O}_{\mathcal{X}}) = O_{\mathbf{K}}$.

PROOF. The abelian scheme $\mathcal{X}$ is automatically of finite presentation over S [**2**, Remark 1.2(a) on p. 3]. Since S is normal and $f : \mathcal{X} \to S$ is smooth, it follows from [**7**, Prop. 5.5 on pp. 235–236] that $\mathcal{X}$ is *normal* and therefore *reduced*. It follows from [**7**, Criterion 5.5.2 on pp. 234–235] that the *irreducible components* of the topological space $\mathcal{X}$ are *disjoint*. Since $X_{\mathbf{K}}$ is a nonempty open irreducible subset of $\mathcal{X}$, it lies in an irreducible component of $\mathcal{X}$, say Y_1. Since X_s is a nonempty closed irreducible subset of $\mathcal{X}$, it lies in an irreducible component of $\mathcal{X}$, say Y_2. This implies that the topological space $\mathcal{X}$ is a union of Y_1 and Y_2. Since irreducible components are closed subsets, both Y_1 and Y_2 are closed subsets in $\mathcal{X}$. Since f is proper, $f(Y_1)$ is a closed subset of S. By definition, $f(Y_1)$ contains the generic point η of S and therefore contains its closure S. This implies that $f(Y_1) = S$. Hence Y_1 meets X_s and therefore meets Y_2 as well. It follows that $Y_1 = Y_2$ and therefore $\mathcal{X} = Y_1$ is *irreducible*. This implies that $\Gamma(\mathcal{X}, \mathcal{O}_{\mathcal{X}})$ is an $O_{\mathbf{K}}$-algebra without zero divisors. On the other hand, the properness of f implies that the $O_{\mathbf{K}}$-algebra $\Gamma(\mathcal{X}, \mathcal{O}_{\mathcal{X}})$ is a finitely generated $\Gamma(S, \mathcal{O}_S) = O_{\mathbf{K}}$-module.

Recall that $X_{\mathbf{K}}$ is an open nonempty subset of $\mathcal{X}$. The irreducibility of $\mathcal{X}$ implies that $X_{\mathbf{K}}$ is dense in $\mathcal{X}$. It follows that the restriction map

$$\Gamma(\mathcal{X}, \mathcal{O}_{\mathcal{X}}) \to \Gamma(X_{\mathbf{K}}, \mathcal{O}_{X_{\mathbf{K}}}) = \mathbf{K}$$

is *injective*. This implies that the (isomorphic) image of $\Gamma(\mathcal{X}, \mathcal{O}_{\mathcal{X}})$ in $\mathbf{K}$ is an *order* containing $O_{\mathbf{K}}$ and therefore coincides with $O_{\mathbf{K}}$. It follows that $\Gamma(\mathcal{X}, \mathcal{O}_{\mathcal{X}}) = O_{\mathbf{K}}$. □

We write $\Omega^1(X_{\mathbf{K}})$ and $\Omega^1(X_s)$ for the $\mathbf{K}$-vector space of differentials of the first kind on $X_{\mathbf{K}}$ and the κ-vector space of differentials of the first kind on X_s respectively (both of dimension g).

THEOREM 5.3. *Let q be a positive integer that is not divisible by* $\operatorname{char}(\kappa)$ *such that κ contains a primitive qth root of unity.*

Suppose that δ_S is an automorphism of the abelian scheme $\mathcal{X}/S$ such that δ_S^q is the identity map. Let $\delta_{\mathbf{K}} : X_{\mathbf{K}} \to X_{\mathbf{K}}$ and $\delta_s : X_s \to X_s$ be the automorphisms of abelian varieties $X_{\mathbf{K}}$ and X_s induced by δ_S. Let us consider the linear automorphisms

$$\delta_{\mathbf{K}}^* : \Omega^1(X_{\mathbf{K}}) \to \Omega^1(X_{\mathbf{K}}),\ \delta_s^* : \Omega^1(X_s) \to \Omega^1(X_s)$$

induced by $\delta_{\mathbf{K}}$ and δ_s respectively.

Let $\operatorname{spec}(\delta_{\mathbf{K}}^*) \subset \bar{\mathbf{K}}$ *and* $\operatorname{spec}(\delta_s^*) \subset \bar{\kappa}$ *be the sets of eigenvalues of $\delta_{\mathbf{K}}^*$ and δ_s^* respectively.*

Then:

(i) *Both $\delta_{\mathbf{K}}^*$ and δ_s^* are diagonalizable linear operators and*

$$\operatorname{spec}(\delta_{\mathbf{K}}^*) \subset \mu_{q,\mathbf{K}} \subset \mathbf{K},\ \operatorname{spec}(\delta_s^*) \subset \mu_{q,\kappa} \subset \kappa.$$

(ii) *A root of unity $\gamma \in \mu_{q,\mathbf{K}}$ lies in* $\operatorname{spec}(\delta_{\mathbf{K}}^*)$ *if and only if $\bar{\gamma}$ lies in* $\operatorname{spec}(\delta_s^*)$.

(iii) *The multiplicity of each $\gamma \in \operatorname{spec}(\delta_{\mathbf{K}}^*)$ coincides with the multiplicity of $\bar{\gamma} \in \operatorname{spec}(\delta_s^*)$.*

PROOF. Clearly,

$$(\delta_{\mathbf{K}})^q = 1_{X_{\mathbf{K}}}, (\delta_s)^q = 1_{X_s}$$

and therefore both $(\delta_{\mathbf{K}}^*)^q$ and $(\delta_s^*)^q$ are the identity automomorphisms of $\Omega^1(X_{\mathbf{K}})$ and $\Omega^1(X_s)$ respectively. The conditions on q imply that both $\delta_{\mathbf{K}}^*$ and δ_s^* are diagonalizable and their eigenvalues lie in $\mu_{q,\mathbf{K}}$ and $\mu_{q,\kappa}$ respectively. This proves (i).

In order to prove (ii) and (iii), recall [**1**, Cor. 3 on p. 102] that the sheaf $\Omega^1_{\mathcal{X}/S}$ of relative differentials is a free $\mathcal{O}_{\mathcal{X}}$-module of rank g and therefore is generated by certain g sections that remain linearly independent at every point of $\mathcal{X}$. It follows from Lemma 5.2 that the group $\Gamma(\mathcal{X}, \Omega^1_{\mathcal{X}/S})$ of its global sections is a free $O_{\mathbf{K}}$-module of rank g. Let

$$\delta_S^* : \Gamma(\mathcal{X}, \Omega^1_{\mathcal{X}/S}) \to \Gamma(\mathcal{X}, \Omega^1_{\mathcal{X}/S})$$

be the automorphism of $\Gamma(S, \mathcal{O}_S) = \mathcal{O}_K$-module $\Gamma(\mathcal{X}, \Omega^1_{\mathcal{X}/S})$ induced by δ_S. Clearly, $(\delta_S^*)^q$ is the identity map. Our conditions on q imply the existence of a basis $\{\omega_1, \dots, \omega_g\}$ of $\Gamma(\mathcal{X}, \Omega^1_{\mathcal{X}/S})$ such that for each $i = 1, \dots g$ there exists $\gamma_i \in \mu_{q,\mathbf{K}}$ such that

$$\delta_S^* \omega_i = \gamma_i \omega_i.$$

Clearly, $\{\omega_1, \dots, \omega_g\}$ remain linearly independent at every point of $\mathcal{X}$. This implies that their *restrictions* $\{\omega_{1,s}, \dots, \omega_{g,s}\}$ to X_s constitute a basis of the κ-vector space

$$\Gamma(X_s, \Omega^1_{X/\kappa}) = \Omega^1(X_s),$$

their *restrictions* $\{\omega_{1,\mathbf{K}}, \dots, \omega_{g,\mathbf{K}}\}$ to $X_{\mathbf{K}}$ constitute a basis of the $\mathbf{K}$-vector space

$$\Gamma(X_{\mathbf{K}}, \Omega^1_{X_{\mathbf{K}}/\mathbf{K}}) = \Omega^1(X_{\mathbf{K}}).$$

The functoriality of the formation of relative differentials and its compatibility with base change [**1**, Prop. 3 on p. 34 and p. 35] imply that

$$\delta_{\mathbf{K}}^{*}\omega_{i,\mathbf{K}} = \gamma_i \cdot \omega_{i,\mathbf{K}},\ \delta_s^{*}\omega_{i,s} = \bar{\gamma}_i \cdot \omega_{i,s}.$$

for all $i = 1, \ldots g$. This implies readily (ii) and (iii). □

THEOREM 5.4. *We keep the notation and assumptions of Theorem* 5.3. *Suppose that there exist a prime ℓ and a positive integer r such that $q = \ell^r$ (in particular, $\ell \neq$ char(κ), the field κ contains a primitive qth root of unity) and $\delta_S \in \mathrm{End}_S(\mathcal{X})$ satisfies the qth cyclotomic equation*

$$\Phi_q(\delta) = 0,$$

where the qth cyclotomic polynomial

$$\Phi_q(t) = \frac{t^{\ell^r} - 1}{t^{\ell^{r-1}} - 1} = \sum_{j=0}^{l-1} t^{j\ell^{r-1}} \in \mathbb{Z}[t].$$

Let us consider the subgroups of fixed points

$$X_{\mathbf{K}}^{\delta_{\mathbf{K}}} = \{x \in X_{\mathbf{K}}(\bar{\mathbf{K}}) \mid \delta_{\mathbf{K}}(x) = x\} \subset X_{\mathbf{K}}(\bar{\mathbf{K}})$$

and

$$X_s^{\delta_s} = \{y \in X_s(\kappa) \mid \delta_s(y) = y\} \subset X_s(\bar{\kappa}).$$

Then:

(i) $X_{\mathbf{K}}^{\delta_{\mathbf{K}}}$ *is a* Gal(K)*-submodule of* $X_{\mathbf{K}}[\ell]$ *and* $X_s^{\delta_s}$ *is a* Gal(κ)*-submodule of* $X_s[\ell]$.

(ii) $X_s^{\delta_s}$ *viewed as the* Gal(K)*-module is isomorphic to* $X_{\mathbf{K}}^{\delta_{\mathbf{K}}}$.

PROOF. (i) We have

$$\ell x = \sum_{j=0}^{l-1} \delta_{\mathbf{K}}^{j\ell^{r-1}}(x) = 0,\ \ell y = \sum_{j=0}^{l-1} \delta_s^{j\ell^{r-1}}(y) = 0.$$

(ii) It follows readily from Lemma 5.1 applied to $n = \ell$ and $\mathfrak{u} = \delta$. □

REMARK 5.5. Let κ be a perfect field of prime characteristic p, $W(\kappa)$ its ring of Witt vectors and $\mathbf{K}$ the field of fractions of $W(\kappa)$. The field $\mathbf{K}$ is a complete discrete valuation field of characteristic zero with valuation ring $W(\kappa)$ and residue field κ.

Let X_0 be an abelian variety over κ of positive dimension g. If X_0 is *ordinary* then there exists a *canonical* Serre-Tate lifting of X_0 - an abelian scheme $\mathcal{X}$ over $S = \mathrm{Spec}(W(\kappa))$, whose closed fiber coincides with X_0, and the natural ring homomorphism $\mathrm{End}_S(\mathcal{X}) \to \mathrm{End}_\kappa(X_0)$ is an isomorphism [**3**, p. 172, Th. 3.3]. We write $X_{\mathbf{K}}$ for the generic fiber of $\mathcal{X}$; it is a g-dimensional abelian variety over $\mathbf{K}$. The abelian scheme $\mathcal{X}$ is a Néron model of $X_{\mathbf{K}}$ and therefore the natural ring homomorphism $\mathrm{End}_S(\mathcal{X}) \to \mathrm{End}_K(X_{\mathbf{K}})$ is an isomorphism [**1**].

6. Proof of Theorem 3.1

Since $\Phi_q(t)$ divides $t^q - 1$ in $\mathbb{Z}[t]$, it follows from (4) that $\delta_{\mathbf{K}}^q - 1_{\mathbb{Y}} = 0$, which proves (a). The ring homomorphism $\mathrm{i}_{\mathbb{Y}}$ defined in (5) is an embedding, because the qth cyclotomic polynomial is irreducible over $\mathbb{Q}$. This proves (b), which, in turn, implies (c) in light of Remark 4.3. The assertion (d) about the cardinality of the spectrum of $\delta_{\mathbf{K}}^*$ is actually proven on pp. 357–358 of [**24**].

Proof of (e). Let us put

$$E = \mathbb{Q}(\zeta_q),\ O = \mathbb{Z}[\zeta_q] \subset \mathbb{Q}(\zeta_q) = E$$

and write λ for the maximal ideal $(1 - \zeta_q)O$ of O. It is well known that

- λ is the only maximal ideal of O that lies above ℓ;
- the residue field $k(\lambda) = O/\lambda = \mathbb{F}_\ell$.

As in [**24**] we define

$$\mathbb{Y}_\lambda = \{y \in \mathbb{Y}(\bar{\mathbf{K}}) \mid \mathrm{i}_{\mathbb{Y}}(e)y = 0\ \forall e \in \lambda\}.$$

Since λ is generated by $(1 - \zeta_q)O$ and $\mathrm{i}_{\mathbb{Y}}(\zeta_q) = \delta_{\mathbf{K}}$,

$$\mathbb{Y}_\lambda = \{y \in \mathbb{Y}(\bar{\mathbf{K}}) \mid (1_{\mathbb{Y}} - \delta_{\mathbf{K}})y = 0\} = \mathbb{Y}^\delta.$$

Combining Property (2) of Theorem 3.1 and (13), we obtain that

$$n - 1 = \frac{2\mathrm{dim}(\mathbb{Y})}{[E:\mathbb{Q}]},\ 2\mathrm{dim}(\mathbb{Y}) = (n-1)\varphi(q). \tag{19}$$

Let $\mathbb{Q}[\delta_{\mathbf{K}}]$ be the $\mathbb{Q}$-subalgebra of $\mathrm{End}^0(\mathbb{Y})$ generated by $\delta_{\mathbf{K}}$. Actually, $\mathbb{Q}[\delta_{\mathbf{K}}]$ is a (sub)field isomorphic to $\mathbb{Q}(\zeta_q)$. We write $\mathrm{End}^0(\mathbb{Y}, \mathrm{i}_{\mathbb{Y}})$ for the *centralizer* of $\mathbb{Q}[\delta_{\mathbf{K}}]$ in $\mathrm{End}^0(\mathbb{Y})$. Clearly

$$\delta_{\mathbf{K}} \in \mathbb{Q}[\delta_{\mathbf{K}}] \subset \mathrm{End}^0(\mathbb{Y}, \mathrm{i}_{\mathbb{Y}}) \subset \mathrm{End}^0(\mathbb{Y}).$$

It follows from Property (2) and Theorem 4.7 of [**21**] that the $\mathrm{Gal}(\mathbf{K})$-module $\mathbb{Y}_\lambda$ is *very simple* in a sense of [**19**]. It follows from Theorem 3.8 of [**24**] that one of the following conditions holds.

(A) The subring $\mathrm{i}_{\mathbb{Y}}(O)$ coincides with its own centralizer in $\mathrm{End}(\mathbb{Y})$. In other words, every endomorphism in $\mathrm{End}(\mathbb{Y})$ that commutes with $\delta_{\mathbf{K}}$ may be presented as a polynomial in $\delta_{\mathbf{K}}$ with integer coefficients. Equivalently,

$$\mathbb{Q}[\delta_{\mathbf{K}}] = \mathrm{End}^0(\mathbb{Y}, \mathrm{i}_{\mathbb{Y}}).$$

(B) $\mathrm{End}^0(\mathbb{Y}, \mathrm{i}_{\mathbb{Y}})$ is a *central simple* $\mathbb{Q}[\delta_{\mathbf{K}}]$-algebra of dimension $(2\mathrm{dim}(\mathbb{Y})/[E:\mathbb{Q}])^2$. (It follows from (19) that the $\mathbb{Q}[\delta_{\mathbf{K}}]$-dimension of $\mathrm{End}^0(\mathbb{Y}, \mathrm{i}_{\mathbb{Y}})$ is $(n-1)^2$.)

Suppose that the case (B) holds. Every eigenspace of $\delta_{\mathbf{K}}^*$ is invariant under the natural action of $\mathrm{End}^0(\mathbb{Y}, \mathrm{i}_{\mathbb{Y}})$. The properties of $\mathrm{End}^0(\mathbb{Y}, \mathrm{i}_{\mathbb{Y}})$ imply that the multiplicity of every eigenvalue of $\delta_{\mathbf{K}}^*$ is divisible by $(n-1)$. (It is the only place where we use that $\mathrm{char}(\mathbf{K}) = 0$.) Notice that for each positive integer $i < q$ with $(i, \ell) = 1$

$$\left[\frac{ni}{q}\right] + \left[\frac{n(q-i)}{q}\right] = n - 1. \tag{20}$$

This implies that if ζ^{-i} is an eigenvalue of $\delta_{\mathbf{K}}^*$ then its multiplicity $[ni/q]$ is a positive integer divisible by $(n-1)$ and therefore equals $(n-1)$. By (20), $[n(q-i)/q] = 0$, i.e., $\zeta^{-(q-i)} = \zeta^{-i}$ is *not* an eigenvalue of $\delta_{\mathbf{K}}^*$. This implies that the number of distinct eigenvalues of δ^* does *not* exceed $\varphi(q)/2$, which is not true. This proves that the case (B) does *not* occur.

So, the case (A) holds, i.e., the field $\mathbb{Q}[\delta_{\mathbf{K}}] \cong \mathbb{Q}(\zeta_q)$ coincides with its own centralizer in $\mathrm{End}^0(\mathbb{Y})$. Then $\mathbb{Q}[\delta_{\mathbf{K}}]$ contains the center $\mathfrak{C}_{\mathbb{Y}}$ of $\mathrm{End}^0(\mathbb{Y})$. It follows from Corollary 2.2 of [**24**] that $\mathbb{Q}[\delta_{\mathbf{K}}]$ coincides with $\mathfrak{C}_{\mathbb{Y}}$. This implies that the centralizer of $\mathbb{Q}[\delta_{\mathbf{K}}]$ is the whole $\mathrm{End}^0(\mathbb{Y})$ and, therefore, $\mathrm{End}^0(\mathbb{Y}) = \mathbb{Q}[\delta_{\mathbf{K}}]$. Since $\mathbb{Z}[\delta_{\mathbf{K}}] \cong \mathbb{Z}[\zeta_q]$ and $\mathbb{Z}[\zeta_q]$ is the maximal order in $\mathbb{Q}[\zeta_q]$,

$$\mathrm{End}(\mathbb{Y}) = \mathbb{Z}[\delta_{\mathbf{K}}] \cong \mathbb{Z}[\zeta_q].$$

This ends the proof of Theorem 3.1.

7. Proof of Theorem 3.2

Let $W(K)$ be the ring of Witt vectors over K and $\mathbf{K}$ its field of fractions, which has characteristic 0. Since K contains a primitive qth root of unity, there is a primitive qth root of unity

$$\zeta \in W(K) \subset \mathbf{K},$$

that goes to $\bar{\zeta}$ in K under the reduction map.

Let $S := \mathrm{Spec}(W(K))$ and $f : \mathcal{Y} \to S$ be the canonical Serre-Tate lifting of Y [**3**], which is an abelian scheme over S, whose closed fiber is Y. We write $\mathbb{Y}$ for the generic fiber of f, which is an abelian variety over $\mathbf{K}$ with the same dimension as Y. We know (see Remark 5.5) that the natural ring homomorphisms

$$\mathrm{End}_K(Y) \leftarrow \mathrm{End}_S(\mathcal{Y}) \to \mathrm{End}_{\mathbf{K}}(\mathbb{Y}) \tag{21}$$

are isomorphisms. Let $\delta_S \in \mathrm{End}_S(\mathcal{Y})$ be the preimage (lifting) of δ in $\mathrm{End}_S(\mathcal{Y})$ and $\delta_{\mathbf{K}} \in \mathrm{End}_{\mathbf{K}}(\mathbb{Y})$ be the image (restriction to the generic fiber) of δ_S in $\mathrm{End}_{\mathbf{K}}(\mathbb{Y})$. Isomorphisms (21) induce ring isomorphisms

$$\mathbb{Z}[\delta_{\mathbf{K}}] \leftarrow \mathbb{Z}[\delta_S] \to \mathbb{Z}[\delta]. \tag{22}$$

In light of (3.2),

$$\Phi_q(\delta_S) = 0, \ \Phi_q(\delta_{\mathbf{K}}) = 0; \ \delta_S^q = 1_{\mathcal{Y}}, \ \delta_{\mathbf{K}}^q = 1_{\mathbb{Y}}. \tag{23}$$

Then the ring homomorphism

$$i_{\mathbb{Y}} : \mathbb{Z}[\zeta_q] \to \mathbb{Z}[\delta_{\mathbf{K}}] \subset \mathrm{End}_{\mathbf{K}}(\mathbb{Y})$$

that sends 1 to $1_{\mathbb{Y}}$ and ζ_q to $\delta_{\mathbf{K}}$ is a ring embedding. This proves Theorem 3.2(b), which, in turn, implies Theorem 3.2(a,c). In order to prove Theorem 3.2(d,e), we are going to apply Theorem 3.1 to $\mathbb{Y}/\mathbf{K}$. Theorems 5.4 and 5.3 applied to $\mathcal{Y}/S$ imply that all the conditions of Theorem 3.1 are fulfilled. Applying Theorem 3.1(d,e) to $\mathbb{Y}/\mathbf{K}, \delta_{\mathbf{K}}, \zeta$, we conclude that the spectrum of δ^* consists of more than $\varphi(q)/2$ eigenvalues and

$$i_{\mathbb{Y}}(\mathbb{Z}[\zeta_q]) = \mathrm{End}_{\mathbf{K}}(\mathbb{Y}) = \mathrm{End}(\mathbb{Y}).$$

It remains to recall that there is a canonical ring isomorphism between $\mathrm{End}(Y)$ and $\mathrm{End}(\mathbb{Y})$ under which $\delta = i_Y(\zeta_q)$ goes to $\delta_{\mathbf{K}} = i_{\mathbb{Y}}(\zeta_q)$.

8. Superelliptic Jacobians

We are heading for the proof of Theorem 2.9. We keep the notation of Section 2. In particular, ℓ is a prime, K is a field of characteristic $p \neq \ell$ that contains ζ, which is a primitive $q = \ell^r$th root of unity; $f(x) \in K[x]$ is a separable polynomial of degree $n \geq 4$.

We write $C_{f,q}$ for the superelliptic K-curve $y^q = f(x)$ and $J(C_{f,q})$ for its Jacobian. Recall that $J(C_{f,q})$ is an abelian variety that is defined over K. In **[24]** I constructed an abelian positive-dimensional K-subvariety $J^{(f,q)} \subset J(C_{f,q})$ that enjoys the following properies.

(i) There exists a K-isogeny of abelian varieties

$$\prod_{j=1}^{r} J^{(f,\ell^i)} \to J(C_{f,q}).$$

(ii) There is a ring embedding

$$e_q : \mathbb{Z}[\zeta_q] \hookrightarrow \mathrm{End}_K(J^{(f,q)})$$

that sends 1 to $1_{J(f,q)}$ and enjoys the following properies.

(1) $\delta := e_q(\zeta_q) \in \mathrm{End}_K(J^{(f,q)})$ is an automorphism of $J^{(f,q)}$ that satisfies

$$\sum_{i=0}^{\ell-1} \delta^{i\ell^{r-1}} = 0, \ \delta^q = 1_{J(f,q)}.$$

(2) If $l \nmid n$ then the $\mathrm{Gal}(K)$-(sub)module $\left(J^{(f,q)}\right)^{\delta}$ of δ-invariants is isomorphic to $\left(\mathbb{F}_\ell^{\mathfrak{R}_f}\right)^0$.

(3) If $l \nmid n$ then the eigenvalues of δ^* in $\Omega^1(J^{(f,q)})$ are primitive qth roots of unity. In addition, for each positive integer i with

$$1 \leq i < q, \ (i,\ell) = 1$$

the multiplicity of ζ^{-i} as an eigenvalue of δ^* is $[ni/q]$. In particular, ζ^{-i} is an eigenvalue of δ^* if and only if $[ni/q] > 0$.

PROOF OF THEOREM 2.9. In light of Remarks 2.10, we may assume that ℓ does not divide n, the field K is perfect and contains a primitive qth root of unity, and $\mathrm{Gal}(f) = \mathbf{A}_n$. Since $J(C_{f,q})$ is ordinary, it follows from Property (i) that $J^{(f,\ell^j)}$ is ordinary if $1 \leq j \leq r$. Applying Theorem 3.2 to

$$Y = J^{(f,\ell^j)}, \delta = e_{\ell^j}(\zeta_{\ell^j})$$

for all such j, we obtain that

$$\mathrm{End}(J^{(f,\ell^j)}) \cong \mathbb{Z}[\zeta_{\ell^j}], \ \mathrm{End}^0(J^{(f,\ell^j)}) \cong \mathbb{Q}(\zeta_{\ell^j}).$$

In particular, all $J^{(f,\ell^j)}$ are absolutely simple and their endomorphism algebras are non-isomorphic for distinct j. Now Property (i) implies that

$$\mathrm{End}^0(J(C_{f,q})) \cong \oplus_{j=1}^r \mathrm{End}^0(J^{(f,\ell^j)}) \cong \oplus_{j=1}^r \mathbb{Q}(\zeta_{\ell^j}).$$

This implies that

$$\mathrm{End}^0(J(C_{f,q})) \cong \oplus_{j=1}^r \mathbb{Q}(\zeta_{\ell^j}).$$

On the other hand, $\mathrm{End}^0(J(C_{f,q}))$ contains the $\mathbb{Q}$-subalgebra $\mathbb{Q}[\delta_q]$ and the latter is isomorphic to $\oplus_{j=1}^r \mathbb{Q}(\zeta_{\ell^j})$ (see Subsect. 2.6). Now $\mathbb{Q}$-dimension arguments imply that $\mathrm{End}^0(J(C_{f,q}) = \mathbb{Q}[\delta_q]$. □

Acknowledgments

The author is grateful to Noriko Yui and Daqing Wan for stimulating discussions. The author's special thanks go to Tatiana Bandman and the referees, whose comments helped to improve the exposition.

References

[1] Siegfried Bosch, Werner Lütkebohmert, and Michel Raynaud, *Néron models*, Ergebnisse der Mathematik und ihrer Grenzgebiete (3) [Results in Mathematics and Related Areas (3)], vol. 21, Springer-Verlag, Berlin, 1990. MR1045822

[2] Gerd Faltings and Ching-Li Chai, *Degeneration of abelian varieties*, Ergebnisse der Mathematik und ihrer Grenzgebiete (3) [Results in Mathematics and Related Areas (3)], vol. 22, Springer-Verlag, Berlin, 1990. With an appendix by David Mumford. MR1083353

[3] William Messing, *The crystals associated to Barsotti-Tate groups: with applications to abelian schemes*, Lecture Notes in Mathematics, Vol. 264, Springer-Verlag, Berlin-New York, 1972. MR0347836

[4] Brian Mortimer, *The modular permutation representations of the known doubly transitive groups*, Proc. London Math. Soc. (3) **41** (1980), no. 1, 1–20, DOI 10.1112/plms/s3-41.1.1. MR579714

[5] David Mumford, *Abelian varieties*, Tata Institute of Fundamental Research Studies in Mathematics, No. 5, Published for the Tata Institute of Fundamental Research, Bombay; Oxford University Press, London, 1970. MR0282985

[6] D. Mumford, J. Fogarty, and F. Kirwan, *Geometric invariant theory*, 3rd ed., Ergebnisse der Mathematik und ihrer Grenzgebiete (2) [Results in Mathematics and Related Areas (2)], vol. 34, Springer-Verlag, Berlin, 1994. MR1304906

[7] David Mumford and Tadao Oda, *Algebraic geometry. II*, Texts and Readings in Mathematics, vol. 73, Hindustan Book Agency, New Delhi, 2015. MR3443857

[8] Bjorn Poonen and Edward F. Schaefer, *Explicit descent for Jacobians of cyclic covers of the projective line*, J. Reine Angew. Math. **488** (1997), 141–188. MR1465369

[9] Kenneth A. Ribet, *Galois action on division points of Abelian varieties with real multiplications*, Amer. J. Math. **98** (1976), no. 3, 751–804, DOI 10.2307/2373815. MR457455

[10] Edward F. Schaefer, *Computing a Selmer group of a Jacobian using functions on the curve*, Math. Ann. **310** (1998), no. 3, 447–471, DOI 10.1007/s002080050156. MR1612262

[11] Jean-Pierre Serre and John Tate, *Good reduction of abelian varieties*, Ann. of Math. (2) **88** (1968), 492–517, DOI 10.2307/1970722. MR236190

[12] Jean-Pierre Serre, *Topics in Galois theory*, Research Notes in Mathematics, vol. 1, Jones and Bartlett Publishers, Boston, MA, 1992. Lecture notes prepared by Henri Damon [Henri Darmon]; With a foreword by Darmon and the author. MR1162313

[13] Goro Shimura, *Introduction to the arithmetic theory of automorphic functions*, Publications of the Mathematical Society of Japan, No. 11. Iwanami Shoten, Publishers, Tokyo; Princeton University Press, Princeton, N.J., 1971. Kanô Memorial Lectures, No. 1. MR0314766

[14] Goro Shimura, *Abelian varieties with complex multiplication and modular functions*, Princeton Mathematical Series, vol. 46, Princeton University Press, Princeton, NJ, 1998. MR1492449

[15] Jiangwei Xue, *Endomorphism algebras of Jacobians of certain superelliptic curves*, J. Number Theory **131** (2011), no. 2, 332–342, DOI 10.1016/j.jnt.2010.07.013. MR2910215

[16] Jiangwei Xue and Chia-Fu Yu, *Endomorphism algebras of factors of certain hypergeometric Jacobians*, Trans. Amer. Math. Soc. **367** (2015), no. 11, 8071–8106, DOI 10.1090/tran/6330. MR3391910

[17] Noriko Yui, *On the Jacobian variety of the Fermat curve*, J. Algebra **65** (1980), no. 1, 1–35, DOI 10.1016/0021-8693(80)90236-7. MR578793

[18] Yuri G. Zarhin, *Hyperelliptic Jacobians without complex multiplication*, Math. Res. Lett. **7** (2000), no. 1, 123–132, DOI 10.4310/MRL.2000.v7.n1.a11. MR1748293

[19] Yuri G. Zarhin, *Hyperelliptic Jacobians and modular representations*, Moduli of abelian varieties (Texel Island, 1999), Progr. Math., vol. 195, Birkhäuser, Basel, 2001, pp. 473–490. MR1827030

[20] Yu. G. Zarhin, *On the endomorphism rings of some Jacobians in finite characteristic* (Russian, with Russian summary), Mat. Sb. **193** (2002), no. 8, 39–48, DOI 10.1070/SM2002v193n08ABEH000673; English transl., Sb. Math. **193** (2002), no. 7-8, 1139–1149. MR1934543

[21] Yuri G. Zarhin, *Cyclic covers of the projective line, their Jacobians and endomorphisms*, J. Reine Angew. Math. **544** (2002), 91–110, DOI 10.1515/crll.2002.030. MR1887891

[22] Yuri G. Zarhin, *The endomorphism rings of Jacobians of cyclic covers of the projective line*, Math. Proc. Cambridge Philos. Soc. **136** (2004), no. 2, 257–267, DOI 10.1017/S0305004103007102. MR2040573

[23] Yuri G. Zarhin, *Non-supersingular hyperelliptic Jacobians* (English, with English and French summaries), Bull. Soc. Math. France **132** (2004), no. 4, 617–634, DOI 10.24033/bsmf.2477. MR2131907

[24] Yuri G. Zarhin, *Endomorphism algebras of superelliptic Jacobians*, Geometric methods in algebra and number theory, Progr. Math., vol. 235, Birkhäuser Boston, Boston, MA, 2005, pp. 339–362, DOI 10.1007/0-8176-4417-2_15. MR2166091

[25] Yuri G. Zarhin, *Abelian varieties without homotheties*, Math. Res. Lett. **14** (2007), no. 1, 157–164, DOI 10.4310/MRL.2007.v14.n1.a13. MR2289628

[26] Yuri G. Zarhin, *Superelliptic Jacobians*, Diophantine geometry, CRM Series, vol. 4, Ed. Norm., Pisa, 2007, pp. 363–390. MR2349666

[27] Yuri G. Zarhin, *Endomorphisms of superelliptic Jacobians*, Math. Z. **261** (2009), no. 3, 691–707, DOI 10.1007/s00209-008-0342-5. MR2471095

[28] Yu. G. Zarhin, *Endomorphisms of superelliptic Jacobians.* arXiv:math/0605028v6 [math.AG].

[29] Yuri G. Zarhin, *Endomorphism algebras of abelian varieties with special reference to superelliptic Jacobians*, Geometry, algebra, number theory, and their information technology applications, Springer Proc. Math. Stat., vol. 251, Springer, Cham, 2018, pp. 477–528, DOI 10.1007/978-3-319-97379-1_22. MR3880401

Department of Mathematics, Pennsylvania State University, University Park, Pennsylvania 16802

Email address: zarhin@math.psu.edu

Contemporary Mathematics
Volume **767**, 2021
https://doi.org/10.1090/conm/767/15398

An introduction to Deuring's theory of constant reductions

Barry Green and Peter Roquette

ABSTRACT. This note presents a brief introduction to Deuring's theory of reduction of function fields. We assume that the reader is familiar with the elementary theory of function fields (or curves) and their divisors, up to the Theorem of Riemann-Roch. One of the new features here is the definition and use of the notion of "inert" extensions of valued fields, which works for extensions of arbitrary degree of transcendency. We study the behavior of the genus of a function field under reduction (section 4). In the case of "good" reduction, where the genus does not change, we obtain Deuring's reduction map for divisors (section 5). Any function field admits good reduction with respect to almost all valuations of the base field (section 6). In a follow-up paper it will be shown that the valuation for good reduction, if it exists, is unique.

Introduction

In this note, a "**function field**" is a finitely generated field extension $F|K$ of transcendence degree 1. Thus there exists an element $x \in F$ which is transcendent over K, and F appears as a finite algebraic extension of the field $K(x)$ of rational functions. We assume that the reader is familiar with the elementary theory of function fields and their divisors, up to the Theorem of Riemann-Roch.

A "**valued function field**" is a function field equipped with a valuation v and the property that the residue field $\overline{F}$ is transcendental over the residue field $\overline{K}$ of K. Thus there exists an element $x \in F$ such that its residue $\overline{x}$ is transcendental over $\overline{K}$. We shall see in section 1 that this condition uniquely characterizes the valuation on the rational subfield $K(x) \subset F$, with residue field $\overline{K}(\overline{x})$. The residue field $\overline{F}$ of F is a finite extension of $\overline{K}(\overline{x})$ and so $\overline{F}|\overline{K}$ is also a function field. It is called the "**reduction**" of $F|K$ modulo the given valuation v.

In general we have $[\overline{F} : \overline{K}(\overline{x})] \leq [F : K(x)]$. The reduction is called "**regular**" if there exists $x \in F$ such that

$$[\overline{F} : \overline{K}(\overline{x})] = [F : K(x)] \,.$$

The main object of this note is to study regular reduction of function fields. In particular we will be interested in the corresponding reduction of divisors, as well as the behavior of the genus under reduction. The theory of reduction of function fields has been developed by Deuring in the year 1942, but for discrete valuations only; see [**1**].

2020 *Mathematics Subject Classification.* Primary 12J20; Secondary 11R58, 14H05.

NOTATION. v is to be a (non-trivial) valuation in the sense of Krull; there is no restriction concerning its value group or its residue field. Given the valuation v of F we denote by $\mathcal{O}_F$ its valuation ring, and similarly $\mathcal{O}_K$ its valuation ring in K. Thus we have $\mathcal{O}_K = K \cap \mathcal{O}_F$. Similarly let $\mathcal{M}_F, \mathcal{M}_K$ be the maximal ideals of $\mathcal{O}_F$ and $\mathcal{O}_K$ respectively, so that $\mathcal{M}_K = K \cap \mathcal{M}_F$. For the residue fields we use the notations already explained above: $\overline{F} = \mathcal{O}_F/\mathcal{M}_F$ and $\overline{K} = \mathcal{O}_K/\mathcal{M}_K$. If $x \in \mathcal{O}_F$ then $\overline{x}$ is its residue in $\overline{F}$. If $x \in F$ and $x \notin \mathcal{O}_F$ then, as usual, we write $\overline{x} = \infty$. For any $0 \neq x \in F$ we denote by (x) its principal divisor. $(x)_\infty$ is the pole divisor of x and $(x)_0$ its zero divisor, so that $(x) = (x)_0 - (x)_\infty$.

1. The Gauss valuation

Let K be a field equipped with a valuation v. Consider the rational function field $F = K(x)$ over K. There are many ways to extend the valuation v from K to F. Some of them, called "Gauss valuations", are particularly easy to describe.

THEOREM 1 (Gauss Theorem). *Let x be a generator of the rational function field $F = K(x)$. There is one and only one extension of v to F for which $\overline{x}$ is transcendental over $\overline{K}$. The value $v(f)$ of a polynomial*

$$f = a_o + a_1x + \cdots + a_nx^n \in K[x]$$

is given by

$$v(f) = \min_{0\le i\le n} v(a_i)\,. \tag{1}$$

The value group and residue field are:

$$v(F) = v(K) \qquad \textit{and} \qquad \overline{F} = \overline{K}(\overline{x})\,.$$

PROOF. If such an extension of v to F exists then its valuation ring $\mathcal{O}_F$ contains the polynomial ring $\mathcal{O}_K[x]$, and its reduction map $\mathcal{O}_F \to \overline{F}$ induces a homomorphism

$$\mathcal{O}_K[x] \to \overline{K}[\overline{x}] \tag{2}$$

of the respective polynomial rings. The homomorphism is obtained coefficientwise, i.e. the image of a polynomial

$$f = f(x) = a_o + a_1x + \cdots + a_nx^n \in \mathcal{O}_K[x] \tag{3}$$

is given by

$$\overline{f} = \overline{f}(\overline{x}) = \overline{a}_o + \overline{a}_1\overline{x} + \cdots + \overline{a}_n\overline{x}^n \in \overline{K}[\overline{x}]. \tag{4}$$

This homomorphism extends naturally and uniquely to its ring of quotients, which consists of all quotients fg^{-1} with $f, g \in \mathcal{O}_K[x]$ and $\overline{g} \neq 0$. We shall see below that this ring of quotients equals $\mathcal{O}_F$. This gives uniqueness.

As to the existence, consider the rational function field of one variable over $\overline{K}$ and one of its generators; denote this generator by $\overline{x}$. Define the homomorphism (2) coefficient wise by (4). Extend that homomorphism to its ring of quotients as above. We claim that this ring of quotients is a valuation ring. Let us denote this ring of quotients already by $\mathcal{O}_F$, so that we have a homomorphism

$$\mathcal{O}_F \to \overline{K}(\overline{x})\,. \tag{5}$$

We are now going to show that, indeed, $\mathcal{O}_F$ is a valuation ring of F. To this end we will verify the characteristic property of a valuation ring: for $h \in F$ we have to show:

$$h \notin \mathcal{O}_F \Longrightarrow h^{-1} \in \mathcal{O}_F\,. \tag{6}$$

We write $h = fg^{-1}$ with $f, g \in K[x]$. We define the function $f \mapsto v(f)$ on $K[x]$ by formula (1). It will turn out that this function is a valuation but for the moment we just have to know that it is a function defined on $K[x]$ with values in the value group $v(K)$.

Let $a \in K^\times$ be an element whose value $v(a) = v(f)$. For instance, a may be a coefficient a_i of f in (3) for which the minimum on the right hand side of (1) is assumed. We write

$$f = a \cdot f_o \qquad \text{with} \qquad v(f_o) = 0\,. \tag{7}$$

The polynomial f_o with $v(f_o) = 0$ is called *primitive.* Thus every polynomial can be normalized by a factor from K to become primitive. The condition $v(f_o) = 0$ is equivalent to $\overline{f}_o \neq 0$.

Similarly we write $g(x) = b \cdot g_o(x)$ with $b \in K$ and g_o primitive. Thus

$$h = fg^{-1} = c \cdot f_o g_o^{-1} \qquad \text{with} \qquad c = ab^{-1} \in K^\times\,. \tag{8}$$

Now, if $v(c) \geq 0$ then $c \in \mathcal{O}_K$ and thus $c \cdot f_o \in \mathcal{O}_K[x]$; hence h is contained in the ring of quotients $\mathcal{O}_F$ as defined above. Consequently, if $h \notin \mathcal{O}_F$ then $v(c) < 0$; this implies $c^{-1} \in \mathcal{O}_K$ and therefore

$$h^{-1} = c^{-1} \cdot g_o f_o^{-1} \in \mathcal{O}_F\,.$$

This establishes our contention (6) and so, indeed, $\mathcal{O}_F$ is a valuation ring.

By construction, $K \cap \mathcal{O}_F = \mathcal{O}_K$. Thus the valuation of F belonging to $\mathcal{O}_F$ is an extension of the valuation v in K. Next we show that this valuation of F is given by (1).

First we claim that the function on $K[x]$ defined by (1) is multiplicative:

$$v(fg) = v(f) + v(g) \qquad \text{for} \qquad f, g \in K[x]\,. \tag{9}$$

(This statement is usually called **Gauss' Lemma.**) To see this, write $f = a \cdot f_o$ as in (7), and similarly $g = b \cdot g_o$, so that

$$fg = ab \cdot f_o g_o\,.$$

Here $\overline{f}_o \neq 0$ and $\overline{g}_o \neq 0$ and thus $\overline{f_o g_o} = \overline{f}_o \overline{g}_o \neq 0$ (because the polynomial ring $\overline{K}[\overline{x}]$ has no zero divisors). Consequently $v(f_o g_o) = 0$ and thus

$$v(fg) = v(ab) = v(a) + v(b) = v(f) + v(g)$$

as contended.

The formula (1) defines the function v on the polynomial ring $K[x]$. We now extend this function to the whole field F by multiplicativity: If $h = fg^{-1} \in F$ with $f, g \in K[x]$ we put $v(h) = v(f) - v(g)$. Because of (9) this is well defined, and it gives a multiplicative function from $F^\times$ to $v(K)$.

We have seen above that any $h \in F$ may be written in the form (8), and so $v(h) = v(c)$. We have also seen that $v(c) \geq 0$ if and only if $h \in \mathcal{O}_F$. We conclude:

$$v(h) \geq 0 \Longleftrightarrow h \in \mathcal{O}_F.$$

This implies by multiplicativity the triangle inequality:

$$v(f+g) \geq \min(v(f), v(g)) \,. \tag{10}$$

To see this, assume $v(g) \leq v(f)$ and write

$$f + g = (h+1)g \qquad \text{with} \qquad h = fg^{-1} \,.$$

By assumption $v(h) \geq 0$, hence $h \in \mathcal{O}_F$ and therefore $h+1 \in \mathcal{O}_F$. Hence $v(h+1) \geq 0$ and $v(f+g) \geq v(g)$.

We have shown that the function v on F is a valuation and that $\mathcal{O}_F$ is its valuation ring. The statements in Theorem 1 concerning value group and residue field are immediate from (1) and (5). □

By construction, the Gauss valuation depends on the choice of the generator x of F. If we wish to indicate this then we speak of the x-Gauss valuation of $K(x)$, and we write v_x. If y is another generator, then the question arises under which condition $v_y = v_x$.

Now, if $K(y) = K(x)$ then

$$y = \frac{ax+b}{cx+d} \qquad \text{with} \qquad a, b, c, d \in K \quad \text{and} \quad ad - bc \neq 0 \tag{11}$$

Note that the coefficients $a, b, c, d \in K$ are not uniquely determined by y. They are determined up to a common factor in $K^\times$.

PROPOSITION 2. *We have $v_y = v_x$ if and only if the coefficients a, b, c, d in (11) are in $\mathcal{O}_K$ and $v_x(ad - bc) = 0$ – after division by a suitable common factor in $K^\times$ if necessary.*

PROOF. Let the $\overline{y}$ denote the residue of y with respect to the valuation v_x. If $v_y = v_x$ then

$$\overline{y} \neq 0, \infty \qquad \text{and} \qquad \overline{K}(\overline{y}) = \overline{K}(\overline{x}). \tag{12}$$

The first condition implies $0 = v_x(y) = v_x(ax+b) - v_x(cx+d)$. We take $u \in K$ such that $v_x(u) = v_x(ax+b) = v_x(cx+d)$; after dividing a, b, c, d by u we may assume that $v_x(ax+b) = 0 = v_x(cx+d)$. We then have

$$\overline{y} = \frac{\overline{a}\,\overline{x} + \overline{b}}{\overline{c}\,\overline{x} + \overline{d}} \,.$$

The second condition in (12) implies $\overline{a}\,\overline{d} - \overline{b}\,\overline{c} \neq 0$ which is to say that $v_x(ad - bc) = 0$. □

REMARK 3. Theorem 1 and the above proof remain valid without change for rational function fields $F = K(x)$ in several variables $x = (x_1, x_2, \ldots)$. In this general situation the condition that "$\overline{x}$ is transcendental over $\overline{K}$" has to be replaced by: "$\overline{x}$ is algebraically free over $\overline{K}$".

When we try to treat power series fields $K((x))$ in the same way then there arises the problem that in formula (1) there are infinitely many coefficients a_i involved and hence the minimum on the right hand side may not exist. Hence one has to restrict the discussion to the case of the field $K((x))_o$ of "convergent" power series, i.e.,

$$f(x) = \sum_{0 \leq i < \infty} a_i x^i \qquad \text{with} \qquad \lim_{i \to \infty} a_i = 0$$

where the limit is meant in the sense of the v-adic valuation of K; this means $v(a_i) \to \infty$. In this situation Theorem 1 and its proof remain valid.

2. Regular reduction

Now let $F|K, v$ be an arbitrary valued function field, not necessarily rational. By definition, there exists $x \in F$ such that $\overline{x}$ is transcendental over $\overline{K}$. From Theorem 1 we see that the valuation in the subfield $K(x) \subset F$ coincides with the Gauss valuation with respect to x. Thus, in order to investigate valued function fields, we have to study the extensions of the Gauss valuation of $K(x)$ to the finite extension F.

General valuation theory tells us that there are only finitely many such extensions. If these are $v_1, \ldots, v_r$ with ramification degrees e_i and residue degrees f_i then

$$\sum_{1 \le i \le r} e_i f_i \le [F : K(x)]. \tag{13}$$

If, say, $v_1 = v$ is the given valuation of $F|K$ then $f_1 = [\overline{F} : \overline{K}(\overline{x})]$ and we conclude

$$[\overline{F} : \overline{K}(\overline{x})] \le [F : K(x)] .$$

As said in the introduction, here we are mainly interested in the case when

$$[\overline{F} : \overline{K}(\overline{x})] = [F : K(x)] . \tag{14}$$

In this case $\overline{F}|\overline{K}$ is said to be a **regular reduction** of $F|K$, and x is called a **regular element**. From (13) we conclude that in the regular case, v is the *only* valuation of F which extends the Gauss valuation v_x of $K(x)$. And v is *unramified* over $K(x)$. Since the Gauss valuation v_x of $K(x)$ is unramified over K (compare (1)) we conclude that $F|K$ is an unramified field extension, provided we have regular reduction.

Suppose for example that $F = K(x, y)$ and there is an irreducible equation of the form

$$\Phi(x, y) = 0 \tag{15}$$

where

$$\Phi(X, Y) = Y^n + f_{n-1}(X)Y^{n-1} + \cdots + f_o(X) \in K[X, Y] \tag{16}$$

with $f_i(X) \in \mathcal{O}_K[X]$. (This implies $n = [F : K(x)]$.) Then y is integral over $\mathcal{O}_K[x]$ and hence $\overline{y} \neq \infty$. The equation

$$\overline{\Phi}(\overline{x}, \overline{y}) = 0 \tag{17}$$

is obtained by reducing the coefficients of $\Phi(X, Y)$ modulo v, where

$$\overline{\Phi}(X, Y) = Y^n + \overline{f}_{n-1}(X)Y^{n-1} + \cdots + \overline{f}_o(X) \in \overline{K}[X, Y] \tag{18}$$

Now, *if* $\overline{\Phi}(X, Y)$ is an *irreducible* polynomial over $\overline{K}$ then we have

$$n = [\overline{K}(\overline{x}, \overline{y}) : \overline{K}(\overline{x})] \le [\overline{F} : \overline{K}(\overline{x})] \le [F : K(x)] = n$$

and we see that (14) holds, i.e., $F|K$ has regular reduction. Thus, the irreducibility of the reduced polynomial $\overline{\Phi}(X, Y)$ implies regularity.

However, this sufficient criterion is not necessary. If $\overline{\Phi}(X, Y)$ is reducible there may be other generators of F over K whose defining equation stays irreducible after reduction.

But if we start from an arbitrary function field $\overline{E}|\overline{K}$ over the residue field of K, then it is possible to construct a "regular lift", i.e., a function field $F|K$ with a valuation extending the given valuation v on K, such that the residue field $\overline{F} = \overline{E}$, and that F admits a regular element.

Indeed: for example we assume $\overline{E} = \overline{K}(\overline{x}, \overline{y})$ with a defining irreducible equation (17) of the form (18). For each non-zero coefficient of $\overline{\Phi}(X, Y)$ we choose a foreimage in K and and obtain a polynomial $\Phi[X, Y] \in K[X, Y]$, of the same degree in Y, which is irreducible since its reduction $\overline{\Phi}(X, Y) \in \overline{K}[X, Y]$ is irreducible. Now let $F = K(x, y)$ be the function field over K defined by the irreducible equation $\Phi(x, y) = 0$. By construction, the substitution $(x, y) \mapsto (\overline{x}, \overline{y})$ defines a homomorphism $K[x, y] \to \overline{K}[\overline{x}, \overline{y}]$ extending the residue map $K \to \overline{K}$ modulo v. That homomorphism can be extended to a place of $F = K(x, y)$, i.e., to the residue map of a valuation of F, again denoted by v, extending the given valuation v of K. In this way $F|K$ becomes a valued function field and, we claim, it has regular reduction, the residue field being the given field $\overline{E}$. Indeed: On the one hand, since x is mapped onto $\overline{x}$ which is transcendental over $\overline{K}$ it follows that in $K(x)$ the valuation v is the Gauss valuation v_x. (See Theorem 1.) The residue field of v_x is $\overline{K}(\overline{x})$. Now by construction, $[F : K(x)] = n$ and the residue field of F contains $\overline{F}(\overline{x}, \overline{y}) = \overline{E}$ which is of degree n over $\overline{K}(\overline{x})$. In view of (13) we conclude that, indeed, $\overline{F} = \overline{E}$, and that x is a regular element in F.

In the situation discussed above it is said that $F|K$ arises by **lifting** the given function field $\overline{E}|\overline{K}$. Note that the lifting process as described above is *not unique.* There are several liftings, in fact infinitely many, of any given function field $\overline{E}|\overline{K}$.

3. The inertia theorem

The notion of an "inert" extension of valued fields had been introduced by Hilbert. But he considered extensions $L|K$ of *finite degree* only. [1] According to Hilbert, $L|K$ is called "inert" if $[\overline{L} : \overline{K}] = [L : K]$. Hence the definition of "regular element" in (14) can be phrased by saying that F is inert over $K(x)$, with respect to the Gauss valuation of $K(x)$.

Hilbert's definition, however, does not make sense for valued field extensions $F|K$ of infinite degree. In that case we have to consider finite dimensional K-submodules $M \subset F$. We denote by $\overline{M}$ the reduction of M. More precisely, if $\mathcal{O}_F$ denotes the valuation ring of F then $\overline{M}$ is the image of $\mathcal{O}_F \cap M$ under the residue map. $\overline{M}$ is a $\overline{K}$-module and we note:

LEMMA 4. *For any finite K-module $M \subset F$ we have* $\dim_{\overline{K}} \overline{M} \le \dim_K M$. *More generally, if $N \subset M$ are K-submodules of F and if* $\dim_K M/N$ *is finite then* $\dim_{\overline{K}} \overline{M}/\overline{N} \le \dim_K M/N$.

PROOF. Let $\overline{u}_1, \ldots, \overline{u}_r \in \overline{M}$ be $\overline{K}$-linearly independent modulo $\overline{N}$, and let the $u_i \in M$ be foreimages of the $\overline{u}_i$. We claim that the u_i are K-linearly independent modulo N. For, assume that there would be a nontrivial linear relation

$$a_1 u_1 + \cdots + a_r u_r \equiv 0 \bmod N \tag{19}$$

[1] In Hilbert's situation $L|K$ was an extension of finite number fields, and the valuation belonged to a prime ideal of the ring of integers.

with $a_i \in K$. Let $a \in K$ such that $v(a) = \min_i v(a_i)$. After dividing the above relation by a and changing notation we may assume

$$\min_{1 \le i \le r} v(a_i) = 0 .$$

In other words, not all $\overline{a}_i = 0$. Applying the residue homomorphism to (19) we obtain a nontrivial linear relation

$$\overline{a}_1 \, \overline{u}_1 + \cdots + \overline{a}_r \, \overline{u}_r \equiv 0 \bmod \overline{N}$$

contradicting the $\overline{K}$-linear independence of the $\overline{u}_i$ modulo $\overline{N}$. □

COROLLARY 5. *Suppose $M \subset F$ is finite dimensional over K. If $\dim \overline{M} = \dim M$ then for every K-submodule $N \subset M$ we also have $\dim \overline{N} = \dim N$.*[2]

PROOF. Let $m, n, \overline{m}, \overline{n}$ be the dimensions of $M, N, \overline{M}, \overline{N}$. By assumption we have $\overline{m} = m$. By Lemma 4 we have $\overline{m} - \overline{n} \le m - n$ and so

$$n \le m - (\overline{m} - \overline{n}) = \overline{n} \le n$$

hence $\overline{n} = n$. □

Corollary 5 permits to generalize Hilbert's notion of "inert" field extensions to valued field extensions $F|K$ of infinite degree and, more generally, for K-submodules $M \subset F$. As follows:

DEFINITION 1. Let $F|K$ be an extension of valued fields and $M \subset F$ any K-module, of finite or infinite dimension. M is said to be **inert** if for every finite-dimensional K-module $N \subset M$ we have $\dim \overline{N} = \dim N$.

We are now able to formulate the Inertia Theorem which is announced in the title of this section. It will be the basis of all that follows.

THEOREM 6 (Inertia Theorem). *Let $F|K$ be a valued function field with regular reduction $\overline{F}|\overline{K}$. Then $F|K$ is inert in the sense as defined above.*

REMARK 7. The definition of "regular" reduction as given in the foregoing section says that $F|K(x)$ is inert (in the sense of Hilbert). The above Theorem now asserts that this implies the whole function field $F|K$ to be inert.

The proof of the inertia Theorem rests on certain general properties of the notion of inertia. Let us state these properties in two Lemmas. In these Lemmas $F|K$ may be an arbitrary extension of valued fields.

If $M \subset F$ is a K-module of finite dimension and $\dim \overline{M} = \dim M$ then the proof of Lemma 4 (for $N = 0$) shows that there exists a K-basis $u_1, \ldots, u_m$ of M such that their residues $\overline{u}_1, \ldots, \overline{u}_m$ form a $\overline{K}$-basis of $\overline{M}$. If $z \in M$ is represented as a linear combination

$$z = \sum_{1 \le i \le m} a_i u_i \qquad \text{with} \qquad a_i \in K$$

then

$$v(z) = \min_{1 \le i \le m} v(a_i) . \tag{20}$$

[2]Here and in the following, to simplify notation we shall write $\dim M$ and $\dim \overline{M}$, thus omitting the reference to K and $\overline{K}$ respectively, whenever it is clear from the context which base field is referred to.

Such a basis is called a **valuation basis** of M. Hence $\dim M = \dim \overline{M}$ if and only if M admits a valuation basis over K.

The notion of "valuation basis" can also be defined for infinite dimensional modules M, as a K-basis u_i such that the $\overline{u}_i$ form a $\overline{K}$-basis of $\overline{M}$ (where i ranges over a suitable index set I). We do not claim that every infinite inert K-module $M \subset F$ admits a valuation basis. However:

LEMMA 8. *Let M be any K-submodule of F, of finite or infinite dimension. If M admits a valuation basis u_i $(i \in I)$ then M is inert.*

PROOF. Every finite dimensional submodule $N \subset M$ is contained in a finite dimensional submodule of the form $M_1 = \sum_{i \in I_1} Ku_i$ with a suitable finite index set $I_1 \subset I$. Since M_1 is inert (because it admits a valuation basis) we may apply Corollary 5 to conclude that N is inert. □

LEMMA 9. *Suppose $R \subset F$ is a K-algebra and F is the field of quotients of R. If R is inert over K then so is F.*

PROOF. Let $M \subset F$ be a finite dimensional K-submodule. We choose a K-basis of M. Every basis element is a quotient of two elements from R. We can choose a common denominator $z \in R$ for these finitely many quotients (because R is an algebra, hence closed with respect to products). Hence $zM \subset R$, and therefore zM is inert. Since $v(z) \in v(K)$ we may assume $v(z) = 0$, after dividing z by a suitable element $a \in K$. It follows $\overline{zM} = \overline{z}\overline{M}$ and therefore

$$\dim \overline{M} = \dim \overline{z}\overline{M} = \dim \overline{zM} = \dim zM = \dim M\,.$$

This being the case for every finite-dimensional module $M \subset F$, it follows that F is inert. □

COROLLARY 10. *Let $F = K(x)$ be a rational function field, equipped with the Gauss valuation with respect to x. Then F is inert over K.*

For, F is the field of quotients of the polynomial ring $R = K[x]$ and this admits the powers x^i $(i = 0, 1, 2, \ldots)$ as a valuation basis in view of (1).

REMARK 11. The corollary holds also for function fields of several variables and for the field of convergent power series; see Remark 3 at the end of section 1.

Now consider again an arbitrary valued field extension $F|K$.

LEMMA 12. *Let $K \subset F_o \subset F$. If $F_o|K$ is inert and $F|F_o$ is inert then $F|K$ is inert.*

PROOF. Let $M \subset F$ be a finite dimensional K-module. Then F_oM is a finite dimensional F_o-module. By the hypothesis of the Theorem, F_oM is inert over F_o. Hence there exists a valuation-F_o-basis u_i of F_oM.

Every element of a K-basis of M can be written as a linear combination of the u_i with coefficients in F_o. Let $M_o \subset F_o$ denote the K-module generated by those finitely many coefficients. M_o is inert over K and so there exists a valuation-K-basis w_j of M_o. By construction, M is contained in the K-module M^* generated by the w_ju_i. These elements form a valuation basis of M^*. For, the $\overline{u}_i$ are $\overline{F}_o$-linearly independent and the $\overline{w}_j \in \overline{F}_o$ are $\overline{K}$-linearly independent. Hence M^* is inert over K and so is M as a K-submodule of M^*. □

PROOF OF THE INERTIA THEOREM. Let $x \in F$ be a regular element. Then the subfield $K(x) \subset F$ is valued by the Gauss valuation and, hence, it is inert over K by corollary 10. Since $[F : K(x)] = [\overline{F} : \overline{K}(\overline{x})]$ we see that F is inert over $K(x)$. Hence $F|K$ is inert by Lemma 12. □

4. Reduction of genus

In dealing with the genus of a function field $F|K$ it is usually assumed that K is relatively algebraically closed in F. Sometimes this is expressed by saying that K is the "exact field of constants" of F. Then the genus g can be defined as the number occurring in the Riemann-Roch Theorem, as follows: Every divisor A of $F|K$ defines a K-module consisting of all elements $y \in F^\times$ whose pole divisor divides A. This module is denoted by $\mathcal{L}(A)$. Its K-dimension is finite and denoted by $\dim A$. The **Riemann-Roch Theorem** implies that $\dim A$ is related to the degree of A by the formula

$$\dim A \geq \deg A + (1-g)\,. \tag{21}$$

More precisely, we have

$$\dim A = \deg A + (1-g) + \dim(W - A) \tag{22}$$

where W is a canonical divisor of the function field $F|K$. Here we do not need to go into the details of definition of canonical divisor. The only fact which we have to know is that

$$\dim W = g \qquad \text{and} \qquad \deg W = 2g-2\,,$$

which is immediate from (22). Since divisors of negative degree have vanishing dimension we conclude

$$\deg A > 2g-2 \qquad \Longrightarrow \qquad \dim A = \deg A + (1-g) \tag{23}$$

Since $\dim A \leq 1 + \deg A$ (proof by induction on $\deg A$) we also have

$$\dim A > 2g-1 \qquad \Longrightarrow \qquad \dim A = \deg A + (1-g) \tag{24}$$

The number $\chi = 1 - g$ which appears in these formulas is called the **arithmetic genus** of the function field. Sometimes g is then called the "geometric genus" of $F|K$.

The above formulas hold if K is relatively algebraically closed in F. But we do not wish to make this assumption, simply because it is not necessary for the general theory. So let K' denote the algebraic closure of K in F and $d = [K' : K]$ its degree. K' is called the "field of constants" of F and d the "degree of constants". For a divisor A of $F|K$ its degree over K is d times the degree over K', and same for $\dim A$. Hence, if the arithmetic genus χ of $F|K$ is now defined to be $d(1-g)$ then the Riemann-Roch Theorem (22) for $F|K$ appears in the form

$$\dim A = \deg A + \chi + \dim(W - A) \tag{25}$$

without the assumption that K is relatively algebraically closed in F. If in (25) we set $A = 0$ and then $A = W$ we obtain [3]

$$\dim W = -\chi + d \qquad \text{and} \qquad \deg W = -2\chi\,. \tag{26}$$

Note that $\chi < 0$ except if $g = 1$ when $\chi = 0$, or when $g = 0$ in which case $\chi = d > 0$.

[3] Observe that for the zero divisor $A = 0$ we have $\deg(0) = 0$ and $\dim(0) = d$.

The formulas (23) and (24) now appear in the form:

$$\text{(27)} \qquad \deg A > -2\chi \qquad \Longrightarrow \qquad \dim A = \deg A + \chi$$

$$\text{(28)} \qquad \dim A > -2\chi + d \qquad \Longrightarrow \qquad \dim A = \deg A + \chi\,.$$

In order to simplify our formulas we shall write $A \gg 0$ if $A > 0$ and $\deg A$ and $\dim A$ are sufficiently large, so that we have in particular

$$\text{(29)} \qquad A \gg 0 \qquad \Longrightarrow \qquad \dim A = \deg A + \chi\,.$$

If $A > 0$ then for a sufficiently large integer $t > 0$ we have $tA \gg 0$.

Now we consider a valued function field $F|K$ with regular reduction $\overline{F}|\overline{K}$. Then we have the same formulas as above for divisors $\overline{A}$ of $\overline{F}|\overline{K}$, where $\deg \overline{A}$ and $\dim \overline{A}$ are to be understood over $\overline{K}$: [4]

$$\text{(30)} \qquad \overline{A} \gg 0 \qquad \Longrightarrow \qquad \dim \overline{A} = \deg \overline{A} + \overline{\chi}\,.$$

Here, $\overline{\chi}$ denotes the arithmetic genus of $\overline{F}|\overline{K}$, and the condition $\overline{A} \gg 0$ refers to the bounds given by $\overline{F}|\overline{K}$.

In general, if we consider at the same time both $F|K$ and its reduction $\overline{F}|\overline{K}$ then the condition $A \gg 0$ refers to the above degree conditions for both $F|K$ and $\overline{F}|\overline{K}$.

Our aim is to compare the arithmetic genus χ of $F|K$ with the arithmetic genus $\overline{\chi}$ of the reduced field $\overline{F}|\overline{K}$.

THEOREM 13 (Genus Reduction Theorem). *Let $F|K$ be a valued function field with regular reduction $\overline{F}|\overline{K}$. Let χ be the arithmetic genus of $F|K$ and $\overline{\chi}$ the arithmetic genus of $\overline{F}|\overline{K}$. Then $\chi \le \overline{\chi}$. Specifically, let $x \in F$ be a regular element and let R_x denote the integral closure of $K[x]$ in F. Similarly $R_{\overline{x}}$ denotes the integral closure of $\overline{K}[\overline{x}]$ in $\overline{F}$. Then*

$$\text{(31)} \qquad \overline{\chi} - \chi = \dim R_{\overline{x}} / \overline{R_x}.$$

Note that since $\dim R_{\overline{x}}/\overline{R_x}$ is finite it follows that $R_{\overline{x}}$ is the integral closure of $\overline{R_x}$ in $\overline{F}$. In particular, we have $\overline{\chi} = \chi$ if and only if $\overline{R_x}$ is integrally closed.

REMARK 14. By definition we have $\chi = d(1-g)$ where g is the geometric genus of $F|K$ and d is the degree of constants. Similarly $\overline{\chi} = \overline{d}(1 - \overline{g})$. In general we have $\overline{d} \ge d$. If the reduction preserves the degree of constants, i.e., if $\overline{d} = d$ then the statement $\chi \le \overline{\chi}$ means $g \ge \overline{g}$.

PROOF OF THEOREM 13. For a divisor $A \gg 0$ of $F|K$ define the divisor $\overline{A}$ of $\overline{F}|\overline{K}$ by

$$\text{(32)} \qquad \overline{A} := \{\sup(\overline{y})_\infty \mid 0 \ne \overline{y} \in \overline{\mathcal{L}(A)}\,\}.$$

Let us call $\overline{A}$ the divisor "associated to A".[5] By definition we have

$$\text{(33)} \qquad \overline{\mathcal{L}(A)} \subset \mathcal{L}(\overline{A}) \qquad \text{hence} \qquad \dim A \le \dim \overline{A}$$

where we have used the Inertia Theorem. In particular it follows $\overline{A} \gg 0$.

---

[4]More precisely we should have written $\deg_{\overline{F}|\overline{K}}(\overline{A})$ and $\dim_{\overline{F}|\overline{K}}(\overline{A})$. But for simplicity of notation we skip the index $\overline{F}|\overline{K}$ if it is clear from the context which function field is referred to. Similarly we write $\mathcal{L}(\overline{A})$ instead of $\mathcal{L}_{\overline{F}|\overline{K}}(\overline{A})$ etc.

[5]Until further notice $\overline{A}$ is defined for $A \gg 0$ only. But see Theorem 17 below.

If there exists $\overline{y} \in \overline{\mathcal{L}(A)}$ with $(\overline{y})_\infty = \overline{A}$ let $y \in \mathcal{L}(A)$ be a foreimage. We compute:

$$\deg \overline{A} = \deg(\overline{y})_\infty = [\overline{F} : \overline{K}(\overline{y})] \leq [F : K(y)] = \deg(y)_\infty \leq \deg A\,, \tag{34}$$

the last relation since $y \in \mathcal{L}(A)$ and hence $(y)_\infty \leq A$.

If the base field $\overline{K}$ of $\overline{F}$ is infinite then indeed there exists such $\overline{y} \in \overline{\mathcal{L}(A)}$. (See Lemma 15 below.) If $\overline{K}$ is finite we have to use some other argument conclude that

$$\deg \overline{A} \leq \deg A\,. \tag{35}$$

(See Lemma 16 below.) From (29) it follows in view of (33) and (35)

$$\begin{aligned} \overline{\chi} - \chi &= (\dim \overline{A} - \deg \overline{A}) - (\dim A - \deg A) \\ &= (\dim \overline{A} - \dim A) + (\deg A - \deg \overline{A}) \geq 0\,. \end{aligned} \tag{36}$$

Now let $A = (x)_\infty$ be the pole divisor of a regular element x of $F|K$. After replacing x by x^t with sufficiently large t we may assume $A \gg 0$. We have $\deg(\overline{x})_\infty = \deg A$. Since $\overline{x} \in \overline{\mathcal{L}(A)}$ it follows $\deg A \leq \deg \overline{A}$, hence in view of (35)

$$\deg \overline{A} = \deg A \tag{37}$$

and therefore by (36)

$$\overline{\chi} - \chi = \dim \overline{A} - \dim A = \dim \mathcal{L}(\overline{A})/\overline{\mathcal{L}(A)}\,. \tag{38}$$

If we replace x by x^t with $t > 0$ then A has to be replaced by tA and we obtain in the same way

$$\overline{\chi} - \chi = \dim \mathcal{L}(\overline{tA})/\overline{\mathcal{L}(tA)}\,. \tag{39}$$

Here we have $\overline{tA} = t\overline{A}$. This can be seen as follows: We have

$$\begin{aligned} \mathcal{L}(A)^t &\subset \mathcal{L}(tA)\,, \\ \overline{\mathcal{L}(A)}^{\,t} &\subset \overline{\mathcal{L}(tA)}\,, \\ t\overline{A} &\leq \overline{tA} && \text{by definition (32)}\,, \\ t\overline{A} &= \overline{tA}\,, \end{aligned}$$

the last relation holds since both sides have the same degree by (37) when applied to tA. Note that x^t is also regular, and it has pole divisor tA.

Now, R_x consists of all elements y whose pole divisor $(y)_\infty$ is composed by prime divisors contained in A. Thus

$$R_x = \bigcup_{t>0} \mathcal{L}(tA) \qquad \text{hence} \qquad \overline{R_x} = \bigcup_{t>0} \overline{\mathcal{L}(tA)}\,.$$

Similarly

$$R_{\overline{x}} = \bigcup_{t>0} \mathcal{L}(t\overline{A})\,,$$

We conclude from (39) for $t \to \infty$ that $\overline{\chi} - \chi = \dim R_{\overline{x}}/\overline{R_x}$ as announced in the Genus Theorem.

We now turn to the two lemmas which we have used in the foregoing proof.

LEMMA 15. *Let $F|K$ be a function field and $M \neq 0$ a finitely generated K-module in F. Define the divisor*

$$M_\infty := \{\sup(y)_\infty \mid 0 \neq y \in M\}. \tag{40}$$

If K is infinite then there exists $y \in M$ such that $(y)_\infty = M_\infty$.

PROOF. The assertion is trivial if $M_\infty = 0$. Otherwise, decompose M_∞ into prime divisors:

$$M_\infty = \sum_i a_i P_i$$

with coefficients $a_i \geq 1$. For each i there exists $y_i \in M$ with $v_{P_i}(y_i) = -a_i$, according to the definition (40). (v_{P_i} denotes the valuation of $F|K$ belonging to P_i.) Let $N_i \subset M$ be the submodule consisting of those $y \in M$ with $v_{P_i}(y) > -a_i$. Then $y_i \notin N_i$, hence N_i is a proper submodule of M. Now, if K is infinite there exists $y \in M$ which is not contained in any of these finitely many proper submodules N_i. (Proof by induction on the dimension of M.) □

We apply Lemma 15 to the function field $\overline{F}|\overline{K}$ and the $\overline{K}$-module $\overline{\mathcal{L}(A)}$. We obtain an element $\overline{y} \in \overline{\mathcal{L}(A)}$ with $(\overline{y})_\infty = \overline{A}$, as we have used in the computation (34) – provided the base field $\overline{K}$ is infinite.

If $\overline{K}$ is finite we have to argue by means of base field extensions. Again, as in Lemma 15 we formulate the next Lemma for any function field $F|K$, but then we will have to apply it for $\overline{F}|\overline{K}$ in order to obtain (35).

In the situation of Lemma 15 let $K' = K(z)$ be a rational function field with a new variable z. Consider the base field extension $F' = FK'$, i.e., the field of quotients of the integral domain $F \otimes_K K'$. The divisor group of $F|K$ embeds naturally into the divisor group of $F'|K'$:

$$\mathrm{Div}(F|K) \subset \mathrm{Div}(F'|K'). \tag{41}$$

Under this embedding the degree of divisors is stable, i.e., for $A \in \mathrm{Div}(F|K)$ we have $\deg_{F|K}(A) = \deg_{F'|K'}(A)$.

LEMMA 16. *Let $0 \neq M$ be a finite dimensional submodule of $F|K$ and $M' = M \otimes_K K'$. Then $M_\infty = M'_\infty$ according to the embedding* (41). *Hence* $\deg_{F|K} M_\infty = \deg_{F'|K'} M'_\infty$.

PROOF. Since $M \subset M'$ we have $M_\infty \leq M'_\infty$. On the other hand, M' consists of the elements y' of the form $y' = \sum_j c_j y_j$ with $c_j \in K'$ and $y_j \in M$. It follows $(y')_\infty \leq \sup_j (y_j)_\infty \leq M_\infty$, hence $M'_\infty \leq M_\infty$. □

We apply this Lemma in the proof of Theorem 13 as follows: We consider the function field $F' = F(z)$ over $K' = K(z)$. We extend the given valuation of F to F' by means of the Gauss valuation with respect to z. Then $F'|K'$ becomes a valued function field with reduction $\overline{F'} = \overline{F}(\overline{z})$ over $\overline{K'} = \overline{K}(\overline{z})$. For any divisor A we have

$$\mathcal{L}_{F'|K'}(A) = \mathcal{L}_{F|K}(A) \otimes_K K'.$$

(This is well known from the general theory of base field extensions.) It follows

$$\overline{\mathcal{L}_{F'|K'}(A)} = \overline{\mathcal{L}_{F|K}(A)} \otimes_{\overline{K}} \overline{K'}.$$

We apply Lemma 16 to $\overline{F}|\overline{K}$ and the module $\overline{\mathcal{L}_{\overline{F}|\overline{K}}(A)}$ and conclude that the divisor $\overline{A} = \overline{\mathcal{L}_{\overline{F}|\overline{K}}(A)}_\infty$ of $\overline{F}|\overline{K}$ (defined by (32)) coincides with the divisor which is defined similarly for $\overline{F'}|\overline{K'}$. Since $\overline{K'}$ is infinite we obtain

$$\deg_{\overline{F}|\overline{K}}(\overline{A}) = \deg_{\overline{F'}|\overline{K'}}(\overline{A}) \leq \deg_{F'|K'}(A) = \deg_{F|K}(A)$$

as required for (35). □

5. Good reduction

$F|K$ denotes a valued function field with regular reduction $\overline{F}|\overline{K}$, arithmetic genus χ and reduced arithmetic genus $\overline{\chi}$.

DEFINITION 2. $F|K$ is said to have **good reduction** if: $\chi = \overline{\chi}$.

Formula(36) shows that in the case of good reduction we have for any divisor $A \gg 0$:

$$\deg \overline{A} = \deg A \qquad \text{and} \qquad \dim \overline{A} = \dim A\,. \tag{42}$$

This implies

$$\overline{\mathcal{L}(A)} = \mathcal{L}(\overline{A}) \tag{43}$$

in view of (33). Here $\overline{A}$ is the divisor which we have defined in (32) for $A \gg 0$, and which we had called there "associated to A". The following Theorem shows that this definition can be extended to all divisors of $F|K$, leading to a homomorphism of the divisor groups.

THEOREM 17 (Divisor Reduction Theorem). *Suppose the valued function field $F|K$ admits good reduction. Then there exists one and only one homomorphism* $\mathrm{Div}(F|K) \to \mathrm{Div}(\overline{F}|\overline{K})$ *of the divisor groups which preserves the degree, the order relation and principal divisors.*

More precisely, the homomorphism $\mathrm{Div}(F|K) \to \mathrm{Div}(\overline{F}|\overline{K})$ *has the following properties, where $\overline{A}$ denotes the image of $A \in \mathrm{Div}(F|K)$:*

$$\deg \overline{A} = \deg A \tag{44}$$

$$A \le B \implies \overline{A} \le \overline{B} \tag{45}$$

$$\overline{(x)} = (\overline{x}) \qquad (\textit{if } \overline{x} \neq 0, \infty)\,. \tag{46}$$

PROOF. For any divisor $A \gg 0$ of $F|K$ we define $\overline{A}$ as in (32). If $B \gg 0$ is another divisor then we conclude successively:

$$\begin{aligned} \mathcal{L}(A) \cdot \mathcal{L}(B) &\subset \mathcal{L}(A+B) \\ \overline{\mathcal{L}(A)} \cdot \overline{\mathcal{L}(B)} &\subset \overline{\mathcal{L}(A+B)} \\ \overline{A} + \overline{B} &\le \overline{A+B} \qquad \text{by definition (32)} \\ \overline{A} + \overline{B} &= \overline{A+B}\,, \end{aligned} \tag{47}$$

the last conclusion holds since both sides have the same degree in view of (42). This holds for $A, B \gg 0$. Now, if A is an arbitrary divisor of $F|K$ then we write

$$A = B - C \qquad \text{with} \qquad B, C \gg 0 \tag{48}$$

and define:

$$\overline{A} := \overline{B} - \overline{C}\,. \tag{49}$$

Because of (47) this is well defined, i.e., does not depend on the choice of $B, C \gg 0$, and it leads to a homomorphism of the divisor groups.. By (42) the degree formula (44) holds if $A \gg 0$, hence it remains true for arbitrary A because of (49).

If $A > 0$ then $tA \gg 0$ for sufficiently large t, hence $\overline{tA} = t\overline{A} \gg 0$ and therefore $\overline{A} > 0$, i.e. the map $A \mapsto \overline{A}$ preserves the order relation for divisors.

If $\overline{x} \neq 0, \infty$ then we write $(x) = B - A$ with $A, B \gg 0$ and have

$$\begin{aligned}(x) + A &= B\\ x \cdot \mathcal{L}(A) &= \mathcal{L}(B)\\ \overline{x} \cdot \overline{\mathcal{L}(A)} &= \overline{\mathcal{L}(B)}\\ \overline{x} \cdot \mathcal{L}(\overline{A}) &= \mathcal{L}(\overline{B})\\ (\overline{x}) + \overline{A} &= \overline{B}\\ (\overline{x}) = \overline{B} - \overline{A} &= \overline{B - A} = \overline{(x)}\,.\end{aligned}$$

This proves the existence of the divisor reduction map. To see that it is unique, assume that there is another such map $A \mapsto A^*$. It suffices to show that $A^* = \overline{A}$ for $A \gg 0$.

If $0 \neq x \in \mathcal{L}(A))$ then $(x) \geq -A$. From (45) and (46) (applied to A^*) it follows $(\overline{x}) \geq -A^*$, i.e., $(\overline{x})_\infty \leq A^*$. In view of (32) it follows $\overline{A} \leq A^*$, hence $\overline{A} = A^*$ since both have the same degree by (44). □

COROLLARY 18. *Suppose $F|K$ admits good reduction. If W is a canonical divisor of $F|K$ then its image $\overline{W}$ is a canonical divisor of $\overline{F}|\overline{K}$.*

PROOF. Until now we have used the Riemann-Roch Theorem in the short version (29) only, which applies to divisors $A \gg 0$. But now, since $W \not\gg 0$ we have to go back to its original version (25). For a canonical divisor W we have seen in (26) that

$$\dim W = -\chi + d \qquad \text{and} \qquad \deg W = -2\chi\,. \tag{50}$$

Canonical divisors are characterized by these values of $\dim(\cdot)$ and $\deg(\cdot)$. For, if A is any divisor with $\deg A = -2\chi$ then by the Riemann-Roch Theorem:

$$\dim(A) = -2\chi + \chi + \dim(W - A)$$

The divisor $W - A$ is of degree 0. A divisor of degree 0 has dimension 0 except if it is a principal divisor in which case its dimension is d. Now, if $W - A$ is a principal divisor then A belongs to the same divisor class as W, i.e., A is in the canonical class. We conclude:

$$\deg A = -2\chi \implies \dim(A) = \begin{cases} -\chi + d \text{ if } A \text{ is a canonical divisor} \\ -\chi \ \text{ if } A \text{ is not canonical.} \end{cases}$$

Now, for the reduced function field $\overline{F}|\overline{K}$ we have the same situation, with d to be replaced by $\overline{d}$, the degree of the relative algebraic closure of $\overline{K}$ in $\overline{F}$. From (44) we know that $\deg \overline{W} = \deg W = -2\chi = -2\overline{\chi}$. Thus, if $\overline{W}$ would not be canonical then $\dim \overline{W} = -\overline{\chi} = -\chi$. But this is not the case, as can be seen as follows: $\mathcal{L}(W)$ consists of those $x \in F$ for which $(x) \geq -W$. Hence for $\overline{x} \in \overline{\mathcal{L}(W)}$ we have $(\overline{x}) \geq -\overline{W}$ by (45) and (46). It follows $\overline{\mathcal{L}(W)} \subset \mathcal{L}(\overline{W})$. Using the Inertia Theorem it follows $\dim W = -\chi + d \leq \dim(\overline{W})$, hence $\dim(\overline{W}) > -\chi$. □

REMARK 19. If the reduction preserves the degree of constants, i.e., if $d = \overline{d}$, then from (50), applied to W and $\overline{W}$, we see that $\dim W = \dim \overline{W}$. Hence $\overline{\mathcal{L}(W)} = \mathcal{L}(\overline{W})$ in this case.

Proposition 20 (Regular elements). *Let $F|K$ be a valued function field with good reduction. For every divisor $A > 0$ and sufficiently large $t > 0$ the divisor tA is the pole divisor of a regular element.*

Proof. First let us choose t such that $tA \gg 0$. Changing notation, we write again A instead of tA. In view of (43) we have $\overline{\mathcal{L}(A)} = \mathcal{L}(\overline{A})$. If $\overline{K}$ is infinite then we have seen in the proof of the Genus Theorem that there exists $\overline{y} \in \overline{\mathcal{L}(A)}$ with $(\overline{y})_\infty = \overline{A}$. (See Lemma 15.) Let $y \in \mathcal{L}(A)$ be a foreimage. We apply the computation (34) and conclude since $\deg \overline{A} = \deg A$:

$$[\overline{F} : \overline{K}(\overline{y})] = [F : K(y)] .$$

Hence y is regular and $(y)_\infty = A$. If $\overline{K}$ is finite then we apply the following Lemma to $\overline{F}|\overline{K}$.

Lemma 21. *Let $F|K$ be a function field and $A \gg 0$ a divisor. For every $t \geq 2$ there exists $y \in \mathcal{L}(tA)$ with denominator $(y)_\infty = tA$.*

Proof. Decompose A into prime divisors:

$$A = \sum_{1 \leq i \leq r} a_i P_i$$

with multiplicities $a_i \geq 1$. We define $B_i := A + (ta_i - 1)P_i \geq A$. Then $B_i \gg 0$. Put $C_i := B_i + P_i = A + ta_i P_i$. Then also $C_i \gg 0$ and therefore

$$\dim(C_i) - \dim(B_i) = \deg(C_i) - \deg(B_i) = \deg P_i > 0 .$$

Therefore there exists $y_i \in \mathcal{L}(C_i)\,,\ y_i \notin \mathcal{L}(B_i)\ (1 \leq i \leq r)$. We have

$$v_{P_i}(y_i) = -ta_i \qquad \text{and} \qquad v_{P_j}(y_i) \geq -a_j > -ta_j \quad \text{for} \quad j \neq i .$$

(v_{P_i}, v_{P_j} are the valuations belonging to the prime divisors P_i, P_j respectively.) The sum $y = \sum_{1 \leq i \leq r} y_i$ has $v_{P_i}(y) = -ta_i$ for $1 \leq i \leq r$, hence $(y)_\infty = tA$. □

Corollary 22. *Suppose $F|K$ has good reduction. Let $R \subset F$ be any subring containing K and with quotient field F. If R is integrally closed then $\overline{R}$ is integrally closed.*

Proof. Let $x \in R$ be a transcendental element and $A = (x)_\infty$. Since R is integrally closed it contains R_x, the integral closure of $K[x]$. The latter consists of all elements $y \in F$ whose denominator $(y)_\infty$ contains those primes only which are contained in A. Let t be sufficiently large so that tA, by Proposition 20, is the denominator of a regular element, say z. Then $R_z = R_x \subset R$. By the Genus Theorem the reduction $\overline{R_z}$ is integrally closed, hence a Dedekind ring. Every overring of a Dedekind ring in its quotient field is again a Dedekind ring. Hence $\overline{R}$ is integrally closed in $\overline{F}$. □

6. Almost all reductions

Consider a function field $F|K$. This time we assume that the base field K is algebraically closed. Our aim is the following

Theorem 23. *For almost all valuations v of K there exists an extension of v to F such that $F|K$ has good reduction with respect to v.*

Here, the expression "almost all" means that there are finitely many non-zero elements $c_1, \dots, c_r \in K$ (depending on F) such that the contention holds for those valuations v of K for which $\overline{c}_i \neq \infty$ for $i = 1, \dots, r$.

PROOF. Let $x \in F$ be transcendental over K and $n = [F : K(x)]$. Let $F = K(x, y)$ and

$$\Phi(x, y) = y^n + f_{n-1}(x)y^{n-1} + \cdots + f_0(x) = 0 \tag{51}$$

the corresponding irreducible equation; we may assume that the coefficients $f_i(x) \in K[x]$. Thus $[F : K(x)] = n$.

Let v be any valuation of K. We use the so-called

THEOREM 24 (Theorem of Bertini-Noether). *Let $\Phi[X, Y] \in K[X, Y]$ be any irreducible polynomial. Then for almost all valuations v of K the reduced polynomial $\overline{\Phi}(X, Y) \in \overline{K}[X, Y]$ is defined and is irreducible.*

A proof of the Bertini-Noether Theorem can be found in the book "Field Arithmetic" by Jarden and Fried, third edition (2008), page 184; see [**3**]. Recall that here we assume the base field K to be algebraically closed. Without this assumption the Bertini-Noether Theorem is valid for *absolutely irreducible* polynomials, i.e., polynomials which stay irreducible over the algebraic closure of K.

When we say that "$\overline{\Phi}(X, Y)$ is defined" then this means that for all coefficients $c \neq 0$ appearing in $\Phi[X, Y]$ we have $\overline{c} \neq \infty$.

We apply this to the polynomial (51) and obtain for almost all v an irreducible polynomial equation

$$\overline{\Phi}(\overline{x}, \overline{y}) = \overline{y}^n + \overline{f}_{n-1}(\overline{x})\overline{y}^{n-1} + \cdots + \overline{f}_0(\overline{x}) = 0$$

which defines a function field $\overline{F}|\overline{K}$ with $[\overline{F} : \overline{K}(\overline{x})] = n$. For those v the map $x, y \mapsto \overline{x}, \overline{y}$ defines a valuation of $F|K$ extending the given valuation v of K, having x as a regular element. $F|K$ is inert with respect to v. In particular we have $\overline{\chi} \geq \chi$ for the genera of $\overline{F}|\overline{K}$ and $F|K$. Let V denote the set of those valuations v. Our claim is that for almost all $v \in V$ we have $\overline{\chi} = \chi$.

Using Theorem 13 we have to show that $\overline{R_x}$ is integrally closed for almost all $v \in V$. It is well known that R_x is a finite $K[x, y]$-module. In fact, there exists one element $z \neq 0$ such that $R_x = K[x, y, z]$. For almost all v we have $\overline{z} \neq 0, \infty$. For, writing

$$z = \sum_{0 \leq i \leq r-1} \frac{f_i(x)}{g_i(x)} y^i$$

with polynomials $f_i(x), g_i(x) \in K[x]$ then for almost all v, every nonzero coefficient c of these polynomials has $\overline{c} \neq 0, \infty$, so that

$$\overline{z} = \sum_{0 \leq i \leq r-1} \frac{\overline{f}_i(\overline{x})}{\overline{g}_i(\overline{x})} \overline{y}^i .$$

Consider the affine curve Γ over K with generic point (x, y, z). The fact that R_x is integrally closed can be expressed in geometric terms by saying that every K-rational point (a, b, c) of Γ is simple. From algebraic geometry we will use the following well known necessary and sufficient criterion for a point (a, b, c) on Γ to be simple.

Let $\varphi_i(X, Y, Z)$ be a defining set of polynomials for Γ, i.e., the φ_i generate the ideal in the polynomial ring $K[X, Y, Z]$ which is the kernel of the homomorphism given by $(X,Y,Z) \mapsto (x,y,z)$. The index i ranges over a finite index set, say $1 \leq i \leq r$.

Consider the matrix of partial derivatives

$$\Delta(X,Y,Z) := \begin{pmatrix} \partial_X \varphi_i(X,Y,Z) \\ \partial_Y \varphi_i(X,y,Z) \\ \partial_Z \varphi_i(X,Y,Z) \end{pmatrix} \qquad (i = 1, \dots r)$$

which determines the tangent space. It has 3 rows and r columns. A point (a,b,c) of Γ is simple if and only if the matrix $\Delta(a,b,c)$ has rank 2, i.e., there is at least one 2×2 non-vanishing subdeterminant. If this is to hold for all K-rational points (a,b,c) of Γ then the ideal generated by all 2×2 subdeterminants $d_\nu(x,y,z)$ of $\Delta(x,y,z)$ is not contained in any maximal ideal of $K[x,y,z] = R_x$, i.e., there exists a relation of the form

$$1 = \sum_\nu g_\nu(x,y,z)\, d_\nu(x,y,z)$$

with polynomials $g_\nu(X,Y,Z) \in K[X,Y,Z]$. (The index ν ranges over a suitable finite index set.) Now, for almost all v every nonzero coefficient c appearing in these polynomials has $\overline{c} \neq 0, \infty$. Hence

$$1 = \sum_\nu \overline{g}_\nu(\overline{x},\overline{y},\overline{z})\, \overline{d}_\nu(\overline{x},\overline{y},\overline{z})$$

for these v. This implies that each K-rational point $\overline{a},\overline{b},\overline{c}$ of the reduced curve $\overline{\Gamma}$ is simple. The reduced curve $\overline{\Gamma}$ is defined by the generic point $(\overline{x},\overline{y},\overline{z})$. Since $\overline{K}[\overline{x},\overline{y},\overline{z}] = \overline{R_x}$ we conclude that $\overline{R_x}$ is integrally closed. □

COROLLARY 25. *The statement of the theorem holds also if the base field K is arbitrary infinite, provided the function field $F|K$ is conservative, i.e., the genus of $F|K$ coincides with the genus of every constant field extension.*

PROOF. Take K' to be the algebraic closure of K and let $F' = FK'$ the corresponding constant field extension. By Theorem 23 almost all valuations $v_{K'}$ of K' can be extended to a valuation $v_{F'}$ such as to give good reduction of $F'|K'$, i.e., $\chi_{F'} = \chi_{\overline{F'}}$. On the other hand, $v_{F'}$ induces in F a valuation v_F and in K a valuation v_K. In this way $F|K$ becomes a valued function with regular reduction, and from the Genus Reduction Theorem 13 we see that $\chi_F \leq \chi_{\overline{F}}$. Since $F|K$ is conservative we have $\chi_F = \chi_{F'}$. Together we conclude $\chi_{\overline{F}} \geq \chi_{\overline{F'}}$. Since $F' = FK'$ we have $\overline{F'} = \overline{F}\,\overline{K'}$, i.e., $\overline{F'}|\overline{K'}$ is a constant field extension of $\overline{F}|\overline{K}$. From the general theory it is well known that the arithmetic genus χ can only increase after constant field extensions. (The geometric genus g can only decrease.) It follows $\chi_{\overline{F}} = \chi_{\overline{F'}} = \chi_F$. Hence v_F defines good reduction on $F|K$.

This holds "for almost all valuations $v_{K'}$". By definition this means that there is finite set $S_{K'} \subset K'$ such that the contention holds for those $v_{K'}$ for which $v_{K'}(c) \neq \infty$ for all $c \in S_{K'}$. For every such c let $f_c(X)$ denote its irreducible polynomial over K. Let $S_K \subset K$ denote the set of coefficients of all those polynomials $f_c(X)$. If for every $a \in S_K$ we have $v_K(a) \neq \infty$ then $v_{K'}(c) \neq \infty$ for every $c \in K'$ and every extension $v_{K'}$ of v_K to K'. Hence, as shown above, v_K admits an extension v_F to $F|K$ with good reduction. □

REMARK 26. *A sufficient criterium for conservativity of a function field $F|K$ is that the base field K is perfect and is relatively algebraically closed in F.* As a reference we mention Deuring's Lectures on the theory of algebraic functions, published 1973; see[**2**].

References

[1] Max Deuring, *Reduktion algebraischer Funktionenkörper nach Primdivisoren des Konstantenkörpers* (German), Math. Z. **47** (1942), 643–654, DOI 10.1007/BF01180977. MR15055

[2] Max Deuring, *Lectures on the theory of algebraic functions of one variable*, Lecture Notes in Mathematics, Vol. 314, Springer-Verlag, Berlin-New York, 1973. MR0344231

[3] Michael D. Fried and Moshe Jarden, *Field arithmetic*, 3rd ed., Ergebnisse der Mathematik und ihrer Grenzgebiete. 3. Folge. A Series of Modern Surveys in Mathematics [Results in Mathematics and Related Areas. 3rd Series. A Series of Modern Surveys in Mathematics], vol. 11, Springer-Verlag, Berlin, 2008. Revised by Jarden. MR2445111

African Institute for Mathematical Sciences, 6 Melrose Road, Muizenberg, Cape Town 7945, South Africa

Email address: `bwg@aims.ac.za`

Mathematisches Institut, Universitaet Heidelberg, Heidelberg, Germany

Email address: `roquette@uni-hd.de`

Contemporary Mathematics
Volume **767**, 2021
https://doi.org/10.1090/conm/767/15399

Purely inseparable points on curves

Damian Rössler

To Gerhard Frey on the occasion of his 75th birthday

ABSTRACT. We give effective upper bounds for the number of purely inseparable points on non isotrivial curves over function fields of positive characteristic and of transcendence degree one. These bounds depend on the genus of the curve, the genus of the function field and the number of points of bad reduction of the curve.

1. Introduction

I am very happy to have the possibility to dedicate this article to Gerhard Frey, who made so many interesting contributions to number theory. An electronic correspondence with him on the subject of Kim's article [**6**] prompted me to look for effective upper bounds for the number of purely inseparable points on curves and the present text summarises what I found.

Let k be an algebraically closed field of characteristic $p > 0$. Let K be the function field of a smooth, proper and geometrically connected curve B over k. Let C be a smooth, proper and geometrically connected curve over K. Let g_B (resp. g_C) be the genus of B (resp. C). We shall denote by A the Jacobian of C over K. We let $\Sigma \subseteq B$ be the set of closed points of bad reduction of A and we let $N = \#\Sigma$. Suppose that $g_C > 0$ and that $C(K) \neq \emptyset$. We suppose that C is not isotrivial over K. By definition, this means that there is no finite field extension $L|K$, such that $C_L \simeq C_{0,L}$, where C_0 is a smooth curve over k. We choose an algebraic closure $\bar{K}$ of K and we denote by $K_{\text{ins}} := \bigcup_{i\geq 0} K^{p^{-i}} \subseteq \bar{K}$ the maximal purely inseparable extension of K in $\bar{K}$. This is also called the perfection of K. The elements of $C(K_{\text{ins}})$ will be called the *perfect points* or *purely inseparable points* of C.

We shall prove

THEOREM 1.1. (a) *If $g_C > 1$ then $C(K_{\text{ins}})$ is a finite set.*

(b) *Suppose that the Jacobian A of C has semiabelian reduction over B. Let $\ell_0 \geq 1$ be such that*

$$p^{\ell_0} - p^{\ell_0-1} g_C(g_C-1)\max\{2g_B-2, 0\} - g_C(2g_B-2) - g_C N > 0. \tag{1.1}$$

Let ℓ be the smallest integer such that $\ell \geq \ell_0$ and $p^\ell > 2g_C - 2$. Then:

- the set $C(K_{\text{ins}}) \smallsetminus C(K^{p^{-\ell+1}})$ is empty;

2020 *Mathematics Subject Classification.* Primary 12F15, 11G30.

Key words and phrases. Function fields, curves, inseparability, rational points.

- *we have* $\#(C(K_{\mathrm{ins}}) \setminus C(K^{p^{-\ell_0+1}})) \leq 2g_C - 2$.

Note that inequality (1.1) is a condition on both p and ℓ_0. For a fixed ℓ_0 (for example $\ell_0 = 1$) it is satisfied if p is large enough. On the other hand, if

$$g_C(g_C - 1)\max\{2g_B - 2, 0\} < p$$

then it will be satisfied for any sufficiently large ℓ_0. In particular, we have the

COROLLARY 1.2. *Suppose that the Jacobian A of C has semiabelian reduction over B. If $g_B > 0$ and*

$$p > \max\{2g_C - 2, g_C^2(2g_B - 2) + g_C N\}$$

then the set $C(K_{\mathrm{ins}}) \setminus C(K)$ is empty. If $g_B = 0$ and

$$p > \max\{2g_C - 2, g_C(N - 2)\}$$

then the set $C(K_{\mathrm{ins}}) \setminus C(K)$ is empty.

REMARK 1.3. Theorem 1.1 (a) also follows from the Claim on page 665 of [**6**].

REMARK 1.4. Let $L|K$ be a finite field extension. If $g_C > 1$ then $C(L)$ is finite. This is a consequence of (a strong form of) the Mordell conjecture over function fields (see [**9**]).

REMARK 1.5. The Jacobian A of C has semiabelian reduction over B if and only if the minimal regular model of C over B has semistable reduction. See eg [**1**, 4.] for a lucid exposition.

Outline of the proof of Theorem 1.1. The basic idea behind the proof of Theorem 1.1 comes from M. Kim's article [**6**]. However, unlike him we make use of the connected component of the Néron model of A and of the theory of semistable sheaves in positive characteristic. By contrast Kim uses the theory of heights and the fact that the relative dualising sheaf of the minimal regular model of C over B is big over k. This follows from [**10**, p. 53, after Th. 2] (private communication from M. Kim to the author). The proof is in several steps.

STEP (1). We associate with a purely inseparable point of degree p^ℓ over K a non vanishing map of K-vector spaces $F_K^{\circ\ell,*}(\omega_K) \to \Omega_{K/k}$. Here ω is the Hodge bundle of the connected component of the Néron model of A, ie its sheaf of differentials restricted to the zero section. This map is basically just a restriction map (see details below).

STEP (2). We quote a result from [**8**] stating that the poles of this map only have logarithmic singularities along the points of bad reduction of A, so that it extends to a morphism of vector bundles $F_B^{\circ\ell,*}(\omega) \to \Omega_{B/k}(\Sigma)$.

STEP (3). From a result of Szpiro (see below for references), we know that $\deg_B(\omega) > 0$ and thus ω contains a semistable subsheaf ω_1 of positive slope. If the semistability of ω_1 were preserved under iterated Frobenius pull-backs, we could conclude that if ℓ is large enough, the image of $F_B^{\circ\ell,*}(\omega_1)$ in $\Omega_{B/k}$ would vanish. Choosing a non-zero differential form $\lambda \in \omega_{1,K}$, we could conclude that λ vanishes on all the purely inseparable points of sufficiently large degree. This would immediately show that there is only finitely many such points. However, we cannot assume that the semistability of ω_1 is preserved under iterated Frobenius pull-backs.

STEP (4). To deal with this last issue, we use a result of Langer, which gives a numerical measure for the lack of semistability of Frobenius pull-backs of semistable sheaves. Using this, up to making assumption on p and ℓ, we can basically pull through Step (3).

The structure of the paper is as follows. In section 2 we recall the results from the theory of semistability of vector bundles on curves that we shall need. In section 3, we prove Theorem 1.1.

NOTATION. If S is a scheme of characteristic p, we write F_S for the absolute Frobenius endomorphism of S. If T is a scheme over S and $n \geq 0$, we write $T^{(p^n)}$ for the base-change of T by $F_S^{\circ n}$. For any integral scheme W we write $\kappa(W)$ for the local ring at the generic point of W (which is a field). If $f : X \to Y$ is a morphism of schemes, we write Ω_f or $\Omega_{X/Y}$ for the sheaf of differentials of the morphism f.

2. Semistable sheaves

We shall need some terminology and results from the theory of semistability of vector bundles over curves.

If V be a non zero vector bundle (ie a coherent locally free sheaf) on B, we write as is customary

$$\mu(V) := \deg(V)/\mathrm{rk}(V)$$

where $\deg(V)$ is the degree of the line bundle $\det(V)$ and $\mathrm{rk}(V)$ is the rank of V. Recall that the degree $\deg(L)$ of a line bundle L on B can be computed as follows. Choose any rational section σ of L; the degree of L is the number of zeroes of σ minus the number of poles of σ (both counted with multiplicities); this quantity does not depend on the choice of σ. See [**4**, II, par. 6, Cor. 6.10] for details.

The quantity $\mu(V)$ is called the *slope* of V. Recall that a non zero vector bundle V on B is called semistable if for any non zero coherent subsheaf $W \subseteq V$, we have $\mu(W) \leqslant \mu(V)$ (recall that a coherent subsheaf of V is automatically locally free, because it is torsion free and B is a Dedekind scheme).

Let V be a non zero vector bundle on B. There is a unique filtration by coherent subsheaves

$$0 = V_0 \subsetneq V_1 \subsetneq V_2 \subsetneq \cdots \subsetneq V_{\mathrm{hn}(V)} = V$$

such that all the sheaves V_i/V_{i-1} ($1 \leqslant i \leqslant \mathrm{hn}(V)$) are (vector bundles and) semistable and such that the sequence $\mu(V_i/V_{i-1})$ is strictly decreasing. This filtration is called the *Harder-Narasimhan filtration* of V (short-hand: HN filtration). One then defines

$$V_{\min} := V/V_{\mathrm{hn}(V)-1},\ V_{\max}(V) := V_1$$

and

$$\mu_{\max}(V) := \mu(V_1),\ \mu_{\min}(V) := \mu(V_{\min}).$$

One basic property of semistable sheaves that we shall use is the following. If V and W are non zero vector bundles on B and $\mu_{\min}(V) > \mu_{\max}(W)$ then

$$\mathrm{Hom}_B(V, W) = 0.$$

This follows from the definitions.

See [**3**, chap. 5] (for instance) for all these notions.

If V is a non zero vector bundle on B we say that V is *Frobenius semistable* if $F_B^{\circ r,*}(V)$ is semistable for all $r \geq 0$. The terminology *strongly semistable* also appears in the literature.

THEOREM 2.1. *Let V be a non zero vector bundle on B. There is an $\ell_0 = \ell_0(V) \in \mathbb{N}$ such that the quotients of the Harder-Narasimhan filtration of $F_B^{\circ\ell_0,*}(V)$ are all Frobenius semistable.*

PROOF. See eg [**7**, Th. 2.7, p. 259]. □

Theorem 2.1 shows in particular that the following definitions :

$$\bar{\mu}_{\min}(V) := \lim_{\ell\to\infty} \mu_{\min}(F_B^{\circ\ell,*}(V))/p^\ell,$$

$$\bar{\mu}_{\max}(V) := \lim_{\ell\to\infty} \mu_{\max}(F_B^{\circ\ell,*}(V))/p^\ell,$$

make sense if V is a non zero vector bundle on B.

Note that since F_B is faithfully flat, this implies that if V and W are non zero vector bundles on B and $\bar{\mu}_{\min}(V) > \bar{\mu}_{\max}(W)$ then we also have $\mathrm{Hom}_B(V, W) = 0$.

For use below, we note that one can show that the sequence $\mu_{\min}(F_B^{\circ\ell,*}(V))/p^\ell$ (resp. $\mu_{\max}(F_B^{\circ\ell,*}(V))/p^\ell$) is weakly decreasing (resp. increasing). See [**7**, p. 258, before Cor. 2.5].

We shall need the following numerical estimate. To formulate it, let V be a non zero vector bundle on B and let

$$\alpha(V) := \max\{\mu_{\min}(V) - \bar{\mu}_{\min}(V), \bar{\mu}_{\max}(V) - \mu_{\max}(V)\}.$$

By the preceding remark, we have $\alpha(V) \geq 0$.

THEOREM 2.2 (Langer). *We have*

$$\alpha(V) \leqslant \frac{\mathrm{rk}(V) - 1}{p} \max\{\bar{\mu}_{\max}(\Omega_{B/k}), 0\}.$$

PROOF. See [**7**, Cor. 6.2]. □

3. Proof of Theorem 1.1

We let $\mathcal{A}$ be the connected component of the identity of the Néron model of the Jacobian A of C over B. We suppose throughout the proof that $\mathcal{A}$ is a semiabelian scheme. In (b), this is an assumption and in (a), we may assume it without restriction of generality, since A will acquire semiabelian reduction on a finite extension of K by a classical result of Grothendieck.

We shall write $\omega = \omega_{\mathcal{A}}$ for the restriction of $\Omega_{\mathcal{A}/B}$ to B by the zero section of $\mathcal{A}$. Under the assumptions of Theorem 1.1, we have $\deg_B(\omega) > 0$. This is a consequence of [**10**, Th. 1] and [**2**, 9.5, Th. 4 (b) and 8.4, Th. 1], together with Grothendieck duality.

Let us write $f : A \to \mathrm{Spec}(K)$ (resp. $\pi : \mathcal{A} \to B$) for the structural morphism. In particular $f = \pi_K$.

Choose once and for all an element of $C(K)$ to determine an embedding of C into its Jacobian and write $\iota : C \hookrightarrow A$ for this embedding. Write $g : C \to \mathrm{Spec}(K)$ for the structural morphism.

Fix $\ell_0 \geq 1$.

Let $P \in C(K_{\mathrm{ins}}) \setminus C(K^{p^{-\ell_0+1}})$. Let $\ell \geq 0$ be such that $[\kappa(P) : K] = p^\ell$ (in other words we have $P \in C(K^{p^{-\ell}}) \setminus C(K^{p^{-\ell+1}})$). Note that by construction we have $\ell \geq \ell_0$. The point P corresponds by definition to an element $\widetilde{P} \in C^{(p^\ell)}(K)$ and thus we obtain a commutative diagram

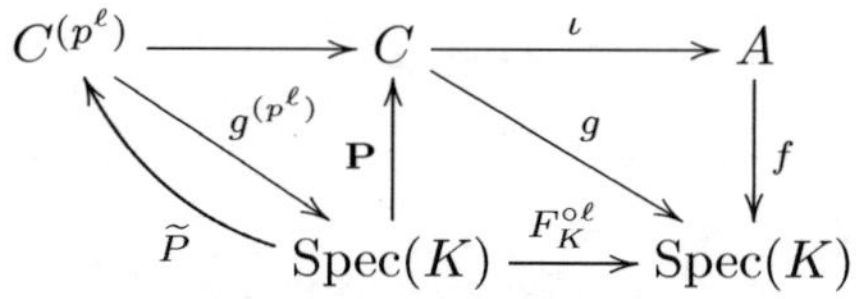

where $\mathbf{P}$ is a closed immersion and the parallelogram on the left is cartesian. We used the short-hand $F_K^{\circ\ell}$ for $F_{\mathrm{Spec}(K)}^{\circ\ell}$. This diagram induces morphisms of differentials

$$\mathbf{P}^*(\iota^*(\Omega_f)) \to \mathbf{P}^*(\Omega_g) \to \Omega_{F_K^{\circ\ell}} \tag{3.1}$$

where the morphism $\mathbf{P}^*(\iota^*(\Omega_f)) \to \mathbf{P}^*(\Omega_g)$ is induced by the natural morphism

$$\iota^*(\Omega_f) \to \Omega_g.$$

In particular, both arrows in (3.1) are surjective maps of K-vector spaces, since ι and $\mathbf{P}$ are closed immersions.

Recall that we have a canonical isomorphism $\Omega_\pi \simeq \pi^*(\omega_{\mathcal{A}})$, since $\mathcal{A}$ is a group scheme over B. Thus we have a canonical isomorphism

$$F_K^{\circ\ell,*}(\omega_K) \simeq \mathbf{P}^*(\iota^*(\Omega_f)).$$

Composing this isomorphism with the two arrows in (3.1), we obtain a map

$$F_K^{\circ\ell,*}(\omega_K) \to \Omega_{F_K^{\circ\ell}},$$

which obviously depends on P.

We shall now make use of the following result.

Proposition 3.1. *The map*

$$F_K^{\circ\ell,*}(\omega_K) \to \Omega_{F_K^{\circ\ell}}$$

extends to a morphism of vector bundles

$$\phi_P : F_B^{\circ\ell,*}(\omega) \to \Omega_{F_B^{\circ\ell}}(\Sigma).$$

Proof. See [**8**, Lemma B.2]. □

Remark 3.2. Note that the proof of Proposition 3.1 is not elementary. It makes use of logarithmic differentials and of the semistable compactifications of semiabelian schemes constructed by Faltings, Chai and Mumford.

Write $b : B \to \mathrm{Spec}(k)$ for the structural morphism. Note that we have an exact sequence

$$F_B^{\circ\ell,*}(\Omega_b) \stackrel{F_B^{\circ\ell,*}}{\to} \Omega_{F_k^{\circ\ell}\circ b} \to \Omega_{F_B^{\circ\ell}} \to 0$$

and a standard computation shows that the first arrow in the sequence vanishes. Hence we have a canonical isomorphism

$$\Omega_{F_k^{\circ\ell}\circ b} \simeq \Omega_{F_B^{\circ\ell}}$$

and since $F_k^{\circ\ell}$ is an isomorphism (because k is perfect), we have a canonical isomorphism $\Omega_{F_k^{\circ\ell}\circ b} \simeq \Omega_b$ so that we finally get a canonical isomorphism

$$\Omega_{B/k} = \Omega_b \simeq \Omega_{F_B^{\circ\ell}}.$$

In particular, $\Omega_{F_B^{\circ\ell}}$ is a line bundle. This implies in particular that the arrow

$$\mathbf{P}^*(\Omega_g) \to \Omega_{F_K^{\circ\ell}}$$

is an isomorphism because both $\mathbf{P}^*(\Omega_g)$ and $\Omega_{F_K^{\circ\ell}}$ are K-vector spaces of dimension one and the arrow is surjective.

We shall now try to exhibit a differential form on C, which vanishes on the non rational purely inseparable points of C.

Recall that we know that $\deg_B(\omega) > 0$. In particular, we must have $\mu_{\max}(\omega) > 0$. Let ω_1 be the first step of the HN filtration of ω. By definition, we have $\mu(\omega_1) = \mu_{\max}(\omega)$. Suppose now until further notice that the inequality

$$p^{\ell_0}(\mu(\omega_1) - \alpha(\omega_1)) > \bar{\mu}_{\max}(\Omega_{F_B^{\circ\ell}}(\Sigma)) \tag{3.2}$$

holds. Since $\Omega_{F_B^{\circ\ell}} \simeq \Omega_{B/k}$, we have

$$\bar{\mu}_{\max}(\Omega_{F_B^{\circ\ell}}(\Sigma)) = 2g_B - 2 + N.$$

Thus inequality (3.2) is equivalent to the inequality

$$p^{\ell_0}(\mu(\omega_1) - \alpha(\omega_1)) > 2g_B - 2 + N. \tag{3.3}$$

We then have

$$p^{\ell}\bar{\mu}_{\min}(\omega_1) \geq p^{\ell_0}\bar{\mu}_{\min}(\omega_1) > \bar{\mu}_{\max}(\Omega_{F_B^{\circ\ell}}(\Sigma))$$

(by the definition of $\alpha(\bullet)$) and thus

$$\bar{\mu}_{\min}(F_B^{\circ\ell,*}(\omega_1)) = p^{\ell}\bar{\mu}_{\min}(\omega_1) > \bar{\mu}_{\max}(\Omega_{F_B^{\circ\ell}}(\Sigma)).$$

This implies that $\phi_P(F_B^{\circ\ell,*}(\omega_1)) = 0$ (see before Theorem 2.2).

On the other hand, if $\phi_P(F_B^{\circ\ell,*}(\omega_1)) = 0$ then from the fact that the arrow $\mathbf{P}^*(\Omega_g) \to \Omega_{F_K^{\circ\ell}}$ is an isomorphism, we can conclude that the arrow

$$\mathbf{P}^*(\iota^*(f^*(\omega_{1,K})) \to \mathbf{P}^*(\Omega_g) \tag{3.4}$$

vanishes.

Now fix $\lambda \in \omega_{1,K}$ with $\lambda \neq 0$. Recall that the canonical map

$$H^0(A, \Omega_f) \to H^0(C, \Omega_g)$$

induced by ι is an isomorphism, because A is the Jacobian of C. Via the identification $\omega_K \simeq H^0(A, \Omega_f)$, the element λ corresponds an element of $H^0(A, \Omega_f)$ and thus to a non zero element λ_C of $H^0(C, \Omega_g)$. Since the arrow (3.4) vanishes, we conclude that P is contained in the vanishing locus $Z(\lambda_C)$ of λ_C. We have

$$\deg(\Omega_g) \geq \sum_{z \in Z(\lambda_C)} [\kappa(z) : K] \geq \#Z(\lambda_C)$$

In particular, since $P \in C(K_{\text{ins}}) \smallsetminus C(K^{p^{-\ell_0+1}})$ was arbitrary, we see that

$$\#(C(K_{\text{ins}}) \smallsetminus C(K^{p^{-\ell_0+1}})) \leq \#Z(\lambda_C) \leq \deg(\Omega_g) = 2g_C - 2.$$

Furthermore, we see that

$$p^{\ell} = [\kappa(P) : K] \leq 2g_C - 2.$$

To summarise:

Let $\ell_0 \geq 1$.

(A) If $p^{\ell_0}(\mu(\omega_1) - \alpha(\omega_1)) > 2g_B - 2 + N$ then

$$\#(C(K_{\text{ins}}) \smallsetminus C(K^{p^{-\ell_0+1}})) \leq 2g_C - 2.$$

(B) If $\ell \geq \ell_0$, $P \in C(K^{p^{-\ell}}) \smallsetminus C(K^{p^{-\ell+1}})$ and $p^{\ell_0}(\mu(\omega_1)-\alpha(\omega_1)) > 2g_B - 2 + N$ then $p^\ell \leq 2g_C - 2$.

We now prove (a) in Theorem 1.1. Note that to prove (a), we may by definition replace C by $C^{(p^c)}$ for any $c \geq 0$. By Theorem 2.1, we may thus assume that the quotients of the HN filtration of $\omega_{\mathcal{A}}$ are Frobenius semistable. In particular, we may assume that ω_1 is Frobenius semistable and thus that $\alpha(\omega_1) = 0$. Now choose $\ell_0 \geq 1$ so that

$$p^{\ell_0} > (2g_B - 2 + N)/\mu(\omega_1).$$

Then (A) above implies that $C(K_{\mathrm{ins}}) \smallsetminus C(K^{p^{-\ell_0+1}})$ is finite. By the Mordell conjecture over function fields (see remark 1.4) $C(K^{p^{-\ell_0+1}})$ is also finite and this concludes the proof of (a).

We now turn to the proof of (b). We are going to seek numerical conditions ensuring that inequality (3.3) holds, ie that

$$p^{\ell_0}(\mu(\omega_1) - \alpha(\omega_1)) > 2g_B - 2 + N.$$

Note that by Theorem 2.2 we have

$$\alpha(\omega_1) \leqslant \frac{\mathrm{rk}(\omega_1) - 1}{p} \max\{\bar{\mu}_{\max}(\Omega_{B/k}), 0\}.$$

Furthermore, we clearly have $\mathrm{rk}(\omega_1) \leq \mathrm{rk}(\omega) = g_C$ and

$$\bar{\mu}_{\max}(\Omega_{B/k}) = \deg(\Omega_{B/k}) = 2g_B - 2.$$

Thus we have

$$\alpha(\omega_1) \leqslant \frac{(g_C - 1)\max\{2g_B - 2, 0\}}{p}$$

Also we have the simple lower bound $\mu(\omega_1) \geq 1/g_C$.

For (3.3) to hold it is thus sufficient to have

$$p^{\ell_0}\Big(1/g_C - \frac{(g_C - 1)\max\{2g_B - 2, 0\})}{p}\Big) > 2g_B - 2 + N$$

or in other words, to have

$$p^{\ell_0} - p^{\ell_0 - 1} g_C(g_C - 1)\max\{2g_B - 2, 0\} - g_C(2g_B - 2) - g_C N > 0. \tag{3.5}$$

From (A) above we see that if $\ell_0 \geq 1$ and inequality (3.5) holds then

$$\#(C(K_{\mathrm{ins}}) \smallsetminus C(K^{p^{-\ell_0+1}})) \leq 2g_C - 2.$$

This is the second statement in (b). The first statement in (b) is now a direct consequence of (B).

Acknowledgments

Many thanks to the anonymous referee for useful remarks and suggestions of improvement. The author is also grateful to Moshe Jarden for his comments.

References

[1] Ahmed Abbes, *Réduction semi-stable des courbes d'après Artin, Deligne, Grothendieck, Mumford, Saito, Winters, ...* (French), Courbes semi-stables et groupe fondamental en géométrie algébrique (Luminy, 1998), Progr. Math., vol. 187, Birkhäuser, Basel, 2000, pp. 59–110. MR1768094

[2] Siegfried Bosch, Werner Lütkebohmert, and Michel Raynaud, *Néron models*, Ergebnisse der Mathematik und ihrer Grenzgebiete (3) [Results in Mathematics and Related Areas (3)], vol. 21, Springer-Verlag, Berlin, 1990. MR1045822

[3] Holger Brenner, Jürgen Herzog, and Orlando Villamayor, *Three lectures on commutative algebra*, University Lecture Series, vol. 42, American Mathematical Society, Providence, RI; Real Sociedad Matemática Española, Madrid, 2008. Lectures from the Winter School on Commutative Algebra and Applications held in Barcelona, January 30–February 3, 2006; Edited by Gemma Colomé-Nin, Teresa Cortadellas Benítez, Juan Elias and Santiago Zarzuela. MR2404090

[4] Robin Hartshorne, *Algebraic geometry*, Springer-Verlag, New York-Heidelberg, 1977. Graduate Texts in Mathematics, No. 52. MR0463157

[5] Daniel Huybrechts and Manfred Lehn, *The geometry of moduli spaces of sheaves*, Aspects of Mathematics, E31, Friedr. Vieweg & Sohn, Braunschweig, 1997. MR1450870

[6] Minhyong Kim, *Purely inseparable points on curves of higher genus*, Math. Res. Lett. **4** (1997), no. 5, 663–666, DOI 10.4310/MRL.1997.v4.n5.a4. MR1484697

[7] Adrian Langer, *Semistable sheaves in positive characteristic*, Ann. of Math. (2) **159** (2004), no. 1, 251–276, DOI 10.4007/annals.2004.159.251. MR2051393

[8] Damian Rössler, *On the group of purely inseparable points of an abelian variety defined over a function field of positive characteristic, II*, Algebra Number Theory **14** (2020), no. 5, 1123–1173, DOI 10.2140/ant.2020.14.1123. MR4129384

[9] Pierre Samuel, *Compléments à un article de Hans Grauert sur la conjecture de Mordell* (French), Inst. Hautes Études Sci. Publ. Math. **29** (1966), 55–62. MR204430

[10] Lucien Szpiro, *Propriétés numériques du faisceau dualisant relatif* (French), Astérisque **86** (1981), 44–78. Seminar on Pencils of Curves of Genus at Least Two. MR3618571

Mathematical Institute, University of Oxford, Andrew Wiles Building, Radcliffe Observatory Quarter, Woodstock Road, Oxford, OX2 6GG, United Kingdom

Email address: `rossler@maths.ox.ac.uk`

Contemporary Mathematics
Volume **767**, 2021
https://doi.org/10.1090/conm/767/15400

Correspondences on curves in positive characteristic

Akio Tamagawa

On the occasion of Professor Gerhard Frey's seventy-fifth birthday

ABSTRACT. We say that two curves over a fixed algebraically closed field are isogenous, if they admit a common finite étale covering or, equivalently, if they admit a bi-finite-étale correspondence. In the present paper, we prove that any two affine, smooth, connected curves over the algebraic closure of a finite field are isogenous. This is in a striking contrast with S. Mochizuki's finiteness result in characteristic 0.

1. Introduction

Let k be an algebraically closed field, and X, Y connected schemes of finite type over k. Then we say that X and Y are isogenous over k and write $X \sim Y$, if there exist a (non-empty) connected scheme Z of finite type over k and finite étale morphisms $f : Z \to X$ and $g : Z \to Y$ over k. We may also regard (f, g) as giving a (bi-finite-étale) correspondence between X and Y.

For the equivalence relation $\sim$ on curves in characteristic 0, S. Mochizuki proved the following fundamental finiteness result.

THEOREM 1.1 ([**M**, Theorem A]). *Assume that the characteristic of k is 0. Let (g, r) be a pair of non-negative integers satisfying the hyperbolicity condition $2g + r - 2 > 0$. Then each $\sim$-class of smooth, connected curves over k contains only finitely many k-isomorphism classes of curves X with $(g_X, r_X) = (g, r)$, where g_X denotes the genus of the smooth compactification X^{cpt} of X and r_X denotes the cardinality of $X^{\mathrm{cpt}} - X$.*

We note that Mochizuki's proof of Theorem 1.1 is transcendental and relies heavily on certain classical results of Margulis and Takeuchi in the theory of arithmetic Riemann surfaces.

In the present paper, we investigate the equivalence relation $\sim$ on curves in positive characteristic. The main results are as follows.

THEOREM 1.2 (cf. Theorem 3.1). *Assume $k = \overline{\mathbf{F}}_p$. Then for every pair of affine, smooth, connected curves X, Y over k, we have $X \sim Y$.*

COROLLARY 1.3 (cf. Corollary 3.3). *Assume $k = \overline{\mathbf{F}}_p$. Then there exists a k-scheme H (unique up to k-isomorphisms), such that, for every affine, smooth, connected curve X over k, the 'pro-finite-étale universal covering' $\tilde{X}$ of X is isomorphic to H over k.*

2020 *Mathematics Subject Classification.* Primary 14H30; Secondary 11G20, 14G15, 14G17.

These results in positive characteristic are in a striking contrast with Mochizuki's result in characteristic 0. The main ingredient of the proof of Theorem 1.2 is the following consequence of the main result of [**Tam3**] on the 'unramified' Skolem problem in positive characteristic.

THEOREM 1.4. *Assume* $k = \overline{\mathbf{F}}_p$. *Let* C *be an affine, smooth, connected curve over* k, Σ *a (possibly empty) finite set of closed points of* C, K *the function field of* C, K_v *the* v*-adic completion of* K $(v \in \Sigma)$, $f : W \to B \stackrel{\text{def}}{=} C - \Sigma$ *a smooth, surjective morphism with* $W \times_B K$ *geometrically connected over* K, *and* $\Omega_v \subset W(K_v)$ *a non-empty,* v*-adically open subset* $(v \in \Sigma)$. *Then there exist a finite extension* K' *of* K *and a morphism* $s : B' \to W$ *over* B, *where* B' *is the integral closure of* B *in* K', *such that* B' *is (finite) étale over* B, *that, for each* $v \in \Sigma$, $B' \times_B K_v$ *is a (finite) disjoint union of copies of* $\mathrm{Spec}(K_v)$, *and that, for each* $v \in \Sigma$, *the image of* $s \times_B K_v : B' \times_B K_v \to W \times_B K_v$ *is contained in* Ω_v $(\subset W(K_v))$.

PROOF. This is a special case of [**Tam3**, Theorem 3.1]. □

For general (algebraically closed) base fields in positive characteristic, very little is known, but we can prove the following small finiteness result.

PROPOSITION 1.5 (cf. Proposition 4.16(ii)). *Assume that the characteristic of* k *is* $p > 0$, *and let* X *be a proper, smooth, connected curve of genus* ≥ 2 *over* k *which does not descend to any curve over* $\overline{\mathbf{F}}_p$. *Then each* $\sim$*-class of* $\{X - \{P\} \mid P \in X\}$ *is a finite set.*

The main ingredient of the proof of Proposition 1.5 is the 'full' Mordell-Lang conjecture for curves in positive characteristic (cf. [**FP**]).

In section 1, we investigate étale morphisms from affine curves to the affine line. In section 2, we prove the main results Theorem 3.1 and Corollary 3.3, based on Theorem 1.4 and the investigation in section 1. In section 3, we discuss several possible/impossible generalizations of the main results.

NOTE 1.6. A substantial portion of the results in the present paper had been obtained by 2002 and were announced in Kyoto (cf. [**Tam1**, **Tam2**]), Banff, Tokyo, Orsay, etc. around 2002-2004. Since then, the author has had no chance of writing up the contents of the present paper till now.

2. Étale morphisms to the affine line

In this section, let k be an arbitrary algebraically closed field of characteristic $p > 0$, X an affine, smooth, connected curve over k, and X^{cpt} the smooth compactification of X, which is a proper, smooth, connected curve over k. In this section, we consider the following condition on X:

$(*_X)$ There exists a finite étale k-morphism $X \to \mathbf{A}^1_k$.

The following basic observation is peculiar to positive characteristic.

LEMMA 2.1. *The following* (i)–(iii) *are equivalent.*
(i) $(*_X)$ *holds.*
(ii) *There exists an étale* k*-morphism* $f : X \to \mathbf{A}^1_k$.
(iii) *There exists an element* $f \in \Gamma(X, \mathcal{O}_X)$ *such that* $\Omega_X = \mathcal{O}_X df$ *holds.*

PROOF. (i) $\Longrightarrow$ (ii) is trivial. Next, note that considering a k-morphism $f : X \to \mathbf{A}^1_k = \mathrm{Spec}(k[T])$ is equivalent to considering an element $f \in \Gamma(X, \mathcal{O}_X)$.

In this identification, it is easy to see (almost by definition) that $f : X \to \mathbf{A}^1_k$ is étale if and only if $f \in \Gamma(X, \mathcal{O}_X)$ satisfies $\Omega_X = \mathcal{O}_X df$. Thus, (ii) $\iff$ (iii) holds.

Finally, (ii) $\implies$ (i) follows from 'Abhyankar's trick'. More precisely, let f be as in (ii) (or, equivalently, as in (iii)). By the Riemann-Roch theorem (or by Noether's normalization theorem), there exists an element $h \in \Gamma(X, \mathcal{O}_X)$, such that h has poles at all points of $X^{\mathrm{cpt}} - X$ (or, equivalently, $h : X \to \mathbf{A}^1_k$ is finite). Namely, we have $\mathrm{ord}_x(h) < 0$ for each $x \in X^{\mathrm{cpt}} - X$. Now, take a sufficiently large natural number N, such that $\mathrm{ord}_x(f) > Np\,\mathrm{ord}_x(h)$ holds for each $x \in X^{\mathrm{cpt}} - X$. Then $g \overset{\mathrm{def}}{=} f + h^{Np}$ satisfies $\mathrm{ord}_x(g) = Np\,\mathrm{ord}_x(h) < 0$ for each $x \in X^{\mathrm{cpt}} - X$, or, equivalently, g gives a finite k-morphism $X \to \mathbf{A}^1_k$. Since $dg = df$, g is also étale. This completes the proof. □

REMARK 2.2. Lemma 2.1, (i) $\iff$ (iii) is essentially the same as [**Z**, Proposition 1], whose proof is slightly different from ours.

For an $\mathbf{F}_p$-scheme S, the p-th power map $\mathcal{O}_S \to \mathcal{O}_S$, $x \mapsto x^p$ induces the absolute Frobenius morphism $F_S : S \to S$. When S is equipped with a structure of k-scheme and n is a non-negative integer, we set $S_n \overset{\mathrm{def}}{=} S \underset{\mathrm{Spec}(k), F^n_{\mathrm{Spec}(k)}}{\to} \times \mathrm{Spec}(k)$. Then F^n_S induces the n-th relative Frobenius morphism $F^n_{S/k} : S \to S_n$, which is a k-morphism. (We write $F_{S/k}$ for $F^1_{S/k}$.)

It is well-known that the following exact sequence of $\mathcal{O}_{X_1}$-modules exists:

$$0 \to \mathcal{O}_{X_1} \overset{i}{\to} (F_{X/k})_*(\mathcal{O}_X) \overset{d}{\to} (F_{X/k})_*(\Omega_X) \overset{c}{\to} \Omega_{X_1} \to 0.$$

Here, $i = i_X$ is the k-algebra homomorphism associated to $F_{X/k}$, $d = d_X$ is the natural derivation, and $c = c_X$ is the Cartier operator.

Following Raynaud [**Ra**] (cf. [**Tan**]), we introduce:

DEFINITION. We define $B = B_X$ to be the cokernel of i (= the image of d = the kernel of c). (Thus, B_X is a locally free $\mathcal{O}_{X_1}$-module of rank $p-1$.)

Here is another interpretation of $(*_X)$:

LEMMA 2.3. *The following are equivalent:*
(i) $(*_X)$ *holds.*
(ii) *There exists an element* $\omega \in \Gamma(X_1, B_X)$ *such that* $\Omega_X = \mathcal{O}_X \omega$ *holds.*

PROOF. Since X_1 is affine, the natural map

$$\Gamma(X, \mathcal{O}_X) = \Gamma(X_1, (F_{X/k})_*(\mathcal{O}_X)) \to \Gamma(X_1, B_X)$$

is surjective. Accordingly, (ii) is just equivalent to Lemma 2.1(iii). □

COROLLARY 2.4. *If* $p = 2$, *the following are equivalent:*
(i) $(*_X)$ *holds.*
(ii) $B_X \simeq \mathcal{O}_{X_1}$.

PROOF. As $p - 1 = 1$, B_X is an invertible sheaf on X_1. Therefore, the natural surjective homomorphism $(F_{X/k})^*(B_X) \to \Omega_X$ of $\mathcal{O}_X$-modules must be an isomorphism. Now, each $\omega \in \Gamma(X_1, B_X)$ defines a homomorphism $\mathcal{O}_{X_1} \to B_X$, $a_1 \mapsto a_1\omega$, of $\mathcal{O}_{X_1}$-modules. Since $F_{X/k}$ is faithfully flat, this homomorphism is an isomorphism if and only if its pull-back to X, $\mathcal{O}_X \to (F_{X/k})^*(B_X) \overset{\sim}{\to} \Omega_X$, $a \mapsto a\omega$, is an isomorphism (of $\mathcal{O}_X$-modules). The latter is further equivalent to saying that

$\mathcal{O}_X \to (F_{X/k})^*(B_X) \overset{\sim}{\to} \Omega_X$ is surjective, as both $\mathcal{O}_X$ and Ω_X are invertible sheaves on X. Now, we see that (ii) is equivalent to Lemma 2.3(ii). □

We shall reinterpret the above facts more geometrically.

PROPOSITION 2.5. *There exists a smooth, surjective X_1-scheme $W = W_X \to X_1$ whose generic fiber is geometrically irreducible, such that the following holds: For each étale k-morphism $j : Y \to X$ from an affine, smooth, connected curve Y over k to X, the following are equivalent:*

(i) *There exists a finite, étale k-morphism $Y \to \mathbf{A}^1_k$, i.e., $(*_Y)$ holds.*

(ii) *There exists an X_1-morphism $Y_1 \to W$.*

PROOF. By taking the adjoint of the $\mathcal{O}_{X_1}$-homomorphism $d : (F_{X/k})_*(\mathcal{O}_X) \to (F_{X/k})_*(\Omega_X)$, we obtain an $\mathcal{O}_X$-homomorphism

$$(F_{X/k})^*((F_{X/k})_*(\mathcal{O}_X)) \to \Omega_X.$$

In terms of geometric vector bundles, this corresponds to

$$\begin{array}{ccc} \mathbf{V}_X((F_{X/k})^*((F_{X/k})_*(\mathcal{O}_X))^\vee) & \to & \mathbf{V}_X(\Omega_X^\vee), \\ \| & & \\ \mathbf{V}_{X_1}((F_{X/k})_*(\mathcal{O}_X)^\vee) \times_{X_1} X & & \end{array}$$

where, for a locally free sheaf E on a scheme S, we set

$$\mathbf{V}_S(E) \overset{\text{def}}{=} \mathbf{Spec}_S(\oplus_{n\geq 0} \operatorname{Sym}^n(E)).$$

Now, let $\tilde{W}$ be the inverse image in $\mathbf{V}_X((F_{X/k})^*((F_{X/k})_*(\mathcal{O}_X))^\vee)$ of the open subset $\mathbf{V}_X(\Omega_X^\vee) - \mathbf{0}$ of $\mathbf{V}_X(\Omega_X^\vee)$. (Here, $\mathbf{0}$ denotes the zero section of the line bundle $\mathbf{V}_X(\Omega_X^\vee)$.) We define W to be the (set-theoretic) image of $\tilde{W}$ by the natural projection $\mathbf{V}_X((F_{X/k})^*((F_{X/k})_*(\mathcal{O}_X))^\vee) = \mathbf{V}_{X_1}((F_{X/k})_*(\mathcal{O}_X)^\vee) \times_{X_1} X \to \mathbf{V}_{X_1}((F_{X/k})_*(\mathcal{O}_X)^\vee)$. Since this projection is a homeomorphism, W is an open subset of $\mathbf{V}_{X_1}((F_{X/k})_*(\mathcal{O}_X)^\vee)$, which we regard as an open subscheme. In particular, the X_1-scheme W is smooth, and its generic fiber is geometrically irreducible.

Take any $j : Y \to X$ as in the statement of Proposition 2.5, which naturally induces $j_1 : Y_1 \to X_1$. Then, by definition, the set of X_1-morphisms $Y_1 \to \mathbf{V}_{X_1}((F_{X/k})_*(\mathcal{O}_X)^\vee)$ is in one-to-one correspondence with the set of sections $\Gamma(Y_1, (j_1)^*((F_{X/k})_*(\mathcal{O}_X)^\vee)^\vee)$. Moreover, we have

$$(j_1)^*((F_{X/k})_*(\mathcal{O}_X)^\vee)^\vee = (j_1)^*((F_{X/k})_*(\mathcal{O}_X)) = (F_{Y/k})_*(\mathcal{O}_Y),$$

where the point of the second equality is that the diagram

$$\begin{array}{ccc} Y & \overset{F_{Y/k}}{\to} & Y_1 \\ j\downarrow & & \downarrow j_1 \\ X & \overset{F_{X/k}}{\to} & X_1 \end{array}$$

is cartesian, as j is étale. Thus, we conclude that the set of X_1-morphisms $Y_1 \to \mathbf{V}_{X_1}((F_{X/k})_*(\mathcal{O}_X)^\vee)$ is in one-to-one correspondence with $\Gamma(Y_1, (F_{Y/k})_*(\mathcal{O}_Y)) = \Gamma(Y, \mathcal{O}_Y)$.

Moreover, take any $f \in \Gamma(Y, \mathcal{O}_Y)$. Then, by the above various definitions, the image of the X_1-morphism $Y_1 \to \mathbf{V}_{X_1}((F_{X/k})_*(\mathcal{O}_X)^\vee)$ that corresponds to f is contained in W if and only if the image df of f in $j^*(\Omega_X) = \Omega_Y$ admits no zeroes, i.e., generates Ω_Y. This shows that the set of X_1-morphisms $Y_1 \to W$ is in one-to-one correspondence with the set of elements $f \in \Gamma(Y, \mathcal{O}_Y)$ with $\Omega_Y = \mathcal{O}_Y df$. This, together with Lemma 2.1, completes the proof. (Note that the above argument also shows that $W \to X_1$ is surjective, since Ω_X is locally generated by a single element in the form of df.) □

REMARK 2.6. The choice of W in Proposition 2.5 is not unique. For example, by Lemma 2.3, we may also define W to be the image in $\mathbf{V}_{X_1}(B_X^\vee)$ of the inverse image in $\mathbf{V}_X((F_{X/k})^*(B_X)^\vee) = \mathbf{V}_{X_1}(B_X^\vee)\times_{X_1} X$ of $\mathbf{V}_X(\Omega_X^\vee) - \mathbf{0}$. Here, the relative dimension of W over X_1 is $p-1$, while that in the above proof is p.

3. The main results and their proofs

From now on, let k_0 be an algebraic closure of the finite prime field $\mathbf{F}_p = \mathbf{Z}/p\mathbf{Z}$ of characteristic $p > 0$.

THEOREM 3.1. *Let X and X' be affine, smooth, connected curves over k_0. Then we have $X \sim X'$.*

Since $\sim$ is an equivalence relation, Theorem 3.1 follows from the following special case:

THEOREM 3.2. *Let X be an affine, smooth, connected curve over k_0. Then we have $X \sim \mathbf{A}^1_{k_0}$.*

PROOF. In more down-to-earth terms, this claims that there exists a finite, étale, connected covering $Y \to X$ which admits a finite, étale k_0-morphism $Y \to \mathbf{A}^1_{k_0}$. By Proposition 2.5, this is equivalent to saying that there exists a finite, étale, connected covering $Y \to X$ which admits an X_1-morphism $Y_1 \to W_X$, or, equivalently, that there exists a finite, étale, connected covering $Y' \to X_1$ which admits an X_1-morphism $Y' \to W_X$. (Observe that the functor $Y \mapsto Y_1$ (or, more precisely, $(Y \to X) \mapsto (Y_1 \to X_1)$) gives an equivalence from the category of finite, étale coverings of X to the category of finite, étale coverings of X_1.) Now, this last condition just follows from Theorem 1.4, applied to $k = k_0$, $C = X_1$, $\Sigma = \emptyset$ and $f : W_X \to X_1$. □

COROLLARY 3.3. *There exists a k_0-scheme H (unique up to k_0-isomorphisms), such that, for every affine, smooth, connected curve X over k_0, the 'pro-finite-étale universal covering' $\tilde{X}$ of X is isomorphic to H over k_0.*

PROOF. This is immediate from Theorem 3.1 or Theorem 3.2. (Set $H \stackrel{\text{def}}{=} \widetilde{\mathbf{A}^1_{k_0}}$.) □

REMARK 3.4. Corollary 3.3 suggests that the role played by the k_0-scheme H for affine, smooth, connected curves over k_0 should be similar to the role played by the complex upper half-plane $\mathbf{H}$ for hyperbolic Riemann surfaces. In fact, we can prove several properties of H formally analogous to those of $\mathbf{H}$.

LEMMA 3.5. *Let X be an affine, smooth, connected curve over k_0. Let λ be an element of* $\mathrm{Pic}(X)$. *Then there exists a finite, étale, connected covering $\mu : Y \to X$, such that the image of λ by the natural homomorphism $\mu^* : \mathrm{Pic}(X) \to \mathrm{Pic}(Y)$ is* 0.

Moreover, we may assume that μ is an abelian covering and that the covering $Y^{\mathrm{cpt}} \to X^{\mathrm{cpt}}$ induced by μ is also (finite and) étale.

PROOF. This is a consequence of the principal ideal theorem in the class field theory of function fields of one variable over finite fields. More precisely, we may assume that the pair (X, λ) over k_0 comes from a similar pair $(X_{\mathbf{F}}, \lambda_{\mathbf{F}})$ over a finite extension $\mathbf{F}$ of $\mathbf{F}_p$ (in k_0). Now, we can apply [**Ro**, Proposition 2.4]. $\square$

In the case $p = 2$, we have the following strengthening of Theorem 3.2, whose proof does not resort to [**Tam3**] (but resorts to class field theory).

THEOREM 3.6. *Notations being as in Theorem 3.2, assume, moreover, $p = 2$. Then there exists a finite, étale, abelian, connected covering $\mu : Y \to X$, such that the covering $Y^{\mathrm{cpt}} \to X^{\mathrm{cpt}}$ induced by μ is also (finite and) étale and that there exists a finite, étale k_0-morphism $Y \to \mathbf{A}^1_{k_0}$.*

PROOF. By Corollary 2.4, this is equivalent to saying that there exists a finite, étale, abelian, connected covering $\mu : Y \to X$, such that the covering $Y^{\mathrm{cpt}} \to X^{\mathrm{cpt}}$ induced by μ is also (finite and) étale and that $B_Y \simeq \mathcal{O}_{Y_1}$ holds. Now, this last condition just follows from Lemma 3.5. Indeed, observe that the formation of locally free sheaf B commutes with any étale base change. (More precisely, for each étale k_0-morphism $\mu : Y \to X$, we have a canonical identification $B_Y = (\mu_1)^*(B_X)$, where μ_1 denotes the natural k_0-morphism $Y_1 \to X_1$ induced by μ.) $\square$

REMARK 3.7. If the order of λ is prime to p, then the assertion of Lemma 3.5 can be proved in a purely algebraic-geometric fashion, by means of the surjectivity of $J(k_0) = \mathrm{Pic}^0(X^{\mathrm{cpt}}) \to \mathrm{Pic}(X)$ (where J denotes the Jacobian variety of X^{cpt} over k_0) and the correspondence between prime-to-p-torsion points of J and prime-to-p-cyclic, finite, étale coverings of X^{cpt}. On the other hand, the first half of the assertion of Lemma 3.5 also follows from [**Tam3**, Theorem 3.1], (or, more precisely, Theorem 1.4 above, applied to $k = k_0$, $C = X$, $\Sigma = \emptyset$, and $f : \mathbf{V}_X(L^\vee) - \mathbf{0} \to X$, where $\lambda = [L]$), whose proof is rather complicated but purely algebraic-geometric.

It may be an interesting problem to ask if there is a full proof of Lemma 3.5 and Theorem 3.6 that does not resort to class field theory.

4. Possible/impossible generalizations

4.1. A pointed version. We have the following generalization of Theorem 3.1.

THEOREM 4.1. *Let X and X' be affine, smooth, connected curves over k_0, and S (resp. S') a (possibly empty) 0-dimensional closed subscheme of X (resp. X'). Let l be an isomorphism $S \xrightarrow{\sim} S'$ over k_0. Then there exist an affine, smooth, connected curve Y over k_0, finite étale k_0-morphisms $f : Y \to X$ and $f' : Y \to X'$, and a 0-dimensional closed subscheme T of Y, such that f (resp. f') induces an (a unique) isomorphism $f_T : T \xrightarrow{\sim} S$ (resp. $f'_T : T \xrightarrow{\sim} S'$) with $f'_T = l \circ f_T$.*

PROOF. First, we impose the extra assumption that $X' = \mathbf{A}^1_{k_0}$, and prove the weaker assertion in which $f' : Y \to X' = \mathbf{A}^1_{k_0}$ is étale but not necessarily finite étale. Note that l determines a closed immersion $S \to \mathbf{A}^1_{k_0}$ or, equivalently, a generator of the finite-dimensional (artinian) k_0-algebra $\Gamma(S, \mathcal{O}_S)$.

We exploit the X_1-scheme $W = W_X$ which we constructed in the proof of Proposition 2.5. Recall that W is an open subscheme of $\mathbf{V}_{X_1}((F_{X/k_0})_*(\mathcal{O}_X)^\vee)$. Let R be a $\Gamma(X_1, \mathcal{O}_{X_1})$-algebra. Then $W(R)$ can be naturally identified with the set of elements of $R' \stackrel{\text{def}}{=} \Gamma(X, \mathcal{O}_X) \otimes_{\Gamma(X_1, \mathcal{O}_{X_1})} R$ whose image (by d) generates $\Gamma(X, \Omega_X) \otimes_{\Gamma(X_1, \mathcal{O}_{X_1})} R = \Gamma(X, \Omega_X) \otimes_{\Gamma(X, \mathcal{O}_X)} R'$ as an R'-module.

For each $x \in S$, we denote by S_x the connected component of S containing x, so that S is scheme-theoretically the disjoint union of S_x. Let x_1 be the image of x in X_1. Let R_x (resp. R_{1,x_1}) be the completion of the discrete valuation ring $\mathcal{O}_{X,x}$ (resp. $\mathcal{O}_{X_1,x_1}$), $\mathfrak{m}_x$ the maximal ideal of R_x, and K_x (resp. K_{1,x_1}) the field of fractions of R_x (resp. R_{1,x_1}). (Observe that $R_x = \Gamma(X, \mathcal{O}_X) \otimes_{\Gamma(X_1, \mathcal{O}_{X_1})} R_{1,x_1}$ and $K_x = \Gamma(X, \mathcal{O}_X) \otimes_{\Gamma(X_1, \mathcal{O}_{X_1})} K_{1,x_1}$.) Then S_x can be naturally identified with the closed subscheme $\operatorname{Spec}(R_x/\mathfrak{m}_x^{n_x})$ of $\operatorname{Spec}(R_x)$ for some $n_x > 0$. Set $l_x \stackrel{\text{def}}{=} l|_{S_x}$, which is a generator of the finite-dimensional (artinian) local k_0-algebra $\Gamma(S_x, \mathcal{O}_{S_x}) = R_x/\mathfrak{m}_x^{n_x}$. Then

$$\Omega_{l,x} \stackrel{\text{def}}{=} \{\gamma \in R_x \mid \gamma|_{S_x} = l_x\} \cap W(R_{1,x_1})$$

defines an x_1-adically open subset of $W(K_{1,x_1})$. We claim that $\Omega_{l,x}$ is non-empty. Indeed, when $n_x > 1$ (resp. $n_x = 1$), fix any element $\tilde{l}_x \in R_x$ (resp. $\tilde{l}_x \in k_0 \subset R_x$) whose image in $R_x/\mathfrak{m}_x^{n_x}$ coincides with l_x. Then it is easy to see

$$\Omega_{l,x} = \begin{cases} \tilde{l}_x + \mathfrak{m}_x^{n_x}, & n_x > 1, \\ \tilde{l}_x + (\mathfrak{m}_x - \mathfrak{m}_x^2), & n_x = 1, \end{cases}$$

which is clearly non-empty.

Now, Theorem 1.4, applied to $k = k_0$, $C = X_1$, $\Sigma = S_1$, $B = X_1' \stackrel{\text{def}}{=} X_1 - S_1$, $f : W' \stackrel{\text{def}}{=} W \times_{X_1} X_1' \to X_1'$, and $\Omega_{x_1} = \Omega_{l,x}$ ($x_1 \in S_1$), ensures that there exist a finite, étale, connected covering $\mu_1' : Y_1' \to X_1'$ and an X_1'-morphism $s' : Y_1' \to W'$, such that, for each $x_1 \in S_1$, $Y_1' \times_{X_1'} K_{1,x_1}$ is a disjoint union of copies of $\operatorname{Spec}(K_{1,x_1})$ and its image in $W(K_{1,x_1})$ is contained in $\Omega_{l,x}$. In particular, the covering $\mu_1' : Y_1' \to X_1'$ extends (uniquely) to a finite, étale, connected covering $\mu_1 : Y_1 \to X_1$, and the X_1'-morphism $s' : Y_1' \to W'$ extends (uniquely) to an X_1-morphism $s : Y_1 \to W$. (Observe that the latter follows from the fact that $\Omega_{l,x} \subset W(R_{1,x_1})$ for every $x_1 \in S_1$.)

For each $x \in S$, choose $y \in \mu^{-1}(x) \subset Y$, where μ denotes the natural morphism $Y \to X$ induced by $\mu_1 : Y_1 \to X_1$. Since μ is étale, the connected component T_y of $\mu^{-1}(S_x)$ containing y is isomorphically mapped onto S_x by μ. We define T to be the disjoint union of T_y, which is a 0-dimensional closed subscheme of Y. By definition, μ induces an isomorphism $\mu_T : T \xrightarrow{\sim} S$. The (quasi-)section s corresponds to an element $f' \in \Gamma(Y, \mathcal{O}_Y)$, or, equivalently, a k_0-morphism $f' : Y \to \mathbf{A}^1_{k_0}$. By the definition of W, f' is étale, and, by the definition of $\Omega_{l,x}$, $f'|_T = l \circ \mu_T$, as desired.

Next, we shall reduce the full assertion of Theorem 4.1 to what we have just proved. Here, we need the following strengthening of (the essential part (ii) $\Longrightarrow$ (i) of) Lemma 2.1.

LEMMA 4.2. *Let k be an algebraically closed field of characteristic $p > 0$, and X an affine, smooth, connected curve over k. Let $f : X \to \mathbf{A}^1_k$ be an étale k-morphism. Let S be a (possibly empty) 0-dimensional closed subscheme of X. Let C be an affine, smooth, connected curve over k, and $j : C \to \mathbf{A}^1_k$ a quasi-finite k-morphism.*

Then there exists a finite étale k-morphism $g : X \to \mathbf{A}^1_k$, such that $g|_S = f|_S$ holds and that $X \times_{\mathbf{A}^1_k} C$ is (affine, smooth and) connected.

PROOF. Enlarging S if necessary, we may assume that S is non-empty. Let t denote the affine coordinate of $\mathbf{A}_k^1$, i.e., $\mathbf{A}_k^1 = \mathrm{Spec}(k[t])$, and write $A_t = \mathbf{A}_k^1$. As in section 1, let X^{cpt} denote the smooth compactification of X. By the Riemann-Roch theorem, there exists an element $h \in \Gamma(X, \mathcal{O}_X)$, such that h has poles at all points of $X^{\mathrm{cpt}} - X$ and zeros at all points of S. Now, take a sufficiently large natural number N and any element $c \in k^\times$. Then $g = g_c \stackrel{\mathrm{def}}{=} f + ch^{Np}$ gives a finite étale k-morphism $X \to A_t$ as in the proof of Lemma 2.1, and $g|_S = f|_S$ holds since the image of h in $\Gamma(S, \mathcal{O}_S)$ is a nilpotent element.

From now, we shall show that for some choice of $c \in k^\times$, $Z_c \stackrel{\mathrm{def}}{=} X \times_{A_t} C$ is connected (or, equivalently, irreducible). This is equivalent to saying that $k(X) \otimes_{k(t)} k(C)$ is a field. Replacing C by the integral closure of A_t in the normal closure of $k(C)$ over $k(t)$, we may assume that the extension $k(C)/k(t)$ is normal. Moreover, let C' be the maximal (finite) étale subcovering of C/A_t (which is finite, étale and Galois over A_t). Then, since $X \to A_t$ is finite étale, it is sufficient to prove that $X \times_{A_t} C'$ is connected. Thus, replacing C by C', we may assume that $C \to A_t$ is finite étale.

Now, take another affine line $A_u \stackrel{\mathrm{def}}{=} \mathrm{Spec}(k[u])$, and set $A'_u \stackrel{\mathrm{def}}{=} A_u - \{0\}$. Consider the following family of coverings $X \to A_t$, parametrized by A'_u:

$$\gamma : X \times A'_u \to A_t \times A'_u, \ (x, u) \mapsto (g_u(x), u),$$

where $g_u(x) \stackrel{\mathrm{def}}{=} f(x) + uh(x)^{Np}$.

CLAIM 4.3. (i) γ extends uniquely to a finite flat morphism $\gamma^{\mathrm{cpt}} : X^{\mathrm{cpt}} \times A'_u \to P_t \times A'_u$ over A'_u, where $P_t \stackrel{\mathrm{def}}{=} A_t^{\mathrm{cpt}} = A_t \coprod \{\infty\}$ is the projective line.
(ii) γ is finite étale.

PROOF. (i) The uniqueness follows from the facts that $X^{\mathrm{cpt}} \times A'_u$ is reduced, that $P_t \times A'_u$ is separated, and that $X \times A'_u$ is dense in $X^{\mathrm{cpt}} \times A'_u$. To prove the existence, we note that $h, h^{-Np}, fh^{-Np} \in k(X)$ give rise to k-morphisms $h, h^{-Np}, fh^{-Np} : X^{\mathrm{cpt}} \to P_t$. Set $S_0 \stackrel{\mathrm{def}}{=} h^{-1}(0)$ and $S_\infty \stackrel{\mathrm{def}}{=} h^{-1}(\infty)$. Then S_0 and S_∞ are disjoint, and it follows from the choice of h that $S_\infty = X^{\mathrm{cpt}} - X$. Now, we define a map $\gamma^{\mathrm{cpt}} : X^{\mathrm{cpt}} \times A'_u \to P_t \times A'_u$ by

$$\gamma^{\mathrm{cpt}}(x, u) = \begin{cases} ((1 : g_u(x)), u) \ (= \gamma(x, u)), & (x, u) \in (X^{\mathrm{cpt}} - S_\infty) \times A'_u \ (= X \times A'_u), \\ ((h^{-Np}(x) : (fh^{-Np})(x) + u), u), & (x, u) \in (X^{\mathrm{cpt}} - S_0) \times A'_u. \end{cases}$$

Observe that this gives a well-defined k-morphism. (The point is that for each $x \in X - S_0$ (resp. $x \in S_\infty$), we have $\mathrm{ord}_x(h^{-Np}) = -Np\,\mathrm{ord}_x(h) = 0$ (resp. > 0), $\mathrm{ord}_x(fh^{-Np}) = \mathrm{ord}_x(f) - Np\,\mathrm{ord}_x(h) \geq 0$ (resp. > 0), hence $(x, u) \mapsto (h^{-Np}(x), (fh^{-Np})(x) + u))$ gives a k-morphism $(X^{\mathrm{cpt}} - S_0) \times A'_u \to \mathbf{A}_k^2 - \{(0,0)\}$.) It is clear that γ^{cpt} is a morphism over A'_u that extends γ. As both $X^{\mathrm{cpt}} \times A'_u$ and $P_t \times A'_u$ are proper over A'_u, γ^{cpt} is proper. Further, it is easy to see that $\gamma^{\mathrm{cpt}}|_{(X^{\mathrm{cpt}} - S_\infty) \times A'_u} (= \gamma)$ and $\gamma^{\mathrm{cpt}}|_{S_\infty \times A'_u}$ is quasi-finite, hence γ^{cpt} is quasi-finite. As γ^{cpt} is proper and quasi-finite, it is finite. Finally, for each $c \in A'_u$, the fiber of γ^{cpt} at c gives a non-constant $k(c)$-morphism $X^{\mathrm{cpt}}_{k(c)} \to (P_t)_{k(c)}$ of curves, hence, by the fiber-by-fiber criterion for flatness ([**EGA4**, Corollaire (11.3.11)]), γ^{cpt} is flat.

(ii) By (i), γ is finite flat. (For the finiteness, observe $X \times A'_u = (\gamma^{\mathrm{cpt}})^{-1}(A_t \times A'_u)$.) Now, as $\mathrm{Im}((\gamma^{\mathrm{cpt}})^*(\Omega_{A_t \times A'_u}) \to \Omega_{X \times A'_u}) = \mathcal{O}_{X \times A'_u} d(f + uh^{Np}) + \mathcal{O}_{X \times A'_u} du$ coincides with $\Omega_{X \times A'_u} = \mathcal{O}_{X \times A'_u} df + \mathcal{O}_{X \times A'_u} du$, γ^{cpt} is étale, as desired. □

Set

$$Z \stackrel{\text{def}}{=} (X \times A'_u) \times_{A_t} C = (X \times A'_u) \times_{(A_t \times A'_u)} (C \times A'_u).$$

(Here, the morphism $X \times A'_u \to A_t$ in the first fiber product is the composite of γ and the projection $A_t \times A'_u \to A_t$, while the morphism $C \times A'_u \to A_t \times A'_u$ in the second fiber product is the base change of j to A'_u.) Being finite étale over $X \times A'_u$, Z is an affine, smooth, 2-dimensional k-scheme. Moreover, Z admits a natural (smooth, surjective) morphism $Z \to A'_u$, obtained as the composite of the projection $Z \to X \times A'_u$ (which is finite étale) and the projection $X \times A'_u \to A'_u$ (which is smooth, surjective).

CLAIM 4.4. (i) Z is irreducible.
(ii) The generic fiber of $Z \to A'_u$ is geometrically irreducible.

PROOF. (i) Note that we have a morphism $X \times A'_u \to X \times A_t$ over A_t, $(x, u) \mapsto (x, g_u(x))$. This is birational, and its inverse (as a rational map) is given by $(x, t) \mapsto (x, (t - f(x))/h(x)^{Np})$. This, together with the fact that $(X \times A_t) \times_{A_t} C = X \times C$ is an integral scheme, implies that $Z = (X \times A'_u) \times_{A_t} C$ is an integral scheme, as desired.

(ii) By (i), it suffices to prove that the integral closure $\tilde{A}'_u$ of A'_u in Z coincides with A'_u. Let x_0 be an element of $S \subset X$, set $t_0 \stackrel{\text{def}}{=} f(x_0)$, and take any (necessarily closed) point w_0 of $j^{-1}(t_0) \subset C$. Then $u \mapsto ((x_0, u), w_0)$ defines a morphism $A'_u \to Z = (X \times A'_u) \times_{A_t} C$, as $h(x_0) = 0$. This morphism gives a section of $Z \to A'_u$, which implies that $\tilde{A}'_u$ must coincide with A'_u, as desired. $\square$

For each integer $n \geq 0$, let $\phi_n : A'_u \to A'_u$, $u \mapsto u^{p^n}$ denote the p^n-th power endomorphism of $A'_u = \mathbf{G}_{m,k}$, and set $Z^{(n)} \stackrel{\text{def}}{=} Z \times_{A'_u, \phi_n} A'_u$, which fits into the following commutative diagram with all the squares cartesian:

$$\begin{array}{ccccccc}
Z^{(n)} & \to & X \times A'_u & \stackrel{\gamma^{(n)}}{\to} & A_t \times A'_u & \to & A'_u \\
\downarrow & & \downarrow \mathrm{id} \times \phi_n & & \downarrow \mathrm{id} \times \phi_n & & \downarrow \phi_n \\
Z & \to & X \times A'_u & \stackrel{\gamma}{\to} & A_t \times A'_u & \to & A'_u,
\end{array}$$

where $\gamma^{(n)} : X \times A'_u \to A_t \times A'_u$ is defined by $(x, u) \mapsto (g_{u^{p^n}}(x), u)$. Let $(Z^{(n)})^{\text{cpt}}$ (resp. $Z^{\text{cpt}} = (Z^{(0)})^{\text{cpt}}$) be the integral closure of $P_t \times A'_u$ in (the function field of) $Z^{(n)}$ (resp. $Z = Z^{(0)}$), which fits into the following commutative diagram with the middle and right squares cartesian:

$$\begin{array}{ccccccc}
(Z^{(n)})^{\text{cpt}} & \to & X^{\text{cpt}} \times A'_u & \stackrel{(\gamma^{(n)})^{\text{cpt}}}{\to} & P_t \times A'_u & \to & A'_u \\
\downarrow & & \downarrow \mathrm{id} \times \phi_n & & \downarrow \mathrm{id} \times \phi_n & & \downarrow \phi_n \\
Z^{\text{cpt}} & \to & X^{\text{cpt}} \times A'_u & \stackrel{\gamma^{\text{cpt}}}{\to} & P_t \times A'_u & \to & A'_u,
\end{array}$$

where $(\gamma^{(n)})^{\text{cpt}}$ is defined by the cartesian property.

CLAIM 4.5. (i) $(Z^{(n)})^{\text{cpt}} \to A'_u$ is proper, flat with geometrically connected fibers.
(ii) $(Z^{(n)})^{\text{cpt}} \to A'_u$ is generically smooth for $n \gg 0$.

PROOF. (i) As $(Z^{(n)})^{\mathrm{cpt}} \to A'_u$ is the composite of the finite morphism $(Z^{(n)})^{\mathrm{cpt}} \to P_t \times A'_u$ and the projective morphism $P_t \times A'_u \to A'_u$, it is proper. As $(Z^{(n)})^{\mathrm{cpt}} \to A'_u$ is a dominant morphism from an integral scheme to a noetherian normal integral 1-dimensional scheme, it is flat. Finally, to prove that $(Z^{(n)})^{\mathrm{cpt}} \to A'_u$ has geometrically connected fibers, it suffices to prove that its Stein factorization is trivial. Thus, as A'_u is normal, it suffices to prove that $k((Z^{(n)})^{\mathrm{cpt}})/k(A'_u)$ is a regular extension. Further, as $k((Z^{(n)})^{\mathrm{cpt}}) = k(Z^{(n)}) = k(Z) \otimes_{k(A'_u),\phi_n} k(A'_u)$, it suffices to prove that $k(Z)/k(A'_u)$ is a regular extension, which follows from (the proof of) Claim 4.4(ii).

(ii) It suffices to prove that the generic fiber $((Z^{(n)})^{\mathrm{cpt}})_{k(u)} = (Z^{(n)}_{k(u)})^{\mathrm{cpt}} \to \mathrm{Spec}(k(u))$, which is the normal (or, equivalently, regular) compactification of the smooth curve $(Z^{(n)})_{k(u)} = Z^{(n)}_{k(u)} \to \mathrm{Spec}(k(u))$, is smooth for $n \gg 0$. Let $k(u)^{\mathrm{perf}} = \cup_{n\geq 0} k(u)^{p^{-n}}$ denote the perfect closure of $k(u)$. Then the normal compactification $(Z_{k(u)^{\mathrm{perf}}})^{\mathrm{cpt}}$ of the smooth curve $Z_{k(u)^{\mathrm{perf}}}$ is regular, hence smooth over $k(u)^{\mathrm{perf}}$, as $k(u)^{\mathrm{perf}}$ is perfect. The proper smooth curve $(Z_{k(u)^{\mathrm{perf}}})^{\mathrm{cpt}}$ over $k(u)^{\mathrm{perf}}$ must descend to a proper smooth curve $(Z_{k(u)^{p^{-n}}})^{\mathrm{cpt}}$ over $k(u)^{p^{-n}}$ for $n \gg 0$, which gives the (unique) normal compactification of $Z_{k(u)^{p^{-n}}}$. Now, as $Z^{(n)}_{k(u)} \to \mathrm{Spec}(k(u))$ is isomorphic to $Z_{k(u)^{p^{-n}}} \to \mathrm{Spec}(k(u)^{p^{-n}})$, $(Z^{(n)}_{k(u)})^{\mathrm{cpt}} \to \mathrm{Spec}(k(u))$ is isomorphic to $(Z_{k(u)^{p^{-n}}})^{\mathrm{cpt}} \to \mathrm{Spec}(k(u)^{p^{-n}})$. Thus, $(Z^{(n)}_{k(u)})^{\mathrm{cpt}} \to \mathrm{Spec}(k(u))$ is smooth for $n \gg 0$, as desired. □

Fix $n \gg 0$ as in Claim 4.5(ii). Then by Claim 4.5(i)(ii), $(Z^{(n)})^{\mathrm{cpt}} \to A'_u$ is proper, flat, generically smooth with geometrically connected fibers. In particular, there exists a k-rational point c' of A'_u, such that the fiber of $(Z^{(n)})^{\mathrm{cpt}} \to A'_u$ at c' is smooth and connected, hence irreducible. In particular, the fiber of $Z^{(n)} \to A'_u$ at c' is irreducible. This last fiber is identified with the fiber Z_c of $Z \to A'_u$ at $c \stackrel{\mathrm{def}}{=} (c')^{p^n}$, as desired. This completes the proof of Lemma 4.2. □

Return to the proof of Theorem 4.1 in the general case. Observe that we can choose a closed subscheme S'' of $\mathbf{A}^1_{k_0}$ and an isomorphism $i' : S' \xrightarrow{\sim} S''$. i' defines an isomorphism $i : S \xrightarrow{\sim} S''$, i.e., $i = i' \circ l$. Then, as a result of the special case, we see that there exist a finite, étale, connected covering $\mu : Y \to X$ (resp. $\mu' : Y' \to X'$), an étale morphism $g : Y \to \mathbf{A}^1_{k_0}$ (resp. $g' : Y' \to \mathbf{A}^1_{k_0}$), and a 0-dimensional closed subscheme T (resp. T') of Y (resp. Y'), such that $g_T = i \circ \mu_T$ (resp. $g'_{T'} = i' \circ \mu'_{T'}$) holds as isomorphisms $T \to S''$ (resp. $T' \to S''$).

First, applying Lemma 4.2 to

$$(f : X \to \mathbf{A}^1_k, S, j : C \to \mathbf{A}^1_k) = (g' : Y' \to \mathbf{A}^1_{k_0}, T', \mathrm{id} : \mathbf{A}^1_{k_0} \to \mathbf{A}^1_{k_0}),$$

we may assume that g' is finite étale. Next, applying Lemma 4.2 to

$$(f : X \to \mathbf{A}^1_k, S, j : C \to \mathbf{A}^1_k) = (g : Y \to \mathbf{A}^1_{k_0}, T, g' : Y' \to \mathbf{A}^1_{k_0}),$$

we may assume that g is also finite étale and that $Y'' \stackrel{\mathrm{def}}{=} Y \times_{\mathbf{A}^1_{k_0}} Y'$ is connected. Y'' admits a 0-dimensional closed subscheme $T'' \stackrel{\mathrm{def}}{=} T \times_{S''} T'$. Now, it is easy to see that the natural morphisms $Y'' \to X$, $Y'' \to X'$ and the closed subscheme $T'' \subset Y''$ satisfy the desired conditions of Theorem 4.1. This completes the proof. □

4.2. Proper and tame versions. It may be an interesting problem to ask whether the assertion of Theorem 3.1 for proper curves holds or not:

QUESTION 4.6. Does the relation $X \sim X'$ hold for every pair (X, X') of proper, smooth, connected curves of genus ≥ 2 over k_0?

We have a variant of this question that involves the following modification of the relation $\sim$.

DEFINITION. Let k be an algebraically closed field, and let X and X' be smooth, connected curves over k. We denote by X^{cpt} and $(X')^{\mathrm{cpt}}$ the smooth compactifications of X and X', respectively. Then we say that X and X' are tamely isogenous to each other (over k), and write $X \sim^{\mathrm{t}} X'$, if there exist a smooth, connected curve Y (whose smooth compactification is denoted by Y^{cpt}) and finite étale k-morphisms $f : Y \to X$, $f' : Y \to X'$, such that the k-morphism $Y^{\mathrm{cpt}} \to X^{\mathrm{cpt}}$ (resp. $Y^{\mathrm{cpt}} \to (X')^{\mathrm{cpt}}$) induced by f (resp. f') is at most tamely ramified along $X^{\mathrm{cpt}} - X$ (resp. $(X')^{\mathrm{cpt}} - X'$).

QUESTION 4.7. Does the relation $X \sim^{\mathrm{t}} X'$ hold for every pair (X, X') of affine, smooth, connected, hyperbolic curves over k_0? (Here, a (smooth, connected) curve X is hyperbolic, if $2g + r - 2 > 0$ holds, where g is the genus of X^{cpt} and r is the cardinality of $X^{\mathrm{cpt}} - X$.)

Concerning these questions, Theorem 3.6 might be regarded as a result slightly in the positive direction. However, it is completely unknown whether the answers of these questions should be affirmative or negative.

4.3. More general base fields. It may be natural to ask for a generalization of Theorem 3.1 to the case of more general base fields (algebraically closed and of positive characteristic). Thus, let k be an arbitrary algebraically closed field of characteristic $p > 0$. Let X be an affine, smooth, connected curve over k, and X^{cpt} the smooth compactification of X. We denote by g the genus of X^{cpt}.

First, we have the following result for $g \leq 1$.

PROPOSITION 4.8. *If $g \leq 1$, then $X \sim \mathbf{A}^1_k$.*

PROOF. If $g = 0$, X is isomorphic to an open subscheme of $\mathbf{A}^1_k$, hence the assertion follows from Lemma 2.1, (ii) $\implies$ (i). If $g = 1$, write $E = X^{\mathrm{cpt}}$ and fix $O \in X^{\mathrm{cpt}} - X$. Then (E, O) is an elliptic curve and X is an open subscheme of $E - \{O\}$. Consider the maximal (finite) étale subcovering $\mu : E' \to E$ of the 2-multiplication morphism $E \to E$. Thus, $\mu : E' \to E$ can be viewed as a separable isogeny of degree 2^δ with $\delta = 2$ (resp. $\delta = 1$, resp. $\delta = 0$) when $p \neq 2$ (resp. $p = 2$ and E is ordinary, resp. $p = 2$ and E is supersingular). As $\mu^{-1}(O) = E'[2](k)$, the quotient morphism $E' - \mu^{-1}(O) \to (E' - \mu^{-1}(O))/\{\pm 1\}$ is (finite) étale. Further, Observe that $(E' - \mu^{-1}(O))/\{\pm 1\}$ is isomorphic to the projective line minus 2^δ points, hence admits an open immersion $(E' - \mu^{-1}(O))/\{\pm 1\} \hookrightarrow \mathbf{A}^1_k$. Finally, set $Y \stackrel{\mathrm{def}}{=} \mu^{-1}(X)$, which is an open subscheme of $E' - \mu^{-1}(O)$. Now, the diagram

$$X \stackrel{\mu}{\leftarrow} Y \subset E' - \mu^{-1}(O) \to (E' - \mu^{-1}(O))/\{\pm 1\} \hookrightarrow \mathbf{A}^1_k,$$

together with Lemma 2.1, (ii) $\implies$ (i) again, implies $X \sim \mathbf{A}^1_k$, as desired. $\square$

For $g \geq 2$, however, Proposition 4.10 below shows that this is impossible in general. (See Example 4.11 below.) We start with the following general lemma.

Here, for a smooth scheme X over k, ω_X denotes the canonical invertible sheaf (i.e., the determinant invertible sheaf of the locally free sheaf Ω_X. (Thus, if $\dim(X) = 1$, we have $\omega_X = \Omega_X$.)

LEMMA 4.9. *Let X, X' and Y be (non-empty) smooth, connected schemes of finite type over k, and $f : Y \to X$ and $f' : Y \to X'$ finite étale morphisms over k. Then $f_*(f')^*[\omega_{X'}] = \deg(f)[\omega_X]$ and $(f')_*f^*[\omega_X] = \deg(f')[\omega_{X'}]$. Here, the homomorphisms $f^* : \mathrm{Pic}(X) \to \mathrm{Pic}(Y)$, $f_* : \mathrm{Pic}(Y) \to \mathrm{Pic}(X)$ are defined by $f^*[L] \stackrel{\text{def}}{=} [f^*(L)]$ for an invertible sheaf L on X and $f_*[M] \stackrel{\text{def}}{=} [N_{Y/X}(M)]$ for an invertible sheaf M on Y, where $N_{Y/X}$ stands for the norm. (See* [**EGA2**, 6.5] *and* [**EGA4**, 21.5].*)*

PROOF. As f and f' are finite étale, we have $f^*\Omega_X = \Omega_Y = (f')^*(\Omega_{X'})$, hence $f^*\omega_X = \omega_Y = (f')^*(\omega_{X'})$. Now, we have:

$$f_*(f')^*[\omega_{X'}] = f_*[\omega_Y] = f_*f^*[\omega_X] = \deg(f)[\omega_X],$$

and, similarly, $(f')_*f^*[\omega_X] = \deg(f')[\omega_{X'}]$, as desired. □

PROPOSITION 4.10. *Let X and X' be smooth, connected curves over k. If $X \sim X'$, then either both $[\Omega_X] \in \mathrm{Pic}(X)$ and $[\Omega_{X'}] \in \mathrm{Pic}(X')$ are torsion elements or both are non-torsion elements.*

In particular, if $X \sim \mathbf{A}^1_k$, then $[\Omega_X] \in \mathrm{Pic}(X)$ is a torsion element.

PROOF. As $X \sim X'$, there exist a smooth, connected curve Y over k and finite étale k-morphisms $f : Y \to X$, $f' : Y \to X'$. By Lemma 4.9, we have $f_*(f')^*[\Omega_{X'}] = \deg(f)[\Omega_X]$ and $(f')_*f^*[\Omega_X] = \deg(f')[\Omega_{X'}]$, from which the first assertion follows. The second assertion follows from the first, since $[\Omega_{\mathbf{A}^1_k}] = 0$ holds in $\mathrm{Pic}(\mathbf{A}^1_k) = \{0\}$. □

EXAMPLE 4.11. Let k be an algebraically closed field of characteristic $p > 0$, and denote by k_0 the algebraic closure of $\mathbf{F}_p$ in k. We assume $k \neq k_0$. Let X_0 be a proper, smooth, connected curve of genus $g \geq 2$ over k_0, and set $X \stackrel{\text{def}}{=} X_0 \otimes_{k_0} k$. Let P be a closed point of X that does not come from a closed point of X_0. (Or, equivalently, the image of P by the natural projection $X \to X_0$ is the generic point of X_0.) Now, set $Y = X - \{P\}$. Then we have $Y \not\sim \mathbf{A}^1_k$.

Indeed, suppose otherwise. Then, by Proposition 4.10, the class of Ω_Y in $\mathrm{Pic}(Y)$ must be a torsion element, or, equivalently, the class of $\Omega_X((2-2g)P)$ in $J(k) = \mathrm{Pic}^0(X)$ must be a torsion element, where J denotes the Jacobian variety of X. (See Proposition 4.12(i) below.) Consider the k-morphism $\phi : X \to J$ that sends $x \in X$ to the class of $\Omega_X((2-2g)x)$. This k-morphism naturally descends to a k_0-morphism $\phi_0 : X_0 \to J_0$, where J_0 denotes the Jacobian variety of X_0. Since $\phi(P)$ is a torsion element of J, it comes from a closed point of J_0. (Observe that $J(k)_{\mathrm{tors}} = J_0(k_0)_{\mathrm{tors}}$.) Namely, the image of P in J_0 is a closed point. Now, since ϕ_0 is a finite morphism (as $g \geq 2$), the image of P in X_0 must be a closed point. This contradicts our assumption.

Example 4.11 can be made more precise, as in the following Proposition 4.12. From now on, we let k be an algebraically closed field of characteristic $p > 0$, and denote by k_0 the algebraic closure of $\mathbf{F}_p$ in k. (We do not assume $k \neq k_0$.) Let X be a proper, smooth, connected curve of genus $g \geq 2$ over k. For each closed point $P \in X$, we set $X_P \stackrel{\text{def}}{=} X - \{P\}$.

PROPOSITION 4.12. *We consider the following conditions on P:*
$(\mathrm{a})_P$ $X_P \sim \mathbf{A}^1_k$.
$(\mathrm{b})_P$ $[\Omega_{X_P}] \in \mathrm{Pic}(X_P)$ *is a torsion element.*
$(\mathrm{b}')_P$ $[\Omega_X((2-2g)P)] \in \mathrm{Pic}(X)$ *is a torsion element.*
(i) *We have* $(\mathrm{a})_P \implies (\mathrm{b})_P \iff (\mathrm{b}')_P$.
(ii) *Assume, moreover, that X descends to a curve over k_0, i.e., $X \simeq X_0 \times_{k_0} k$ holds for some curve X_0 over k_0. We consider:*
$(\mathrm{c}_0)_P$ *P comes from a closed point of X_0.*
Then we have $(\mathrm{a})_P \iff (\mathrm{b})_P \iff (\mathrm{b}')_P \iff (\mathrm{c}_0)_P$.

PROOF. (i) First, $(\mathrm{a})_P \implies (\mathrm{b})_P$ just follows from Proposition 4.10. Next, observe that the natural homomorphism $\mathrm{Pic}^0(X) \to \mathrm{Pic}(X_P)$, $[L] \mapsto [L|_{X_P}]$ is an isomorphism, whose inverse is given by $\mathrm{Pic}(X_P) \to \mathrm{Pic}^0(X)$, $[M] \mapsto [\tilde{M}(-\deg(\tilde{M})P)]$, where $\tilde{M}$ is any invertible sheaf on X with $\tilde{M}|_{X_P} \simeq M$. From this, $(\mathrm{b})_P \iff (\mathrm{b}')_P$ follows.

(ii) $(\mathrm{c}_0)_P \implies (\mathrm{a})_P$ follows from Theorem 3.1. Combining this and (i), it suffices to prove $(\mathrm{b}')_P \implies (\mathrm{c}_0)_P$. But this has already been shown in Example 4.11. □

REMARK 4.13. In Proposition 4.12(i), suppose that condition $(*_{X_P})$, which is stronger than $(\mathrm{a})_P$, holds. Then the conclusion that $[\Omega_{X_P}] = 0$ in $\mathrm{Pic}(X_P)$, which is stronger than $(\mathrm{b})_P$, and the conclusion that $[\Omega_X((2-2g)P)] = 0$ in $\mathrm{Pic}(X)$, which is stronger than $(\mathrm{b}')_P$, follow immediately. This is contained in [**Z**, Corollary 2].

Further, let $|X|$ denote the set of closed points of X, and consider the following relations on $|X|$:

$$P \simeq Q \overset{\text{def}}{\iff} X_P \simeq X_Q \ (\iff \exists \alpha \in \mathrm{Aut}_k(X), \text{ s.t. } Q = \alpha(P)),$$

$$P \sim Q \overset{\text{def}}{\iff} X_P \sim X_Q,$$

$$P \approx Q \overset{\text{def}}{\iff} \exists \alpha \in \mathrm{End}_k(X), \text{ s.t. } Q = \alpha(P) \text{ or } P = \alpha(Q),$$

where $\mathrm{End}_k(X)$ stands for the monoid of finite k-morphisms $X \to X$. It is clear that $\simeq$ and $\sim$ are equivalence relations on $|X|$. We shall prove that $\approx$ is also an equivalence relation on $|X|$.

LEMMA 4.14. *Let $\alpha, \beta \in \mathrm{End}_k(X)$, then there exist $\gamma, \delta \in \mathrm{End}_k(X)$ such that $\beta = \alpha\gamma = \delta\alpha$ or $\alpha = \beta\gamma = \delta\beta$ holds.*

PROOF. As $g \geq 2$, α, β are purely inseparable, hence the (k-linear) embeddings $\alpha^* : k(X) \hookrightarrow k(X)$ (resp. $\beta^* : k(X) \hookrightarrow k(X)$) of $k(X)$ can be identified with the unique purely inseparable extension $k(X) \subset k(X)^{p^{-m}}$ of degree p^m or the unique purely inseparable subfield $k(X)^{p^m} \subset k(X)$ of degree p^m (resp. $k(X) \subset k(X)^{p^{-n}}$ of degree p^n or $k(X)^{p^n} \subset k(X)$ of degree p^n), where we set $p^m = \deg(\alpha)$, $p^n = \deg(\beta)$ ($m, n \in \mathbf{Z}_{\geq 0}$). It follows from this that there exist (k-linear) embeddings $\gamma^*, \delta^* : k(X) \hookrightarrow k(X)$, purely inseparable of degree p^{n-m} (resp. p^{m-n}) if $m \leq n$ (resp. $m \geq n$), such that $\beta^* = \gamma^*\alpha^* = \alpha^*\delta^*$ (resp. $\alpha^* = \gamma^*\beta^* = \beta^*\delta^*$). Now, the k-morphisms $\gamma, \delta : X \to X$ corresponding to $\gamma^*, \delta^* : k(X) \hookrightarrow k(X)$, respectively, satisfy the desired properties. □

COROLLARY 4.15. *$\approx$ is an equivalence relation on $|X|$.*

PROOF. As it is clear that $\approx$ is reflexive and symmetric, it suffices to prove that $\approx$ is transitive. Take $P, Q, R \in |X|$ such that $P \approx Q$ and $Q \approx R$. Then there exist $\alpha, \beta \in \mathrm{End}_k(X)$, such that either $Q = \alpha(P)$ or $P = \alpha(Q)$, and that either $R = \beta(Q)$ or $Q = \beta(R)$. If $Q = \alpha(P)$ and $R = \beta(Q)$ (resp. $P = \alpha(Q)$ and $Q = \beta(R)$), then $R = \beta\alpha(P)$ (resp. $P = \alpha\beta(R)$), hence $P \approx R$. If $Q = \alpha(P)$ and $Q = \beta(R)$, then $\alpha(P) = \beta(R)$. By Lemma 4.14, there exists $\gamma \in \mathrm{End}_k(X)$, such that $\beta = \alpha\gamma$ or $\alpha = \beta\gamma$. If $\beta = \alpha\gamma$ (resp. $\alpha = \beta\gamma$), we have $\alpha(P) = \alpha\gamma(R)$ (resp. $\beta\gamma(P) = \beta(R)$), hence $P = \gamma(R)$ (resp. $\gamma(P) = R$), as α (resp. β) is bijective. Thus, $P \approx R$. Finally, if $P = \alpha(Q)$ and $R = \beta(Q)$, by Lemma 4.14, there exists $\delta \in \mathrm{End}_k(X)$, such that $\beta = \delta\alpha$ or $\alpha = \delta\beta$. If $\beta = \delta\alpha$ (resp. $\alpha = \delta\beta$), we have $\delta(P) = \delta\alpha(Q) = \beta(Q) = R$ (resp. $P = \alpha(Q) = \delta\beta(Q) = \delta(R)$). Thus, $P \approx R$. □

We have the following finiteness result, which can be viewed as in the direction of Mochizuki's theorem (see Theorem 1.1).

PROPOSITION 4.16. (i) *Every $\simeq$-class of $|X|$ is a finite set.*

(ii) *If X does not descend to any curve over k_0, then every $\sim$-class of $|X|$ is a finite set.*

(iii) *If X descends to a curve X_0 over k_0, let S_0 denote the set of closed points of X that comes from a closed point of X_0. Then S_0 forms a single $\sim$-class of $|X|$, which is an infinite set, and, for $P, Q \in |X|$, we have the following implications:*

$$P \simeq Q \implies P \sim Q \implies \text{either } P, Q \in S_0 \text{ or } P \approx Q.$$

PROOF. (i) This follows from the well-known fact that the automorphism group of a (proper, smooth, connected) curve of genus ≥ 2 is a finite group.

(ii) Assume that X does not descend to any curve over k_0, and let S be a $\sim$-class of $|X|$. To prove the finiteness of S, fix an element P_1 of S. Then, for each $P \in S$, we have $X_P \sim X_{P_1}$. Namely, there exist a smooth, connected curve Y over k and finite étale k-morphisms $f : Y \to X_P$ and $f_1 : Y \to X_{P_1}$, which extend uniquely to finite k-morphisms $f^{\mathrm{cpt}} : Y^{\mathrm{cpt}} \to (X_P)^{\mathrm{cpt}} = X$ and $f_1^{\mathrm{cpt}} : Y^{\mathrm{cpt}} \to (X_{P_1})^{\mathrm{cpt}} = X$, respectively. By Lemma 4.9, we have $f_* f_1^* [\Omega_{X_{P_1}}] = \deg(f)[\Omega_{X_P}]$. This implies, via the natural isomorphism $\mathrm{Pic}^0(X) \to \mathrm{Pic}(X_P)$, $[L] \mapsto [L|_{X_P}]$ used in the proof of Proposition 4.12(i),

$$(f^{\mathrm{cpt}})_*(f_1^{\mathrm{cpt}})^*[\Omega_X((2-2g)P_1)] = \deg(f)[\Omega_X((2-2g)P)].$$

Further, it follows from the divisibility of $\mathrm{Pic}^0(X)$ that there exists an invertible sheaf N on X (which is not unique) such that $N^{\otimes(2g-2)} \simeq \Omega_X$. Fix such an N, then we have

$$(2g-2)(f^{\mathrm{cpt}})_*(f_1^{\mathrm{cpt}})^*[\mathcal{O}(P_1) \otimes N^{-1}] = (2g-2)\deg(f)[\mathcal{O}(P) \otimes N^{-1}].$$

Observe that $(f_1^{\mathrm{cpt}})^* : \mathrm{Pic}^0(X) \to \mathrm{Pic}^0(Y)$ and $(f^{\mathrm{cpt}})_* : \mathrm{Pic}^0(Y) \to \mathrm{Pic}^0(X)$ are identified with k-homomorphisms $J \to J_Y$ and $J_Y \to J$, respectively, between the Jacobian varieties J of X and J_Y of Y. Thus, $(2g-2)(f^{\mathrm{cpt}})_*(f_1^{\mathrm{cpt}})^*$ can be viewed as a k-endomorphism $J \to J$. From this, the above equality implies that $[\mathcal{O}(P) \otimes N^{-1}] \in \Gamma \stackrel{\mathrm{def}}{=} (\mathrm{End}_k(J)[\mathcal{O}(P_1) \otimes N^{-1}])^{\mathrm{div}}$. (Here, for a subgroup B of an abelian group A, we set $B^{\mathrm{div}} \stackrel{\mathrm{def}}{=} \{a \in A \mid na \in A \text{ for some } n \in \mathbf{Z}_{>0}\}$.)

Now, consider the embedding $X \hookrightarrow J$, $Q \mapsto [\mathcal{O}(Q) \otimes N^{-1}]$. Then, from the above arguments, we have $S \subset X \cap \Gamma$. This implies that S is finite. Indeed, suppose otherwise. Since Γ is a subgroup of J of finite rank (i.e., $\dim_{\mathbf{Q}}(\Gamma \otimes_{\mathbf{Z}} \mathbf{Q}) < \infty$), the full Mordell-Lang conjecture for curves in positive characteristic ([**FP**, Proposition 4.2], (see Remark 4.17 below)) then implies that X must descend to a curve over k_0. This contradicts our assumption.

(iii) Assume that X descends to a curve X_0 over k_0. Then the first assertion follows from Proposition 4.12(ii), $(\mathrm{a})_P \iff (\mathrm{c}_0)_P$, together with the fact that S_0 ($\simeq |X_0|$) is an infinite set. The first implication of the second assertion is clear. So, it suffices to prove $P \approx Q$, assuming $P \sim Q$ (i.e., $X_P \sim X_Q$) and $P, Q \notin S_0$. As in the proof of (ii), there exist a smooth, connected curve Y over k and finite étale k-morphisms $f : Y \to X_P$ and $g : Y \to X_Q$, which extend uniquely to finite k-morphisms $f^{\mathrm{cpt}} : Y^{\mathrm{cpt}} \to (X_P)^{\mathrm{cpt}} = X$ and $g^{\mathrm{cpt}} : Y^{\mathrm{cpt}} \to (X_Q)^{\mathrm{cpt}} = X$, respectively. As in (ii), we have $f_* g^*[\Omega_{X_Q}] = \deg(f)[\Omega_{X_P}]$,

$$(f^{\mathrm{cpt}})_*(g^{\mathrm{cpt}})^*[\Omega_X((2-2g)Q)] = \deg(f)[\Omega_X((2-2g)P)],$$

and

$$(2g-2)(f^{\mathrm{cpt}})_*(g^{\mathrm{cpt}})^*[\mathcal{O}(Q) \otimes N^{-1}] = (2g-2)\deg(f)[\mathcal{O}(P) \otimes N^{-1}],$$

where we fix an invertible sheaf N on X such that $N^{\otimes(2g-2)} \simeq \Omega_X$. Further, $(g^{\mathrm{cpt}})^* : \mathrm{Pic}^0(X) \to \mathrm{Pic}^0(Y)$, $(f^{\mathrm{cpt}})_* : \mathrm{Pic}^0(Y) \to \mathrm{Pic}^0(X)$, and $\psi \stackrel{\mathrm{def}}{=} (2g-2)(f^{\mathrm{cpt}})_*(g^{\mathrm{cpt}})^* : \mathrm{Pic}^0(X) \to \mathrm{Pic}^0(X)$ are identified with k-homomorphisms $J \to J_Y$, $J_Y \to J$, and $J \to J$, respectively. Now, consider the embedding $\iota = \iota_{[N]} : X \hookrightarrow J$, $R \mapsto [\mathcal{O}(R) \otimes N^{-1}]$. Set $m \stackrel{\mathrm{def}}{=} (2g-2)\deg(f)$, which can be viewed as the m-multiplication k-homomorphism $J \to J$. Then the above equality asserts that $m(P) = \psi(Q)$ holds in J.

Observe that, as X over k descends to X_0 over k_0, J; Ω_X; N; $\iota : X \hookrightarrow J$; $m : J \to J$; and $\psi : J \to J$ over k descend (uniquely) to J_0, Ω_{X_0}; N_0; $\iota_0 : X_0 \hookrightarrow J_0$; $m_0 : J_0 \to J_0$; and $\psi_0 : J_0 \to J_0$ over k_0. More specifically, J_0 is the Jacobian variety of X_0; N descends uniquely to N_0 as $\mathrm{Pic}^0(X_0)$ is divisible and $\mathrm{Pic}(X_0)[2g-2]$ ($= J_0[2g-2]$) $\stackrel{\sim}{\to} \mathrm{Pic}(X)[2g-2]$ ($= J[2g-2]$); $\iota_0 := \iota_{0,[N_0]} : X_0 \hookrightarrow J_0$, $R_0 \mapsto [\mathcal{O}(R_0) \otimes N_0^{-1}]$; m_0 is the m-multiplication k_0-homomorphism $J_0 \to J_0$; and $\psi : J \to J$ descends uniquely to $\psi_0 : J_0 \to J_0$ as $\mathrm{End}_{k_0}(J_0) \stackrel{\sim}{\to} \to \mathrm{End}_k(J)$. (Note that Y, f, g, Y^{cpt}, f^{cpt}, g^{cpt}, J_Y, $(f^{\mathrm{cpt}})_*$, $(g^{\mathrm{cpt}})^*$ may not descend to objects/morphisms over k_0.)

It follows from our assumption $P, Q \notin S_0$ that the images of P and Q under the natural projection $X = X_0 \times_{k_0} k \to X_0$ both coincide with the generic point ξ_0 of X_0. Thus, the equality $m(P) = \psi(Q)$ implies $m_0(\xi_0) = \psi_0(\xi_0)$. By taking the closures, we have $Z_0 \stackrel{\mathrm{def}}{=} m_0(X_0) = \psi_0(X_0)$, which implies $Z \stackrel{\mathrm{def}}{=} Z_0 \times_{k_0} k = m(X) = \psi(X)$. Equipped with the structure of reduced closed subscheme, Z can be regarded as a proper, integral curve over k. Let $\tilde{Z}$ be the normalization of Z, which is a proper, smooth, connected curve over k. Observe that the k-homomorphisms $m, \psi : J \to J$ restrict to k-morphisms $m_X, \psi_X : X \to Z$, respectively, which induce k-morphisms $\tilde{m}_X, \tilde{\psi}_X : X \to \tilde{Z}$, respectively. Further, by the Albanese functoriality of Jacobian varieties, the k-morphisms $\tilde{m}_X, \tilde{\psi}_X : X \to \tilde{Z}$ induce k-homomorphisms $\tilde{m}, \tilde{\psi} : J \to \tilde{J}$, respectively, where $\tilde{J}$ denotes the Jacobian variety of $\tilde{Z}$ over k, while, by the Albanese property of Jacobian varieties, the k-morphism $\tilde{Z} \twoheadrightarrow Z \hookrightarrow J$ induces a k-homomorphism $q : \tilde{J} \to J$. Then we have $m = q \circ \tilde{m}$ and $\psi = q \circ \tilde{\psi}$.

Let $\tilde{g}$ denote the genus of the (proper, smooth, connected) curve $\tilde{Z}$ over k. As there is a finite k-morphism $X \to \tilde{Z}$ (i.e., $\tilde{m}_X$ or $\tilde{\psi}_X$), we have $g \geq \tilde{g}$. On the other hand, as $m = q \circ \tilde{m} : J \to J$ is surjective, so is $q : \tilde{J} \to J$, hence $\tilde{g} \geq g$. Thus, $\tilde{g} = g$ (≥ 2). This implies that $\tilde{m}_X, \tilde{\psi}_X : X \to \tilde{Z}$ are purely inseparable. In particular, $\tilde{\psi}$, hence $\psi = q \circ \tilde{\psi}$, is surjective. Now, as in the proof of Lemma 4.14, there exists $\gamma \in \mathrm{End}_k(X)$, such that $\tilde{\psi}_X = \tilde{m}_X \gamma$ or $\tilde{m}_X = \tilde{\psi}_X \gamma$. If $\tilde{\psi}_X = \tilde{m}_X \gamma$ (resp. $\tilde{m}_X = \tilde{\psi}_X \gamma$), then $m(P) = \psi(Q) = m(\gamma(Q))$, i.e., $m(P - \gamma(Q)) = 0$ (resp. $\psi(Q) = m(P) = \psi(\gamma(P))$, i.e., $\psi(Q - \gamma(P)) = 0$). If $P = \gamma(Q)$ (resp. $Q = \gamma(P)$), we have $P \approx Q$, as desired. Otherwise, $(P, \gamma(Q)) \in X \times X$ (resp. $(\gamma(P), Q) \in X \times X$) lies in the inverse image T of $J[m] - \{0\} \subset J - \{0\}$ (resp. $\mathrm{Ker}(\psi) - \{0\} \subset J - \{0\}$) under the natural k-morphism $d : X \times X - \Delta(X) \to J - \{0\}$, $(R, S) \mapsto [\mathcal{O}(R - S)]$. Since the k-morphism $d : X \times X - \Delta(X) \to J - \{0\}$ and the closed subscheme $J[m] - \{0\} \subset J - \{0\}$ (resp. $\mathrm{Ker}(\psi) - \{0\} \subset J - \{0\}$) descend to a k_0-morphism $d_0 : X_0 \times X_0 - \Delta(X_0) \to J_0 - \{0\}$ and the closed subscheme $J_0[m_0] - \{0\} \subset J_0 - \{0\}$ (resp. $\mathrm{Ker}(\psi_0) - \{0\} \subset J_0 - \{0\}$); d_0 is a finite k_0-morphism (as $g \geq 2$); and $J_0[m_0] - \{0\}$ (resp. $\mathrm{Ker}(\psi_0) - \{0\}$) is finite over k_0, T descends to a closed subscheme $T_0 \subset X_0 \times X_0 - \Delta(X_0)$ which is finite over k_0. Thus, $(P, \gamma(Q)) \in T \subset X \times X$ (resp. $(\gamma(P), Q) \in T \subset X \times X$) must lie in $S_0 \times S_0$, and, in particular, $P \in S_0$ (resp. $Q \in S_0$), which contradicts our assumption. This completes the proof. □

REMARK 4.17. In the proof of Proposition 4.16(ii), we resorted to the full Mordell-Lang conjecture for curves in positive characteristic. More precisely, we applied [**FP**, Proposition 4.2] to the embedding $X \hookrightarrow J$ and the subgroup $\Gamma \subset J$. Here, the readers may consider that we should be able to apply [**GM**, Corollary 2.6], instead of [**FP**, Proposition 4.2] since J is the Jacobian variety of X. However, the proof of [**GM**, Corollary 2.6], seems to be incomplete as it is. More specifically, it relies on the assertion in [**K**, Theorem 1 and Corollary 1] that the curve in question, defined over a function field of one variable over a field k of characteristic $p > 0$, is $\mathbb{F}_p$-isotrivial, but this admits a counterexample as pointed out in [**FP**, Remark 3.2]. (The gap of the proof of [**K**] seems to be on lines 6-7 from the bottom of page 665, where the non-constancy of the map g does not follow in general, unless k is perfect.) In [**K**, Theorem 1 and Corollary 1], the weaker assertion that the curve is k-isotrivial is correct as it is, together with its proof there. This is re-proved and strengthened in [**Rö**, Theorem 1.1], and is used in the proof of [**FP**, Proposition 4.2].

REMARK 4.18. (i) In Proposition 4.16, the set of $P \in |X|$ such that $(*_{X_P})$ holds is always finite. This is [**Z**, Corollary 3].

(ii) In the situation of Proposition 4.16(iii), we do not know whether a $\sim$-class $S \neq S_0$ of $|X|$ is finite or infinite.

4.4. Higher-dimensional varieties. It may be natural to ask for a generalization of Theorem 3.1 to the case of higher-dimensional (affine) varieties. However, the following proposition, which is just similar to Proposition 4.10, shows that this is impossible in general. (See Example 4.20 below.) Let k be an (arbitrary) algebraically closed field of characteristic $p > 0$.

PROPOSITION 4.19. *Let X and X' be smooth, connected, of finite type over k. If $X \sim X'$, then either both $[\omega_X] \in \mathrm{Pic}(X)$ and $[\omega_{X'}] \in \mathrm{Pic}(X')$ are torsion elements or both are non-torsion elements.*

In particular, if $X \sim \mathbf{A}^n_k$ ($n = \dim(X)$), then $[\omega_X] \in \mathrm{Pic}(X)$ is a torsion element.

PROOF. The proof is just similar to that of Proposition 4.10 for curves. As $X \sim X'$, there exist a connected scheme Y over k and finite étale k-morphisms $f : Y \to X$, $f' : Y \to X'$. By Lemma 4.9, we have $f_*(f')^*[\omega_{X'}] = \deg(f)[\omega_X]$ and $(f')_* f^*[\omega_X] = \deg(f')[\omega_{X'}]$, from which the first assertion follows. The second assertion follows from the first, since $[\omega_{\mathbf{A}_k^n}] = 0$ holds in $\mathrm{Pic}(\mathbf{A}_k^n) = \{0\}$. □

EXAMPLE 4.20. Let k be any algebraically closed field of characteristic $p > 0$ (possibly an algebraic closure of $\mathbf{F}_p$ and possibly not). Set $X \stackrel{\text{def}}{=} \mathbf{P}_k^1 \times \mathbf{P}_k^1$. It is well-known that $\mathrm{Pic}(X) = \mathrm{Pic}(\mathbf{P}_k^1) \oplus \mathrm{Pic}(\mathbf{P}_k^1) = \mathbf{Z}^{\oplus 2}$, and that $(a, b) \in \mathbf{Z}^{\oplus 2}$ corresponds to the class of an ample invertible sheaf if and only if both a and b are positive. Now, take any prime divisor D of X that corresponds to $(a, b) \in \mathbf{Z}^{\oplus 2}$ with $a > 0$, $b > 0$, and $a \neq b$. (For example, consider any finite k-morphism $f : \mathbf{P}_k^1 \to \mathbf{P}_k^1$ of degree ≥ 2, and define D to be the graph Γ_f of f. Then this D corresponds to $(a, b) = (\deg(f), 1)$, which satisfies the desired property.)

Now, set $Y = X - D$. We claim that Y is an affine, smooth, connected variety over k, and that the class of ω_Y in $\mathrm{Pic}(Y)$ is not a torsion element. Indeed, since D is an ample divisor, Y is affine. We have $\mathrm{Pic}(Y) = \mathrm{Pic}(X)/\mathbf{Z}[D] = \mathbf{Z}^{\oplus 2}/\mathbf{Z}(a, b)$, in which the class of ω_Y (which is the image of $[\omega_X] \in \mathrm{Pic}(X)$) corresponds to $(-2, -2) \bmod \mathbf{Z}(a, b)$. As $a \neq b$, this class is not a torsion element. Thus, by Proposition 4.19, we have $Y \not\sim \mathbf{A}_k^2$.

Acknowledgments

The author would like to thank the editors of the present volume very much for giving him an opportunity of completing and contributing the present paper, and Moshe Jarden for informing him of useful pieces of information on the full Mordell-Lang conjecture for curves in positive chracteristic (cf. Remark 4.17).

The author would also like to thank the referees very much for their careful reading of the former version of the present paper and for their various comments on it, which were very useful in improving the presentation of the present paper.

References

[EGA2] Alexander Grothendieck, *Éléments de géométrie algébrique. II. Étude globale élémentaire de quelques classes de morphismes* (French), Inst. Hautes Études Sci. Publ. Math. **8** (1961), 5–222, DOI 10.1007/BF02699291. MR217084

[EGA4] Alexander Grothendieck, *Éléments de géométrie algébrique. IV. Étude locale des schémas et des morphismes de schémas IV* (French), Inst. Hautes Études Sci. Publ. Math. **32** (1967), 5–361, DOI 10.1007/BF02732123. MR238860

[FP] Arno Fehm and Sebastian Petersen, *Ranks of abelian varieties and the full Mordell-Lang conjecture in dimension one*, this volume.

[GM] Dragos Ghioca and Rahim Moosa, *Division points on subvarieties of isotrivial semi-abelian varieties*, Int. Math. Res. Not., posted on 2006, Art. ID 65437, 23 pp., DOI 10.1155/IMRN/2006/65437. MR2264715

[K] Minhyong Kim, *Purely inseparable points on curves of higher genus*, Math. Res. Lett. **4** (1997), no. 5, 663–666, DOI 10.4310/MRL.1997.v4.n5.a4. MR1484697

[M] Shinichi Mochizuki, *Correspondences on hyperbolic curves*, J. Pure Appl. Algebra **131** (1998), no. 3, 227–244, DOI 10.1016/S0022-4049(97)00078-9. MR1637015

[Ra] Michel Raynaud, *Sections des fibrés vectoriels sur une courbe* (French, with English summary), Bull. Soc. Math. France **110** (1982), no. 1, 103–125, DOI 10.24033/bsmf.1955. MR662131

[Ro] Michael Rosen, *The Hilbert class field in function fields*, Exposition. Math. **5** (1987), no. 4, 365–378. MR917350

[Rö] Damian Rössler, *Purely inseparable points on curves*, this volume.

[Tam1] Akio Tamagawa, *Coverings and vector bundles of algebraic curves in positive characteristic (Raynaud's theory and its recent developments and applications)* (Japanese), in Algebraic number theory and related topics (Kyoto, 2000), Sūrikaisekikenkyūsho Kōkyūroku **1200** (2001), 186–194.

[Tam2] Akio Tamagawa, *Arithmetic geometry of coverings of curves in positive characteristic* (Japanese), in Algebraic number theory and related topics (Kyoto, 2002), Sūrikaisekikenkyūsho Kōkyūroku **1324** (2003), 1–6. MR2000760

[Tam3] Akio Tamagawa, *Unramified Skolem problems and unramified arithmetic Bertini theorems in positive characteristic*, in Kazuya Kato's fiftieth birthday, Doc. Math., Extra Vol. (2003), 789–831. MR2046615

[Tan] Hiroshi Tango, *On the behavior of extensions of vector bundles under the Frobenius map*, Nagoya Math. J. **48** (1972), 73–89. MR0314851

[Z] Leonardo Zapponi, *On the 1-pointed curves arising as étale covers of the affine line in positive characteristic*, Math. Z. **258** (2008), 711–727, DOI 10.1007/s00209-007-0192-6. MR2369052

Research Institute for Mathematical Sciences, Kyoto University, Kyoto 606-8502, Japan

Email address: tamagawa@kurims.kyoto-u.ac.jp

Contemporary Mathematics
Volume **767**, 2021
https://doi.org/10.1090/conm/767/15401

The inverse Galois problem over $\mathbb{C}(z)$

Arno Fehm, Dan Haran, and Elad Paran

ABSTRACT. We give a self-contained elementary solution for the inverse Galois problem over the field of rational functions over the complex numbers.

1. Introduction

Since the 19th century one of the foundational open questions of Galois theory is the *inverse Galois problem*, which asks whether every finite group occurs as the Galois group of a Galois extension of the field $\mathbb{Q}$ of rational numbers. In 1892 Hilbert proved his celebrated *irreducibility theorem* and deduced that the inverse Galois problem has a positive answer provided that every finite group occurs as the Galois group of a Galois extension of the field $\mathbb{Q}(z)$ of rational functions over $\mathbb{Q}$. It was known already back then that the analogous problem for the field $\mathbb{C}(z)$ of rational functions over the complex numbers indeed has a positive answer: *Every finite group occurs as the Galois group of a Galois extension of* $\mathbb{C}(z)$. The aim of this work is to give a new and elementary proof of this result which uses only basic complex analysis and field theory. The value of such a proof becomes clear when compared to the previously known ones:

1.1. Classical proof using Riemann's existence theorem. The classical proof (likely the one known to Hilbert) uses the fact that the finite extensions of $\mathbb{C}(z)$ correspond to compact Riemann surfaces realized as branched covers of the Riemann sphere $\hat{\mathbb{C}}$, which in turn are described by the fundamental group $\pi_1(D)$ of domains $D = \hat{\mathbb{C}} \smallsetminus \{z_1, \ldots, z_d\}$, where $z_1, \ldots, z_d$ are the branch points. This correspondence follows from Riemann's existence theorem, which states that every compact Riemann surface admits meromorphic functions with prescribed values at finitely many points. This is a deep theorem proven using potential theory [**Lam09**, Ch. X] or the theory of Hilbert spaces and cohomology [**Voe96**, Ch. 6].

1.2. Modern proofs via ultrametric analysis. Various "patching" approaches that appeared in recent years allow the study of the Galois theory of $K(z)$ for fields K that are complete with respect to an *ultrametric* absolute value,

2020 *Mathematics Subject Classification.* Primary 11R32; Secondary 12E30, 30H50.

This research was supported by the Israel Science Foundation (grant No. 696/13), by the Deutsche Forschungsgemeinschaft, and by the Minerva Minkowski Center for Geometry at Tel Aviv University.

The first and the third author were supported by short-term research grants of the Minerva Foundation of the Max-Planck Society.

like the field $\mathbb{Q}_p$ of p-adic numbers or the field $F((t))$ of formal power series over a base field F. Such approaches can be used to give, somewhat artificially, a proof of the inverse Galois problem over $\mathbb{C}(z)$: For example, one could prove a strengthening of the inverse Galois problem (the so-called *regular inverse Galois problem*) over $\mathbb{Q}_p(z)$ and then, via an embedding of $\mathbb{Q}_p$ into $\mathbb{C}$ (which cannot be continuous but exists by the axiom of choice), deduce the inverse Galois problem over $\mathbb{C}(z)$, cf. [**DD97**, §1]. Alternatively, one could prove the regular inverse Galois problem over $\mathbb{C}((t))(z)$, where $\mathbb{C}((t))$ is the complete field of formal power series over $\mathbb{C}$, and then "push down" this solution to $\mathbb{C}(z)$, cf. [**HJ98**, §6].

Such a proof has the obvious disadvantage that it invokes constructions and methods from the non-archimedean world that are foreign to the object of study – the field of complex numbers – while at the same time completely ignoring the useful analytic structure of $\mathbb{C}$: Early patching approaches, such as Harbater's *formal patching* [**Har87**], as well as *rigid patching* [**Ser92**] [**Liu95**] build on the heavy machinery of formal schemes, respectively Tate's rigid geometry and formal / rigid GAGA theorems – cf. the survey [**Har03**]. In more recent approaches, such as Haran-Völklein's *algebraic patching* [**Jar11**] and Harbater-Hartmann's *field patching* [**HH10**], which abstain from employing heavy geometric theories, one needs algebraic structure theorems for certain rings of non-archimedean holomorphic functions.

1.3. A direct proof. The new proof we propose follows the footsteps of the *algebraic patching* approach of [**HV96**], though here we work directly over the complex numbers with intrinsic rings of holomorphic functions on discs and annuli on the Riemann sphere. The heart of the proof is a matrix factorization result for our rings (Proposition 2.5), which is a variant of the classical Wiener-Hopf factorization theorem in harmonic analysis, cf. [**BS06**] or [**GF74**]. This factorization result allows us to realize groups inductively by "patching" two subgroups at a time (Proposition 3.5), where the "patches" correspond to two discs that intersect in an annulus.

We note that while our approach is inspired by the mentioned lofty methods and ideas, the proof itself is self-contained and elementary, and invokes only undergraduate complex analysis, linear algebra, and field theory. On the other hand, the previously explained approaches, in particular the one using Riemann's existence theorem, allow to prove much stronger statements on the Galois theory of $\mathbb{C}(z)$ than just the inverse Galois problem, and we refer the interested reader to [**Har03**] and [**Voe96**] for more details. For results on the inverse Galois problem over other fields see for example [**DD97**], [**Jar11**].

2. Rings of holomorphic functions

Let $\emptyset \neq I \subseteq [0,\infty]$ be an open interval, like (r_1,r_2), $[0,r_2)$, $(r_1,\infty]$ with $0 < r_1 < r_2 < \infty$, and denote by

$$A_I = \{z \in \hat{\mathbb{C}} \mid |z| \in I\}$$

the corresponding annulus around the origin (where $\hat{\mathbb{C}} = \mathbb{C} \cup \{\infty\}$ is the Riemann sphere). For a domain $\emptyset \neq D \subseteq \hat{\mathbb{C}}$ we denote by $\mathcal{H}(D)$ and $\mathcal{M}(D)$ the ring of holomorphic resp. meromorphic functions on D and let $\mathcal{H}_I = \mathcal{H}(A_I)$, $\mathcal{M}_I =$

$\mathcal{M}(A_I)$. Recall that every $f \in \mathcal{H}_I$ has a unique Laurent series expansion

$$f(z) = \sum_{k=-\infty}^{\infty} f_k z^k, \quad f_k \in \mathbb{C}$$

which converges *pointwise* to f on A_I. In particular, $f_k = 0$ for $k < 0$ if $0 \in I$, and $f_k = 0$ for $k > 0$ if $\infty \in I$. We denote by $\mathcal{W}_I$ the set of those such f for which the convergence is even *normal* on A_I, i.e.

$$\mathcal{W}_I = \{f \in \mathcal{H}_I \mid ||f||_I < \infty\},$$

where

$$||f||_I = \sum_{k=-\infty}^{\infty} \sup_{z\in A_I} |f_k z^k| = \sum_{k=-\infty}^{\infty} |f_k| \sup_{z\in A_I} |z|^k.$$

LEMMA 2.1. *$(\mathcal{W}_I, ||.||_I)$ is a Banach algebra.*

PROOF. For $f, g \in \mathcal{W}_I$, obviously $||f+g||_I \leq ||f||_I + ||g||_I$. Also,

$$||fg||_I = \sum_n \left| \sum_{k+l=n} f_k g_l \right| \sup_{z\in A_I} |z^n| \leq \sum_n \sum_{k+l=n} |f_k| \cdot |g_l| \sup_{z\in A_I} |z|^k \sup_{z\in A_I} |z|^l = ||f||_I \cdot ||g||_I,$$

so $\mathcal{W}_I$ is a $\mathbb{C}$-algebra and $||.||_I$ is a norm on $\mathcal{W}_I$.

If $(f_n)_{n=1}^{\infty}$ is a Cauchy sequence in $\mathcal{W}_I$, then, replacing it with a suitable subsequence, we can assume that $||f_{n+1} - f_n||_I < 2^{-n}$ for all n, hence, writing $f_n = \sum_k f_{n,k} z^k$, we have

$$\sum_{n>N} \sum_k |f_{n+1,k} - f_{n,k}| \sup_{z\in A_I} |z|^k = \sum_{n>N} ||f_{n+1} - f_n||_I < \sum_{n>N} 2^{-n} = 2^{-N}$$

for all N. In particular, $(f_{n,k})_{n=1}^{\infty}$ a Cauchy sequence in $\mathbb{C}$, for every k, hence it converges. Let $g_k = \lim_{n\to\infty} f_{n,k}$. Then $g_k - f_{N,k} = \lim_{n\to\infty} f_{n+1,k} - f_{N,k} = \sum_{n\geq N} f_{n+1,k} - f_{n,k}$, for all N, hence $g = \sum_k g_k z^k$ satisfies

$$||g - f_N||_I = \sum_k |g_k - f_{N,k}| \sup_{z\in A_I} |z|^k = \sum_k \left| \sum_{n\geq N} f_{n+1,k} - f_{n,k} \right| \sup_{z\in A_I} |z|^k < 2^{-N+1} \to 0.$$

From this we see that $f_N \to g$ uniformly on A_I, hence $g \in \mathcal{H}_I$, and that $||g||_I < \infty$, hence $g \in \mathcal{W}_I$. Thus, $\mathcal{W}_I$ is complete. □

LEMMA 2.2. *$\mathcal{W}_I$ is an integral domain.*

PROOF. We have $\mathcal{W}_I \subseteq \mathcal{H}_I$, and $\mathcal{H}_I$ is an integral domain since a non-zero holomorphic function on A_I has only isolated zeros, and so the product of two such functions is again non-zero. □

From now on we always view $\mathcal{W}_I$ as a Banach algebra equipped with the norm $||.||_I$ and identify its quotient field $\mathrm{Quot}(\mathcal{W}_I)$ with a subfield of $\mathcal{M}_I$.

LEMMA 2.3. *Let $0 < r_1 < r_2 < \infty$.*

(a) *The ring of Laurent polynomials $\mathbb{C}[z, z^{-1}]$ is dense in $\mathcal{W}_{(r_1,r_2)}$.*

(b) *The Banach algebras $\mathcal{W}_{[0,r_2)}$ and $\mathcal{W}_{(r_1,\infty]}$ are Banach subalgebras of $\mathcal{W}_{(r_1,r_2)}$.*

(c) *There is a continuous additive homomorphism $\varphi\colon \mathcal{W}_{(r_1,r_2)} \to \mathcal{W}_{[0,r_2)}$ with*

(c1) *$\varphi|_{\mathcal{W}_{[0,r_2)}} = \mathrm{id}_{\mathcal{W}_{[0,r_2)}}$,*

(c2) *$\ker(\varphi) \subseteq \mathcal{W}_{(r_1,\infty]}$, and* (c3) $||\varphi(f)||_{[0,r_2)} \leq ||f||_{(r_1,r_2)}$ *for all* $f \in \mathcal{W}_{(r_1,r_2)}$.

(d) $\mathrm{Quot}(\mathcal{W}_{[0,r_2)}) \cap \mathrm{Quot}(\mathcal{W}_{(r_1,\infty]}) = \mathbb{C}(z)$.

(e) $\mathcal{H}_{[0,r_2)} \subseteq \mathcal{W}_{[0,r_1)}$.

PROOF. (a): If $f = \sum_k f_k z^k \in \mathcal{W}_{(r_1,r_2)}$, then $g_N := \sum_{k=-N}^{N} f_k z^k \in \mathbb{C}[z, z^{-1}]$ converges to f in $\mathcal{W}_{(r_1,r_2)}$ as $N \to \infty$.

(b): This is clear from the definition of the norms $||.||_{[0,r_2)}$, $||\cdot||_{(r_1,\infty]}$, $||\cdot||_{(r_1,r_2)}$, since

$$\sup_{z\in A_{(r_1,r_2)}} |z|^k = \begin{cases} r_2^k = \sup_{z\in A_{[0,r_2)}} |z|^k, \text{ for } k \geq 0 \\ r_1^k = \sup_{z\in A_{(r_1,\infty]}} |z|^k, \text{ for } k \leq 0 \end{cases}.$$

(c): For $f = \sum_k f_k z^k \in \mathcal{W}_{(r_1,r_2)}$ let $\varphi(f) := \sum_{k=0}^{\infty} f_k z^k \in \mathcal{W}_{[0,r_2)}$. Clearly, φ is an additive homomorphism with $\varphi|_{\mathcal{W}_{[0,r_2)}} = \mathrm{id}_{\mathcal{W}_{[0,r_2)}}$ and $\ker(\varphi) \subseteq \mathcal{W}_{(r_1,\infty]}$. By (b), $||\varphi(f)||_{[0,r_2)} = ||\varphi(f)||_{(r_1,r_2)} \leq ||f||_{(r_1,r_2)}$ for all $f \in \mathcal{W}_{(r_1,r_2)}$. In particular, φ is continuous.

(d): It follows from Liouville's theorem (cf. [**Lam09**, 1.6.5(3)]) that

$$\mathrm{Quot}(\mathcal{W}_{[0,r_2)}) \cap \mathrm{Quot}(\mathcal{W}_{(r_1,\infty]}) \subseteq \mathcal{M}_{[0,r_2)} \cap \mathcal{M}_{(r_1,\infty]} = \mathcal{M}(\hat{\mathbb{C}}) = \mathbb{C}(z).$$

(e): For $f = \sum_{k=0}^{\infty} f_k z^k \in \mathcal{H}_{[0,r_2)}$ the convergence is absolute on $A_{[0,r_2)}$, in particular at $r_1 \in A_{[0,r_2)}$, hence $||f||_{[0,r_1)} = \sum_{k=0}^{\infty} |f_k| r_1^k < \infty$. □

Let $n \in \mathbb{N}$. Since $\mathcal{W}_I$ is complete, the ring of $n \times n$-matrices $\mathrm{Mat}_n(\mathcal{W}_I)$ is complete with respect to the sub-multiplicative matrix norm defined for $A = (a_{ij})_{i,j=1,\dots,n}$ by

$$|A| = n \cdot \max_{i,j} ||a_{ij}||_I.$$

Notice that the unit matrix $\mathbb{1}$ in $\mathrm{Mat}_n(\mathcal{W}_I)$ satisfies $|\mathbb{1}| = n$.

LEMMA 2.4. *Let* $A \in \mathrm{Mat}_n(\mathcal{W}_I)$. *If* $|\mathbb{1} - A| < 1$, *then* $A \in \mathrm{GL}_n(\mathcal{W}_I)$.

PROOF. Let $B = \mathbb{1} - A$. Since $\mathrm{Mat}_n(\mathcal{W}_I)$ is complete, the geometric series $\sum_{k=0}^{\infty} B^k$ converges to some $C \in \mathrm{Mat}_n(\mathcal{W}_I)$, and $AC = CA = \mathbb{1}$. □

We now fix $0 < r_1 < r_2 < \infty$ and write $R = \mathcal{W}_{(r_1,r_2)}$, $R_1 = \mathcal{W}_{[0,r_2)}$, $R_2 = \mathcal{W}_{(r_1,\infty]}$. For $M \in \mathrm{Mat}_n(R)$ we denote by $M^+ \in \mathrm{Mat}_n(R_1)$ the matrix obtained by applying the continuous map φ of Lemma 2.3(c) to the entries of M. Thus, $M \mapsto M^+$ is a continuous map.

PROPOSITION 2.5 (Matrix factorization). *For every* $A \in \mathrm{GL}_n(R)$ *there exist matrices* $B_1 \in \mathrm{GL}_n(R_1)$, $B_2 \in \mathrm{GL}_n(R_2)$ *and* $C \in \mathrm{GL}_n(\mathbb{C}(z))$ *with* $A = B_1 B_2 C$.

PROOF. By Lemma 2.3(a), applied to the entries of $A^{-1} \in \mathrm{GL}_n(R)$, there exists $D \in \mathrm{Mat}_n(\mathbb{C}[z, z^{-1}])$ with $|D - A^{-1}| < \frac{1}{n+2}|A|^{-1}$, so $|AD - \mathbb{1}| \leq |A| \cdot |D - A^{-1}| < \frac{1}{n+2}$. Let $A_0 = \mathbb{1} - AD$. Since $|A_0| < \frac{1}{n+2} < \frac{1}{2}$, which implies $\frac{|A_0|}{1-|A_0|} < 1$, the infinite sum

$$A_0^+ + (A_0^+ \cdot A_0)^+ + ((A_0^+ \cdot A_0)^+ \cdot A_0)^+ + \dots$$

converges by Lemma 2.3(c3) to a matrix $A_1 \in \mathrm{Mat}_n(R_1)$ with $|A_1| < \sum_{k=1}^{\infty} |A_0|^k \leq \frac{|A_0|}{1-|A_0|} < 1$. Since the map $M \mapsto M^+$ is additive and continuous,

$$\begin{aligned}((\mathbb{1} + A_1)A_0)^+ &= A_0^+ + (A_1A_0)^+ = \\ &= A_0^+ + \left(A_0^+A_0 + (A_0^+ \cdot A_0)^+A_0 + ((A_0^+ \cdot A_0)^+ \cdot A_0)^+A_0 + \dots\right)^+ = \\ &= A_0^+ + (A_0^+ \cdot A_0)^+ + ((A_0^+ \cdot A_0)^+ \cdot A_0)^+ + \dots = A_1.\end{aligned}$$

Put $A_2 = A_1 - (\mathbb{1} + A_1)A_0$. Then the above computation implies that $A_2^+ = A_1^+ - A_1 = 0$, so $A_2 \in \mathrm{Mat}_n(R_2)$. Moreover,

$$|A_2| \leq \frac{|A_0|}{1-|A_0|} + |A_0| \cdot (n + \frac{|A_0|}{1-|A_0|}) = \frac{|A_0|}{1-|A_0|}(n+1-(n-1)|A_0|) \leq \frac{|A_0|}{1-|A_0|}(n+1) < 1,$$

since $|A_0| < \frac{1}{n+2}$. For $i = 1, 2$ we have $\mathbb{1} + A_i \in \mathrm{GL}_n(R_i)$ by Lemma 2.4. Observe that $\mathbb{1} + A_2 = \mathbb{1} + A_1 - (\mathbb{1} + A_1)A_0 = (\mathbb{1} + A_1)(\mathbb{1} - A_0)$ and let $B_2 = \mathbb{1} + A_2$, $B_1 = (\mathbb{1} + A_1)^{-1}$. Then $B_2 = B_1^{-1}AD$, so $D \in \mathrm{GL}_n(\mathbb{C}(z))$ and $A = B_1B_2C$ with $C = D^{-1}$. □

REMARK 2.6. The rings $\mathcal{W}_I$ may be seen as a variant of the classical *Wiener algebra* of absolutely convergent Fourier series. They play a role analogous to that of Tate algebras in patching over non-archimedean fields. The construction of the latter rings uses the uniform norm for functions on certain sets. However, it appears that working with the uniform norm over $\mathbb{C}$ as well does not work: It is unclear to the authors whether any of the rings so obtained satisfy the essential properties of the rings $\mathcal{W}_I$ established in this section; some of them obviously do not, for example if the mentioned sets chosen are as in [**HV96**], see also [**GF74**, p. 30] for further obstructions.

3. Patching of Galois groups

In this section we show how to "patch" certain Galois extensions of $E := \mathbb{C}(z)$.

DEFINITION 3.1. Let $\emptyset \neq D \subseteq \hat{\mathbb{C}}$ be a domain and F a finite extension of E which is a subfield of $\mathcal{M}(D)$. We say that F/E is **analytic over** D if it is Galois and there exists a primitive element f for F/E such that its $\mathrm{Gal}(F/E)$-conjugates $f_1, \dots, f_n$, as well as $(f_\nu - f_\mu)^{-1}$, $1 \leq \nu < \mu \leq n$, belong to $\mathcal{H}(D)$. We say that a finite group G is **analytically realizable** if there exists an analytic extension of E over some domain D with Galois group isomorphic to G.

EXAMPLE 3.2. Every finite *cyclic* group is analytically realizable. Indeed, let n be a positive integer, let $f(z) \in \mathcal{H}_{[0,1)}$ be an n-th root of $z-1$, and let $F = \mathbb{C}(z, f)$. Since f has no zero in the open unit disc, also $f^{-1} \in \mathcal{H}_{[0,1)}$. Since the polynomial

$$p(T) = T^n - (z-1) = \prod_{k=1}^{n} (T - e^{\frac{2\pi i k}{n}} f) \in \mathbb{C}[z][T]$$

completely decomposes over $\mathbb{C}(z, f)$, the extension $\mathbb{C}(z, f)/\mathbb{C}(z)$ is Galois. By Eisenstein's criterion applied to the prime element $z - 1 \in \mathbb{C}[z]$, p is irreducible over $\mathbb{C}(z)$, hence $\mathrm{Gal}(\mathbb{C}(z, f)/\mathbb{C}(z)) \cong \mathbb{Z}/n\mathbb{Z}$ and the conjugates of f are $f_\nu = e^{\frac{2\pi i\nu}{n}} f$, $\nu = 1, \dots, n$. Thus, both $f_1, \dots, f_n$ and $(f_\nu - f_\mu)^{-1} = (e^{\frac{2\pi i\nu}{n}} - e^{\frac{2\pi i\mu}{n}})^{-1} f^{-1}$ are in $\mathcal{H}_{[0,1)}$, so $\mathbb{C}(z, f)$ is analytic over the open unit disc.

LEMMA 3.3. *If $F = E(f)$ is analytic over D, and $\varphi = \frac{az+b}{cz+d} \in \mathbb{C}(z)$ is a Möbius transformation, then $F' = E(f \circ \varphi)$ is analytic over $\varphi^{-1}(D)$, and* $\mathrm{Gal}(F/E) \cong \mathrm{Gal}(F'/E)$.

PROOF. The composition $g \mapsto g \circ \varphi$ induces an isomorphism $\varphi^*\colon \mathcal{M}(D) \to \mathcal{M}(\varphi^{-1}(D))$ which maps $\mathbb{C}(z)$ onto $\mathbb{C}(\varphi(z))$ and $\mathbb{C}(z,f)$ onto $\mathbb{C}(\varphi(z), f(\varphi(z)))$. Since it maps rational functions onto rational functions, and the same holds for φ^{-1}, it follows that $\mathbb{C}(\varphi(z)) = \mathbb{C}(z)$. Thus, φ^* maps F/E onto F'/E. Since $\varphi^*(\mathcal{H}(D)) = \mathcal{H}(\varphi^{-1}(D))$, F' is analytic over $\varphi^{-1}(D)$. □

REMARK 3.4. Since any domain contains a disc and all open discs $D \subseteq \hat{\mathbb{C}}$ are isomorphic to each other, Lemma 3.3 implies that every group G that is analytically realizable is realized by a suitable analytic extension F/E over any open disc, for example $A_{[0,2r_2)}$, where $r_2 > 0$. By Lemma 2.3(e), $\mathcal{H}_{[0,2r_2)} \subseteq \mathcal{W}_{[0,r_2)}$, so there exists a primitive element f for F/E such that all conjugates $f_1, \dots, f_n$ of f, as well as the $(f_\nu - f_\mu)^{-1}$, belong to $\mathcal{W}_{[0,r_2)}$.

The induction step in our proof of the inverse Galois problem over $\mathbb{C}(z)$ is given by the following proposition:

PROPOSITION 3.5 (Patching). *Let G be a finite group, generated by subgroups G_1, G_2. If both G_1 and G_2 are analytically realizable, then so is G.*

Before proving this proposition, we need some additional constructions and lemmas. Fix $0 < r_1 < r_2 < \infty$ and write $R = \mathcal{W}_{(r_1,r_2)}$, $R_1 = \mathcal{W}_{[0,r_2)}$, $R_2 = \mathcal{W}_{(r_1,\infty]}$, as in the previous section. Put $Q = \mathrm{Quot}(R)$, $Q_1 = \mathrm{Quot}(R_1)$, $Q_2 = \mathrm{Quot}(R_2)$. As $R_1, R_2 \subseteq R$, we have $Q_1, Q_2 \subseteq Q$. Let F_1, F_2 be analytic extensions of E with Galois groups G_1, G_2, respectively. By Remark 3.4 we may assume that F_1 is analytic over $A_{(\frac{1}{2}r_1,\infty]}$ and F_2 is analytic over $A_{[0,2r_2)}$. In particular, $F_1 \subseteq Q_2$ and $F_2 \subseteq Q_1$. By Lemma 2.3(d)

$$Q_1 \cap Q_2 = E, \tag{1}$$

hence for $i = 1,2$ we have $F_i \cap Q_i = E$. Let $P_i = F_iQ_i$ be the compositum of F_i and Q_i inside Q. Since F_i/E is Galois with group G_i and $F_i \cap Q_i = E$, we deduce that also P_i/Q_i is Galois and we may identify its Galois group with G_i by restriction to F_i.

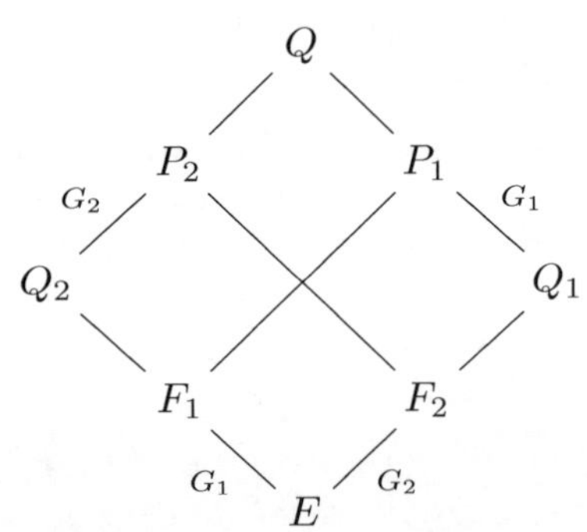

Let $n = |G|$ be the order of G and consider the commutative Q-algebra $N = Q^n$ with Q embedded diagonally. We label the standard basis $\mathcal{S}$ of N by the elements of G, and thus write $\mathcal{S} = (g \mid g \in G)$ and $N = \{\sum_{g\in G} a_g g \mid a_g \in Q\}$. Let G act on

N by $(\sum_g a_g g)^\sigma = \sum_g a_g(\sigma^{-1}g) = \sum_g a_{\sigma g} g$, $\sigma \in G$. Clearly, the fixed ring N^G is Q, where we view Q as a subring of N via the mentioned diagonal embedding. For each $i = 1, 2$ consider the Q_i-subalgebra

$$(2) \qquad N_i = \left\{ \sum_{g \in G} a_g g \in N \;\middle|\; a_g \in P_i,\ a_g^\tau = a_{g\tau},\ \text{ for all } g \in G,\ \tau \in G_i \right\}$$

of N. Put $F = N_1 \cap N_2$.

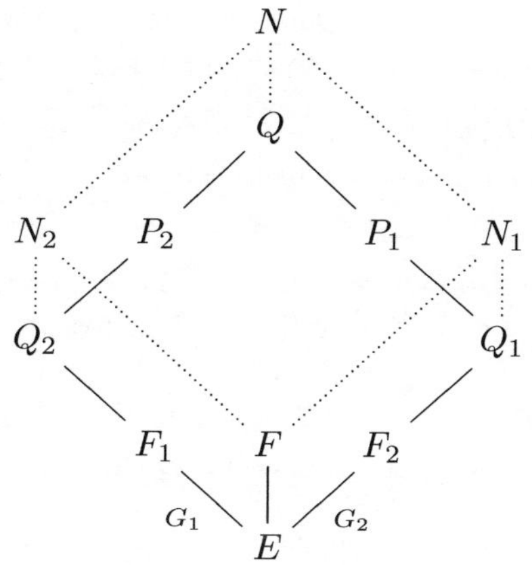

LEMMA 3.6. *For $i = 1, 2$, the Q_i-algebra N_i is G-invariant and $N_i^G = Q_i$.*

PROOF. Let $\sum_{g\in G} a_g g \in N_i$ and $\sigma \in G$. Then $a_g^\tau = a_{g\tau}$ for all $g \in G$ and $\tau \in G_i$, so $a_{\sigma g}^\tau = a_{\sigma g \tau}$. Hence $(\sum_{g\in G} a_g g)^\sigma = \sum_{g\in G} a_{\sigma g} g \in N_i$, so N_i is G-invariant. Furthermore, $N_i^G = N_i \cap N^G = N_i \cap Q \supseteq Q_i$, by (2). If $\sum_{g\in G} ag$ is an element of $N_i \cap Q$, then $a \in P_i$ and (in particular) $a^h = a$ for all $h \in G_i$, thus $a \in P_i^{G_i} = Q_i$. We deduce that $N_i^G = Q_i$. □

LEMMA 3.7. *For $i = 1, 2$, there exists a basis $\mathcal{C}_i$ for N over Q, contained in N_i, such that the transition matrix from $\mathcal{S}$ to $\mathcal{C}_i$ lies in $\mathrm{GL}_n(R)$.*

PROOF. Let $\tau_1, \dots, \tau_r$ be a listing of the elements of G_i and let $\omega_1, \dots, \omega_m$ be a system of representatives for G/G_i. (Thus $n = |G| = mr$.) By our assumptions, there exists a primitive element f for F_i/E, such that its conjugates $f_1 = f^{\tau_1}, \dots, f_r = f^{\tau_r}$ belong to R (since R contains both R_1, R_2), and moreover $d := \prod_{\nu<\mu}(f_\nu - f_\mu)$ is invertible in R. Consider the following n-tuple of elements of N_i

$$\mathcal{C}_i = \left(\sum_{\nu=1}^{r} f_\nu^{\mu-1} \cdot \omega_k \tau_\nu \;\middle|\; 1 \le k \le m,\ 1 \le \mu \le r \right)$$

and let A_i be the transition matrix from $\mathcal{S}$ to $\mathcal{C}_i$ (with respect to any fixed ordering of $\mathcal{S}$ and $\mathcal{C}_i$), i.e. $\mathcal{C}_i = \mathcal{S}A_i$.

By direct computation, A_i consists, up to a permutation of the rows and columns (that depends on the chosen ordering of $\mathcal{S}$ and $\mathcal{C}_i$), of m block matrices on the diagonal, each being the Vandermonde matrix $(f_\nu^{\mu-1})_{\nu,\mu=1,\dots,r}$. By our assumptions, $A_i \in \mathrm{Mat}_n(R)$, and $\det(A_i) = \pm d^m$ is invertible in R. Therefore $A_i \in \mathrm{GL}_n(R) \subseteq \mathrm{GL}_n(Q)$, and in particular $\mathcal{C}_i$ is a basis for N over Q. □

LEMMA 3.8. *The intersection $F = N_1 \cap N_2$ is a field.*

PROOF. The definition (2) of N_i gives an explicit presentation of F as

$$F = \left\{ \sum_{g\in G} a_g g \in N \;\middle|\; a_g \in P_1 \cap P_2,\ a_{g\tau} = a_g^\tau \text{ for all } g \in G \text{ and } \tau \in G_1 \cup G_2 \right\}.$$

It follows that if $\alpha = \sum_{g\in G} a_g g \in F$ is such that $a_g \neq 0$ for some $g \in G$ then $a_{g'} \neq 0$ for all $g' \in G$. Indeed, since $G = \langle G_1, G_2\rangle$, we have $g' = g\tau_1\tau_2 \ldots \tau_l$, where $\tau_1, \ldots, \tau_l \in G_1 \cup G_2$. Now $a_{g'} = a_{g\tau_1\tau_2\ldots\tau_l} = a_{g\tau_1\ldots\tau_{l-1}}^{\tau_l} = \ldots = a_g^{\tau_1\cdots\tau_l} \neq 0$ (since the τ_i are automorphisms). Thus if $\alpha \neq 0$ then $a_g \neq 0$ for all g, and so $\beta = \sum_{g\in G} a_g^{-1} g$ is a multiplicative inverse of α which clearly lies in F. □

Note that since $E \subseteq Q_1, Q_2$, the intersection F is an E-algebra, that is, F/E is a field extension. We conclude the proof of Proposition 3.5 by showing that F/E is Galois with group G:

PROOF OF PROPOSITION 3.5. For $i = 1, 2$ let $\mathcal{C}_i$ be the basis given by Lemma 3.7, and let $A_i \in \mathrm{GL}_n(R)$ be the transition matrix from $\mathcal{S}$ to $\mathcal{C}_i$, so that $\mathcal{C}_i = \mathcal{S}A_i$. By Proposition 2.5 applied to $A_1^{-1}A_2 \in \mathrm{GL}_n(R)$ there exist $B_i \in \mathrm{GL}_n(R_i), i = 1, 2$ and $C \in \mathrm{GL}_n(E)$ such that $A_1^{-1}A_2 = B_1B_2C$. Put

$$\mathcal{C} = \mathcal{C}_1B_1 = \mathcal{S}A_1B_1 = \mathcal{S}A_2C^{-1}B_2^{-1} = \mathcal{C}_2(C^{-1}B_2^{-1}).$$

Since $B_1 \in \mathrm{GL}_n(R_1) \subseteq \mathrm{GL}_n(Q_1)$ and $C^{-1}B_2^{-1} \in \mathrm{GL}_n(E) \cdot \mathrm{GL}_n(R_2) \subseteq \mathrm{GL}_n(Q_2)$ we get that the basis $\mathcal{C}$ for N/Q is contained in $N_1 \cap N_2 = F$.

By Lemma 3.6, $F = N_1 \cap N_2$ is G-invariant and $F^G = N_1^G \cap N_2^G = Q_1 \cap Q_2 = E$, by (1). Since $\mathcal{C}$ consists of n elements that are linearly independent over $E \subseteq Q$, we see that $[F : E] \geq n = |G|$. By Artin's Lemma in basic Galois theory [**Lan02**, VI Thm. 1.8], F/F^G is Galois and its Galois group is a quotient of G. Thus, $[F : E] = n$ and $\mathrm{Gal}(F/E) \cong G$.

Let $\phi\colon N \to Q$ be the homomorphism of Q-algebras $\sum_g a_g g \mapsto a_1$. Since F is a field, ϕ maps F onto a subfield F' of Q. Since ϕ maps E identically onto E (recall that $E \subseteq Q$ embeds into N diagonally), F'/E is Galois with group isomorphic to G. Finally, choose a primitive element $f \in Q$ for F'/E. Then all the conjugates $f_1, \ldots, f_n$ of f over E, as well $(f_\nu - f_\mu)^{-1}$, $\nu < \mu$, belong to $Q = \mathrm{Quot}(R)$. In particular, they are quotients of elements of $R = \mathcal{H}_{(r_1,r_2)}$, and hence are meromorphic functions on $A_{(r_1,r_2)}$. Thus, F' is analytic over some domain $D \subseteq A_{(r_1,r_2)}$. □

We can now easily solve the inverse Galois problem over $\mathbb{C}(z)$:

THEOREM 3.9. *Every finite group G is analytically realizable.*

PROOF. We prove the theorem by induction on the minimal number d of generators of $G = \langle g_1, \ldots, g_d\rangle$.

If $d = 1$, then G is cyclic, so the claim follows from Example 3.2.

If $d > 1$, then $G_1 = \langle g_1\rangle$ and $G_2 = \langle g_2, \ldots, g_d\rangle$ are generated by 1 resp. $d - 1$ elements, hence by the induction hypothesis, G_1 and G_2 are analytically realizable. By Proposition 3.5, so is $G = \langle G_1, G_2\rangle$. □

Acknowledgment

The authors wish to thank Nir Lev for interesting discussions on the Wiener algebra.

References

[BS06] Albrecht Böttcher and Bernd Silbermann, *Analysis of Toeplitz operators*, 2nd ed., Springer Monographs in Mathematics, Springer-Verlag, Berlin, 2006. Prepared jointly with Alexei Karlovich. MR2223704

[DD97] Pierre Dèbes and Bruno Deschamps, *The regular inverse Galois problem over large fields*, Geometric Galois actions, 2, London Math. Soc. Lecture Note Ser., vol. 243, Cambridge Univ. Press, Cambridge, 1997, pp. 119–138, DOI 10.1017/CBO9780511666124.007. MR1653011

[GF74] I. C. Gohberg and I. A. Fel'dman, *Convolution Equations and Projection Methods for Their Solution*, Transl. Math. Monogr., vol. 41. Amer. Math. Soc., 1974.

[HJ98] Dan Haran and Moshe Jarden, *Regular split embedding problems over complete valued fields*, Forum Math. **10** (1998), no. 3, 329–351, DOI 10.1515/form.10.3.329. MR1619723

[HV96] Dan Haran and Helmut Völklein, *Galois groups over complete valued fields*, Israel J. Math. **93** (1996), 9–27, DOI 10.1007/BF02761092. MR1380632

[Har87] David Harbater, *Galois coverings of the arithmetic line*, Number theory (New York, 1984), Lecture Notes in Math., vol. 1240, Springer, Berlin, 1987, pp. 165–195, DOI 10.1007/BFb0072980. MR894511

[Har03] David Harbater, *Patching and Galois theory*, Galois groups and fundamental groups, Math. Sci. Res. Inst. Publ., vol. 41, Cambridge Univ. Press, Cambridge, 2003, pp. 313–424. MR2012220

[HH10] David Harbater and Julia Hartmann, *Patching over fields*, Israel J. Math. **176** (2010), 61–107, DOI 10.1007/s11856-010-0021-1. MR2653187

[Jar11] Moshe Jarden, *Algebraic patching*, Springer Monographs in Mathematics, Springer, Heidelberg, 2011. MR2768285

[Lam09] K. Lamotke, *Riemannsche Flächen*, Second Edition. Springer, 2009.

[Lan02] Serge Lang, *Algebra*, 3rd ed., Graduate Texts in Mathematics, vol. 211, Springer-Verlag, New York, 2002. MR1878556

[Liu95] Qing Liu, *Tout groupe fini est un groupe de Galois sur $\mathbf{Q}_p(T)$, d'après Harbater* (French, with English summary), Recent developments in the inverse Galois problem (Seattle, WA, 1993), Contemp. Math., vol. 186, Amer. Math. Soc., Providence, RI, 1995, pp. 261–265, DOI 10.1090/conm/186/02185. MR1352276

[Ser92] Jean-Pierre Serre, *Topics in Galois theory*, Research Notes in Mathematics, vol. 1, Jones and Bartlett Publishers, Boston, MA, 1992. Lecture notes prepared by Henri Damon [Henri Darmon]; With a foreword by Darmon and the author. MR1162313

[Voe96] Helmut Völklein, *Groups as Galois groups*, Cambridge Studies in Advanced Mathematics, vol. 53, Cambridge University Press, Cambridge, 1996. An introduction. MR1405612

Institut für Algebra, Fakultät Mathematik, Technische Universität Dresden
Email address: arno.fehm@tu-dresden.de

Raymond and Beverly Sackler School of Mathematical Sciences, Tel Aviv University
Email address: haran@tauex.tau.ac.il

Department of Mathematics, Open University of Israel
Email address: paran@openu.ac.il

Contemporary Mathematics
Volume **767**, 2021
https://doi.org/10.1090/conm/767/15402

Explicit Hilbert's irreducibility theorem in function fields

Lior Bary-Soroker and Alexei Entin

Abstract. We prove a quantitative version of Hilbert's irreducibility theorem for function fields: If $f(T_1, \dots, T_n, X)$ is an irreducible polynomial over the field of rational functions in u over a finite field with q elements, then the proportion of n-tuples $(t_1, \dots, t_n)$ of monic polynomials in u of degree d for which $f(t_1, \dots, t_n, X)$ is reducible out of all n-tuples of degree d monic polynomials is $O(dq^{-d/2})$.

1. Introduction

Hilbert's irreducibility theorem says that if $f(T_1, \dots, T_n, X) \in \mathbb{Q}[T, X]$ is a multivariate irreducible polynomial of positive degree in X, then the set

$$H_f = \{(t_1, \dots, t_n) \in \mathbb{Q}^n : f(t_1, \dots, t_n, X) \in \mathbb{Q}[X] \text{ is irreducible}\}$$

is non-empty. We call such a set a (basic) *Hilbert set.*

The original motivation of Hilbert was in the inverse Galois problem, see for example [**8**]. Since then the theorem found numerous applications to different problems in algebra and number theory; e.g., for constructing elliptic curves of high rank, see [**9**].

There are several types of qualitative results that indicate that any Hilbert set H_f is big (see [**3**, §13] for reference): H_f is infinite; H_f is Zariski dense; H_f is S-adic dense, there exist many finite sets of primes S for which H_f is S-adic open.

A more quantitative variant is by counting: for $N \geq 1$, set

$$H_f(N) = \#(H_f \cap \{(t_1, \dots, t_n) \in \mathbb{Z}^n : N \leq t_i < 2N, \ i = 1, \dots, n\})$$

to be the number of integral tuples in H with entries in $[N, 2N)$. S. D. Cohen [**1**] proved that

$$\frac{H_f(N)}{N^n} = 1 + O_f\Big(\frac{\log N}{\sqrt{N}}\Big), \qquad N \to \infty. \tag{1.1}$$

Here the notation O_f means that the implied constant may depend on f, but not on N. The polynomial $f(T_1, \dots, T_n, X) = X^2 - (T_1 + T_2 + \dots + T_n)$ shows that the error term is optimal (up to the logarithmic factor) since obviously the number of n-tuples $(t_1, \dots, t_n)$ with $N \leq t_i < 2N$ for which $\sum t_i$ is a perfect square is $O(N^{n-1/2})$.

2020 *Mathematics Subject Classification.* Primary 11N36, 11R32, 11R58, 12E25.

The first named author was partially supported by grant no. 702/19 of the Israel Science Foundation and the second by grant no. 2507/19 of the Israel Science Foundation.

The qualitative results mentioned above were proved also in the global function field setting. However to the best of our knowledge, (1.1) has not been proved in function fields. An additional challenge arising in this setting is having to deal with inseparable polynomials, a phenomenon that does not occur in number fields.

The goal of this work is to prove a function field analogue. We denote by $\mathbb{F}_q$ the finite field of q elements. We let $\mathbb{F}_q(u)$ be the field of rational functions in u with coefficients in $\mathbb{F}_q$. The subset $M_d \subseteq \mathbb{F}_q(u)$ of monic polynomials of degree d takes the role of the interval $[N, 2N)$ with $q^d = \#M_d$ being the analog of $N = \#[N, 2N) \cap \mathbb{Z}$.

THEOREM 1.1. *Let $K = \mathbb{F}_q(u)$, let $f(T_1, \dots, T_n, X) \in K[T_1, \dots, T_n, X]$ be an irreducible polynomial of positive degree in X. Let $H_f(d)$ be the number of tuples $(t_1(u), \dots, t_n(u)) \in M_d^n$ with $f(t_1, \dots, t_n, X) \in K[X]$ irreducible. Then,*

$$\frac{H_f(d)}{q^{dn}} = 1 + O_{f,q}(dq^{-d/2}), \tag{1.2}$$

where the implied constant may depend on f, q.

To compare with (1.1) we note that q^d corresponds to N and d to $\log N$, so the error term in (1.2) is the perfect analog of (1.1).

REMARK 1.2. In fact we will show that if one fixes $\deg_{\mathbf{T},X} f$ then a similar bound holds uniformly whenever $d \geq 2\log_q \deg_u f + \log_q \lambda + 1$ for some constant $\lambda = \lambda(\deg_{\mathbf{T},X} f)$. See Theorem 4.3 for the precise statement.

When f is separable in the variable X, we will prove (1.2) using the function-field large sieve inequality due to Hsu [**6**], a proof that goes in the same lines of the proof of (1.1). When f is inseparable, those ideas seem to fail. Instead we take an approach inspired from the work of Uchida [**10**].

Theorem 1.1 has many consequences. The first is to Galois theory, which states that if $L/K(T_1, \dots, T_n)$ is a Galois extension, then for most $(t_1, \dots, t_n) \in M_d^n$, the Galois group remains the same under the specialization $T_i \mapsto t_i$. See Corollary 3.5 for a precise statement.

Theorem 1.1 gives a new proof to the following assertions, that also can be derived from [**3**, Theorem 13.3.5].

COROLLARY 1.3. *Let $f(T_1, \dots, T_n, X) \in K[T_1, \dots, T_n, X]$ be irreducible of positive degree in X and H_f the set of $(t_1, \dots, t_n) \in \mathbb{F}_q[u]^n$ for which $f(t_1, \dots, t_n, X) \in \mathbb{F}_q(u)[X]$ is irreducible.*

(1) *H_f contains a Zariski dense set of prime n-tuples; i.e., for every nonzero $g(T_1, \dots, T_n)$, there exists $(t_1, \dots, t_n) \in H_f$ such that $g(t_1, \dots, t_n) \neq 0$ and $t_i \in \mathbb{F}_q[u]$ is irreducible for all i.*
(2) *If S is a finite set of primes of $\mathbb{F}_q(u)$, then H_f is S-adically dense.*

The implication of the corollary from Theorem 1.1 is immediate: the primes have density bigger than $q^{-\epsilon}$, for any $\epsilon > 0$ and any S-adic open set has positive density; hence those sets cannot be contained in the complement of H_f by Theorem 1.1.

2. Set-up and preliminaries

Let $\mathbb{F}_q$ be a finite field with q elements and of characteristic p, let $K = \mathbb{F}_q(u)$ be the field of rational functions in u. For a positive integer d, we denote by M_d the subset of K of monic polynomials of degree d. We denote tuples by bolded letters, e.g. we write $\mathbf{t}$ for $(t_1, \dots, t_n)$ and $\mathbf{T}$ for $(T_1, \dots, T_n)$.

To a polynomial $f \in K[\mathbf{T}, X]$ in $n+1$ variables over the field K we attach the following data. Let $\deg_{\mathbf{T}} f$ (resp. $\deg_{\mathbf{T},X} f$) denote the total degree of f in the variables $T_1, \dots, T_n$ (resp. $T_1, \dots, T_n, X$) and denote $\deg_u(f) = \max\left(\deg_u \tilde{f}, \deg_u a\right)$, where $f = \tilde{f}/a, \tilde{f} \in \mathbb{F}_q[u, \mathbf{T}, X], a \in \mathbb{F}_q[u]$ is a reduced fraction.

It is convenient to work with polynomials in $\mathbb{F}_q[u, \mathbf{T}, X]$ that are monic in X. The following lemmas allow us to reduce to this case.

LEMMA 2.1. *Let $f \in K[\mathbf{T}, X]$ be a polynomial, let $a(u) \in \mathbb{F}_q[u]$ be the common denominator of the coefficients of f, $\tilde{f} = af \in \mathbb{F}_q[u, \mathbf{T}, X]$, $b \in \mathbb{F}_q[u, \mathbf{T}]$ the leading coefficient of $\tilde{f}$ (in the variable X). Then the polynomial*

$$g(\mathbf{T}, X) = b^{\deg_X f - 1} \tilde{f}\left(\mathbf{T}, \frac{X}{b}\right) \in \mathbb{F}_q[u, \mathbf{T}, X]$$

satisfies:

(1) *g is monic in X.*
(2) $\deg_X g = \deg_X f$, $\deg_{\mathbf{T}} g \leq \deg_{\mathbf{T}} f \cdot \deg_X f$ *and* $\deg_u g \leq \deg_u f \cdot \deg_X f$.
(3) *For all $\mathbf{t} \in M_d^n$ with $b(\mathbf{t}) \neq 0$ each of the polynomials $g(\mathbf{t}, X)$ and $f(\mathbf{t}, X)$ is irreducible iff the other is.*

PROOF. Straightforward from the definition of g. □

The following lemma will be useful in several instances.

LEMMA 2.2. *Let $h(\mathbf{T}) \in K[T_1, \dots, T_n]$ be a nonzero polynomial. Then the number of tuples $\mathbf{t} \in M_d^n$ such that $h(\mathbf{t}) = 0$ is at most $\sum_j \deg_{T_j} h \cdot q^{d(n-1)}$.*

PROOF. We apply induction. If $n = 1$, then $h(T)$ has at most $\deg_T h$ solutions in K, hence in M_d. Assume $n \geq 2$ and write $h(\mathbf{T}) = \sum_i h_i(T_1, \dots, T_{n-1}) T_n^i$, where not all of the h_i-s are zero. Then, by induction,

$$\#\{\mathbf{t} \in M_d^n : h(\mathbf{t}) = 0\} \leq \sum_{\substack{\mathbf{t}' \in M_d^{n-1} \\ \forall i : h_i(\mathbf{t}') = 0}} q^d + \sum_{\substack{\mathbf{t}' \in M_d^{n-1} \\ \exists i : h_i(\mathbf{t}') \neq 0}} \deg_{T_n} h \leq \sum_j \deg_{T_j} h \cdot q^{d(n-1)},$$

as needed. □

The last two lemmas immediately imply the following result which allows us to reduce to monic polynomials with coefficients in $\mathbb{F}_q[u, \mathbf{T}]$:

COROLLARY 2.3. *Let $f \in K[\mathbf{T}, X]$ be a polynomial and let $g(\mathbf{T}, X) \in \mathbb{F}_q[u, \mathbf{T}, X]$ and $b \in \mathbb{F}_q[u, \mathbf{T}]$ be as in Lemma 2.1. Then for all $\mathbf{t} \in M_d^n$ but at most $\sum \deg_{T_j} b \cdot q^{d(n-1)}$, each of $f(\mathbf{t}, X)$ and $g(\mathbf{t}, X)$ is irreducible iff the other is.*

3. Separable polynomials

In this section we assume that $f \in \mathbb{F}_q[u, \mathbf{T}, X]$ is irreducible and that it is separable and monic in X. Throughout this section and the next one we fix a bound on $\deg_{\mathbf{T},X} f$ and all asymptotic notation will have implied constants that may depend on this bound. We do not fix the degree in u as we are looking for a result which is uniform in f, q as long as $d \geq 2 \log_q \deg_u f + \log_q \lambda$ for a suitable constant $\lambda = \lambda(\deg_{\mathbf{T},X} f)$.

Let L be the splitting field of f over $K(\mathbf{T})$ and let $\mathbb{F}_{q^r} = \overline{\mathbb{F}}_q \cap L$ be the algebraic closure of $\mathbb{F}_q$ in L. We have the following diagram of fields:

$$\begin{array}{ccccc} & & L & \text{———} & L\overline{\mathbb{F}}_q \\ & & | & & | \\ K(\mathbf{T}) & \text{———} & K\mathbb{F}_{q^r}(\mathbf{T}) & \text{———} & K\overline{\mathbb{F}}_q(\mathbf{T}) \\ | & & | & & | \\ \mathbb{F}_q & \text{———} & \mathbb{F}_{q^r} & \text{———} & \overline{\mathbb{F}}_q \end{array}$$

We denote

$$G = \mathrm{Gal}(L/K(\mathbf{T})) \text{ and } G_{\mathrm{geom}} = \mathrm{Gal}(L/K\mathbb{F}_{q^r}(\mathbf{T})) \cong \mathrm{Gal}(L\overline{\mathbb{F}}_q/K\overline{\mathbb{F}}_q(\mathbf{T})).$$

We have the fundamental exact sequence

$$1 \longrightarrow G_{\mathrm{geom}} \longrightarrow G \longrightarrow \mathbb{Z}/r\mathbb{Z} \longrightarrow 1 \tag{3.1}$$

where $\mathbb{Z}/r\mathbb{Z} \cong \mathrm{Gal}(\mathbb{F}_{q^r}/\mathbb{F}_q)$ and under this identification the map $G \to \mathbb{Z}/r\mathbb{Z}$ is the restriction-of-automorphism map.

REMARK 3.1. By a result of Guralnick [**5**, Theorem 1] any cyclic quotient of a transitive permutation group of degree N has order $\leq N$ and therefore since $\mathbb{Z}/r\mathbb{Z}$ is a quotient of G which acts transitively on the roots of f we must have $r \leq \deg_X f$.

If $\mathbf{t} \in K^n$ is an n-tuple for which $f(\mathbf{t}, X)$ is separable in X, then we denote by $L_{\mathbf{t}}$ the splitting field of $f(\mathbf{t}, X)$ over K and we note that $\mathbb{F}_{q^r} \subseteq L_{\mathbf{t}}$. We further denote $G_{\mathbf{t}} = \mathrm{Gal}(L_{\mathbf{t}}/K)$ and $G_{\mathrm{geom},\mathbf{t}} = \mathrm{Gal}(L_{\mathbf{t}}/K\mathbb{F}_{q^r})$. We have that $G_{\mathbf{t}} \leq G$ and $G_{\mathrm{geom},\mathbf{t}} \leq G_{\mathrm{geom}}$, in a canonical way up to conjugation, and $G/G_{\mathrm{geom}} \cong G_{\mathbf{t}}/G_{\mathrm{geom},\mathbf{t}} \cong \mathbb{Z}/r\mathbb{Z}$. Hence the following observation is immediate:

LEMMA 3.2. *$G_{\mathbf{t}} \cong G$ if and only if $G_{\mathrm{geom},\mathbf{t}} \cong G_{\mathrm{geom}}$.*

Let $P(u) \in \mathbb{F}_q[u]$ be a prime polynomial; that is to say, irreducible and monic. Write $|P| = q^{\deg P}$ for the norm of P. We may reduce the coefficients of f modulo P, and we denote the resulting polynomial by $f_P \in \mathbb{F}_{|P|}[\mathbf{T}, X]$. If in addition f_P is separable in X, then we define L_P, G_P and $G_{\mathrm{geom},P}$ in a similar fashion to the definition of L, G and G_{geom} (with $\mathbb{F}_{|P|}$ taking the role of K). Since f is monic in X, we have that $\deg_X f = \deg_X f_P$, so G_P and $G_{\mathrm{geom},P}$ embed in G and G_{geom}, respectively, and this embedding is canonical, up to conjugation in G.

Consider the set of primes of good reduction in the following sense:

$$\mathcal{P}_f = \{P \in \mathbb{F}_q[u] : P \text{ prime}, \;\; r \mid \deg P, \;\; f_P \text{ separable in } X, \;\; G_{\mathrm{geom}} \cong G_{\mathrm{geom},P}\}.$$

If $P \in \mathcal{P}_f$, then $\mathbb{F}_{q^r} \subseteq \mathbb{F}_{|P|}$, so L_P is regular over $\mathbb{F}_{|P|}$, and we have the diagram of fields:

$$\begin{array}{ccc} L_P & \text{———} & L_P\overline{\mathbb{F}}_q \\ G_{\text{geom}} \Big| & & G_{\text{geom}} \Big| \\ \mathbb{F}_{|P|}(\mathbf{T}) & \text{———} & \overline{\mathbb{F}}_q(\mathbf{T}) \\ \Big| & & \Big| \\ \mathbb{F}_{|P|} & \text{———} & \overline{\mathbb{F}}_q \end{array}$$

LEMMA 3.3. *Let D be a natural number such that $r|D$. The number of (monic) irreducible polynomials $P \in \mathbb{F}_q[u]$ with* $\deg P = D$ *such that $P \notin \mathcal{P}_f$ is $O_{\deg_{\mathbf{T},X}}(\deg_u f/D)$.*

PROOF. Since f is separable in X, there is a nonozero coefficient $g \in \mathbb{F}_q[u, \mathbf{T}]$ of X^i for some $i \not\equiv 0 \mod p$. If f_P is not separable, then P must divide all the coefficients of g (viewed as a polynomial in $\mathbf{T}$). There are at most $\deg_u g/D \leq \deg_u f/D$ such P with $\deg P = D$ (since the same coefficient of g is divisible by all of them).

It remains to show that if $\deg P = D$ and f_P is separable we have $G_{\text{geom}} \cong G_{\text{geom},P}$ with $O(\deg_u f/D)$ many exceptions. For this we choose a monic irreducible polynomial $h \in K\mathbb{F}_{q^r}[X]$ whose root generates L (h is a resolvent of f over $K\mathbb{F}_{q^r}$). We can choose h with coefficients in $\mathbb{F}_{q^r}[u, \mathbf{T}]$ such that $\deg_u h = O(\deg_u f)$. We have $\deg_X h = |G_{\text{geom}}|$. The resolvent h is absolutely irreducible. If $P \in \mathbb{F}_q[u]$ is a prime of degree D such that $r|D$ then over $\mathbb{F}_{q^r}$ it factors as $P = P_1 \cdots P_r$ with $\deg P_i = D/r$. If $h_{P_1} := h \bmod P_1$ (and hence all h_{P_i}) is absolutely irreducible then a standard Galois-theoretic argument shows that $G_{\text{geom},P} \cong G_{\text{geom}}$.

Let $c_1, \ldots, c_m$ be the coefficients of h. By a theorem of Noether (see the beautiful exposition by Geyer [**4**, Theorem 5.3.1]) absolute irreducibility of a polynomial is determined by polynomial non-equations on the coefficients, namely there exist $O_{\deg_{\mathbf{T},X} f}(1)$ polynomials $F_j \in \mathbb{Z}[v_1, \ldots, v_m]$ of degree $O_{\deg_{\mathbf{T},X} f}(1)$ such that $F_j(c_1, \ldots, c_m) \neq 0$ and h_{P_1} is absolutely irreducible unless $F_j(c_1, \ldots, c_m) \equiv 0 \pmod{P_1}$ for some j. Since $\deg_u F_j(c_1, \ldots, c_m) = O(\deg_u f)$ there can only be $O(\deg_u f/D)$ such $P_1 \in \mathbb{F}_{q^r}[u]$ with $\deg P_1 = D/r$ (since $r \leq \deg_X f = O(1)$) and thus only $O(\deg_u f/D)$ corresponding P. For any other P we have $G_{\text{geom},P} \cong G_{\text{geom}}$. $\square$

Take $P \in \mathcal{P}_f$. Every $\mathbf{a} \in \mathbb{F}_{|P|}^n$ with $f_P(\mathbf{a}, X) \in \mathbb{F}_{|P|}[X]$ separable gives rise to the Frobenius conjugacy class

$$C_{\mathbf{a}} \subseteq G_P = G_{\text{geom},P} \cong G_{\text{geom}}.$$

Now fix a conjugacy class $C \subseteq G_{\text{geom}}$ and let

$$\Omega_P = \Omega_P(C) = \{\mathbf{a} \in \mathbb{F}_{|P|}^n : C_{\mathbf{a}} \neq C\}.$$

(The set Ω_P also contains those $\mathbf{a}$ for which $f_P(\mathbf{a}, X)$ is not separable.) The explicit Chebotarev density theorem for function fields (see Theorem 3 in [**2**], for a version with the desired uniformity) gives that

$$\frac{\#\Omega_P}{|P|^n} = \left(1 - \frac{|C|}{|G_{\text{geom}}|}\right) \cdot \left(1 + O_{\deg_{\mathbf{T},X} f}(|P|^{-1/2})\right).$$

In particular, if G_{geom} is non-trivial, then $|C| \neq |G_{\text{geom}}|$, so if we take any $1 - \frac{|C|}{|G_{\text{geom}}|} < c < 1$, then for every $P \in \mathcal{P}_f$ of sufficiently large degree, we have

$$\frac{\#\Omega_P}{|P|^n} \leq c. \tag{3.2}$$

PROPOSITION 3.4. *Let $B_C(d)$ be the number of tuples $\mathbf{t} \in M_d^n$ such that $G_{\text{geom},\mathbf{t}} \cap C = \emptyset$. Then there exists a constant $\lambda = \lambda(\deg_{\mathbf{T},X} f)$ such that whenever $d \geq 2\log_q \deg_u f + \log_q \lambda$ we have*

$$\frac{B_C(d)}{q^{dn}} = O_{\deg_{\mathbf{T},X} f}\left(q^{-d/2+r} d\right)$$

with $r \leq \deg_X f$ as defined above.

PROOF. Let $\Omega = \Omega(d, C, f)$ be the set of tuples $\mathbf{t} \in M_d^n$ such that $G_{\text{geom},\mathbf{t}} \cap C = \emptyset$. Let $\mathbf{t} \in M_d^n$ and $P \in \mathcal{P}_f$. Since $r \mid \deg P$, we have that P is totally split in $K\mathbb{F}_{q^r}$. Moreover, $f(\mathbf{t}, X) \mod P = f_P(\mathbf{a}, X)$, where $\mathbf{a} := \mathbf{t} \mod P$. If we assume that $f_P(\mathbf{a}, X)$ is separable, then we get that the Frobenius class of P in $L_{\mathbf{t}}$ equals the Frobenius class of $\mathbf{a}$ in L_P, which we earlier denoted by $C_{\mathbf{a}}$. Hence if $\mathbf{t} \in \Omega$, then $C_{\mathbf{a}} \neq C$ and so

$$\mathbf{t} \mod P \in \Omega_P$$

for any $P \in \mathcal{P}_f$.

We apply the large sieve inequality for function fields [**6**, Theorem 3.2][1] to conclude that

$$\#\Omega \leq \frac{q^{dn}}{L},$$

where

$$L \geq 1 + \sum_{\substack{P \in \mathcal{P}_f \\ \deg P \leq d/2}} \frac{1-c}{c} \gg \#\{P \in \mathcal{P}_f : \deg P \leq d/2\}.$$

Finally, let s be an integer so that $d/2 - r \leq rs \leq d/2$ (if it is not positive the assertion of the proposition is trivial). By Lemma 3.3, $\mathcal{P}_f$ contains all but $O\left(\frac{\deg_u f}{rs}\right)$ primes of degree rs, so

$$\#\{P \in \mathcal{P}_f : \deg P \leq d/2\} \geq \pi_q(rs) - O\left(\frac{\deg_u f}{rs}\right) \gg \frac{q^{d/2-r}}{d/2}$$

provided that $d \geq 2\log_q \deg_u f + \log_q \lambda$ for a sufficiently large constant $\lambda = \lambda(\deg_{\mathbf{T},X} f)$. Note that this is the step which determines the form of the condition on d. Here $\pi_q(rs)$ denotes the number of primes of degree rs in $\mathbb{F}_q[u]$ and we use the Prime Polynomial Theorem (see e.g. [**7**, Theorem 2.2]) which says that $\pi_q(rs) = \frac{q^{rs}}{rs} + O(q^{rs/2})$. We conclude that

$$\frac{\#\Omega}{q^{dn}} \ll q^{-d/2+r} d.$$

Since $B_C(d) = \#\Omega$ the proof is done. □

[1] In the notation of [**6**, Theorem 3.2] we take $X = \Omega$, $N = d-1$, $f = (t^d, \dots, t^d)$, the variable t is replaced by our u, $K = d/2$, $\alpha_P = c$ if $P \in \mathcal{P}_f$ and $\alpha_P = 1$ otherwise.

COROLLARY 3.5. *Let $B(d)$ be the number of n-tuples $\mathbf{t} \in M_d^n$ such that $f(\mathbf{t}, X)$ is separable and $G_{\mathbf{t}} \not\cong G$. There exists a constant $\lambda = \lambda(\deg_{\mathbf{T},X} f)$ such that whenever $d \geq 2\log_q \deg_u f + \log_q \lambda$ we have*

$$\frac{B(d)}{q^{dn}} = O_{\deg_{\mathbf{T},X} f}(q^{-d/2+r} d),$$

with $r \leq \deg_X f$ as defined above.

PROOF. By Lemma 3.2, if $G_{\mathbf{t}} \not\cong G$, then $G_{\mathrm{geom},\mathbf{t}} \not\cong G_{\mathrm{geom}}$. If $G_{\mathrm{geom},\mathbf{t}}$ is isomorphic to a proper subgroup of G_{geom}, then there exists a conjugacy class C such that $G_{\mathrm{geom},\mathbf{t}} \cap C = \emptyset$ (it is an elementary exercise in group theory to show that every proper subgroup of a finite group is disjoint from some conjugacy class, see for example [**3**, Lemma 13.3.2]). Applying Proposition 3.4, the union bound, and the fact that there are $O_{\deg_X f}(1)$ conjugacy classes in G_{geom}, we conclude that (as long as $d \geq 2\log_q \deg_u f + \log_q \lambda$)

$$\frac{B(d)}{q^{dn}} \ll_{\deg_{\mathbf{T},X} f} q^{-d/2+r} d,$$

as needed. □

THEOREM 3.6. *Let $K = \mathbb{F}_q(u)$, let $f(T_1, \ldots, T_n, X) \in K[T_1, \ldots, T_n, X]$ be an irreducible polynomial that is separable and of positive degree in X. Let $H(d)$ be the number of tuples $\mathbf{t} \in M_d^n$ with $f(\mathbf{t}, X)$ irreducible, separable, and $\deg_X f(\mathbf{t}, X) = \deg_X f(\mathbf{T}, X)$. There exists a constant $\lambda = \lambda(\deg_{\mathbf{T},X} f)$ such that whenever $d \geq 2\log_q \deg_u f + \log_q \lambda$ we have*

$$\frac{H(d)}{q^{dn}} = 1 + O_{\deg_{\mathbf{T},X} f}\left(dq^{-d/2+r}\right),$$

with $r \leq \deg_X f$ as defined above.

PROOF. We fix a bound on $\deg_{\mathbf{T},X} f$. First we observe that by Remark 3.1 we have $r \leq \deg_X f$. By Corollary 2.3 we may assume w.l.o.g. that $f \in \mathbb{F}_q[u, \mathbf{T}, X]$ and that f is monic in X (note that this reduction increases $\deg_u f$ by a constant factor, so the condition on d is preserved, possibly with a different value of λ).

Let Σ be the set of tuples $\mathbf{t} \in M_d^n$ for which $f(\mathbf{t}, X)$ is inseparable; i.e., the zero set of all the coefficients of X^i for $i \not\equiv 0 \mod p$. By Lemma 2.2, $\#\Sigma = O(q^{d(n-1)})$. For $\mathbf{t} \notin \Sigma$, the action of $G_{\mathbf{t}}$ on the roots of $f(\mathbf{t}, X)$ coincides with the action of the decomposition group $G_{\mathbf{t}} \cong D_{\mathbf{t}} \leq G$ on the roots of $f(\mathbf{T}, X)$, hence if $G_{\mathbf{t}} \cong G$, the action is transitive, and $f(\mathbf{t}, X)$ is irreducible. Corollary 3.5 then completes the proof. □

4. Inseparable polynomials

In this part, we let $f(\mathbf{T}, X) \in K[T_1, \ldots, T_n, X]$ be an irreducible polynomial that is inseparable in X.

Our treatment is inspired from Uchida's work [**10**] and its presentation in [**3**]. We will use the following criterion of Uchida for irreducibility of inseparable polynomials, see [**3**, Lemma 12.4.1].

LEMMA 4.1. *Let F be a field of positive characteristic $p > 0$, let $g \in F[X]$ be an irreducible monic polynomial. Assume that at least one of the coefficients of g is not a p-th power in F. Then $g(X^{p^\alpha})$ is irreducible for all $\alpha > 0$.*

We shall need a technical lemma.

LEMMA 4.2. *Let $h(\mathbf{T}) \in \mathbb{F}_q[u, \mathbf{T}]$ be a polynomial which is not a p-th power. Let $A(d)$ denote the number of n-tuples $\mathbf{t} \in M_d^n$ such that $h(\mathbf{t})$ is a p-th power. There exists a constant $\lambda = \lambda(\deg_{\mathbf{T}} h)$ such that as long as $\lfloor d/2 \rfloor \geq \log_q \deg_u f + \log_q \lambda$ we have*

$$\frac{A(d)}{q^{dn}} \ll_{\deg_{\mathbf{T}} h} q^{-d/2+1}.$$

PROOF. We fix a bound on $\deg_{\mathbf{T}} h$. All asymptotic notation will have implied constants depending on this bound. Since every element in $\mathbb{F}_q$ is a p-th power and since h is not a p-th power, h is not a polynomial in v^p for at least one of the variables $v \in \{u, T_1, \dots, T_n\}$. Equivalently, $\frac{\partial h}{\partial v} \neq 0$. Similarly, for $\mathbf{t} \in M_d^n$, the univariate polynomial $h(\mathbf{t})$ is a p-th power if and only if $d(h(\mathbf{t}))/du = 0$.

If $\frac{\partial h}{\partial T_i} = 0$ for all i, hence $\frac{\partial h}{\partial u} \neq 0$, then by the chain rule, for all $\mathbf{t} \in M_d^n$ we have

$$\frac{d(h(\mathbf{t}))}{du} = \frac{\partial h}{\partial u}(\mathbf{t}).$$

By Lemma 2.2,

$$\frac{A(d)}{q^{dn}} = \frac{\#\{\mathbf{t} \in M_d^n : \frac{\partial h}{\partial u}(\mathbf{t}) = 0\}}{q^{dn}} \ll q^{-d},$$

and the proof is done.

Next assume that $\frac{\partial h}{\partial T_i} \neq 0$ for some i. We can reduce this case to the univariate case (i.e. $n = 1$): To ease notation assume $i = n$. For $t_1, \dots, t_{n-1} \in M_d^{n-1}$ consider the polynomial $g(T) = h(t_1, \dots, t_{n-1}, T) \in \mathbb{F}_q[u, T]$. Since

$$\frac{\partial g}{\partial T} = \frac{\partial h}{\partial T_n}(t_1, \dots, t_{n-1}, T)$$

we get by Lemma 2.2 that $\partial g/\partial T \neq 0$ for all $(t_1, \dots, t_{n-1}) \in M_d^{n-1}$ but $O(q^{d(n-2)})$ such $(n-1)$-tuples. So n-tuples $\mathbf{t} \in M_d^n$ for which $\partial g/\partial T = 0$, contribute at most $O(q^{d(n-1)})$ to the total number of n-tuples with $h(\mathbf{t})$ a p-th power. Therefore, we may assume that $\partial g/\partial T \neq 0$ and it suffices to prove that under this assumption

$$\#\{t \in M_d : g(t) \in K^p\} = O(q^{d/2+1}), \tag{4.1}$$

in order to conclude the proof.

Next we want to pick $R(u) \in \mathbb{F}_q[u]$ irreducible of degree $\lfloor \frac{d}{2} \rfloor$ such that $g, \partial g/\partial T \not\equiv 0 \pmod{R}$. We show that under the conditions of the lemma such an R exists. Indeed $\deg_u g, \deg_u(\partial g/\partial T) = O(\deg_u h)$ and so for all but $O(\deg_u h/d)$ irreducible R of degree $\lfloor \frac{d}{2} \rfloor$ one of the coefficients of g (and of $\partial g/\partial T$) is not divisible by R. The condition $\lfloor d/2 \rfloor \geq \log_q \deg_u h + \log_q \lambda$ ensures that $\pi_q(\lfloor \frac{d}{2} \rfloor) - O(\deg_u h/d) > 0$ if $\lambda = \lambda(\deg_{\mathbf{T}} h)$ is chosen sufficiently large and so a prime R as above must exist.

Let $\mathcal{A} = \mathbb{F}_q[u]/R^2$ and let $\mathcal{S} = \{c^p : c \in \mathcal{A}\}$ be the set of p-th powers modulo R^2. We have $\#\mathcal{S} = O\left(q^{\lfloor d/2 \rfloor}\right)$ (a p-th power modulo R^2 is either 0 or an element of the subgroup $(\mathcal{A}^*)^p$ of $\mathcal{A}^*$ which has order $q^{\lfloor d/2 \rfloor} - 1$).

For $t \in M_d$, if $g(t) \in K^p$, then $g(t) \mod R^2 \in \mathcal{S}$. Since $\partial g/\partial T \not\equiv 0 \pmod{r}$ we have

$$\#\left\{t \in M_d : \frac{\partial g}{\partial T}(t) \equiv 0 \pmod{R}\right\} = O\left(q^{d-\deg R}\right) = O\left(q^{d/2+1}\right).$$

Hence it suffices to show that

$$\#\left\{t \in M_d : g(t) \bmod r^2 \in \mathcal{S}, \frac{\partial g}{\partial T}(t) \not\equiv 0 \pmod{r}\right\} = O\left(q^{d/2+1}\right).$$

Fix $s \in \mathcal{S}$. Since $g \not\equiv 0 \pmod{r}$ there are $O(1)$ solutions to $g(\tau) \equiv s \pmod{R}$. If $\tau \in \mathbb{F}_q[u]/r$ is a root of $g(\tau) - s \bmod R$ and $\frac{\partial g}{\partial T}(\tau) \not\equiv 0 \pmod{R}$ then by Hensel's lemma, there are $O(1)$ solutions to $g(\tau') \equiv s \pmod{R^2}$ with $\tau' \equiv \tau \pmod{r}$ and each of them lifts to $q^{d-2\lfloor d/2 \rfloor}$ solutions of $g(t) \equiv s \pmod{R^2}, t \in M_d$. So the total number of solutions to $g(t) \equiv s \pmod{R^2}$, $\frac{\partial g}{\partial T}(t) \not\equiv 0 \pmod{r}, t \in M_d$ is bounded by $\#\mathcal{S} \cdot q = O(q^{d/2+1})$, as required. □

Now we may deduce the quantitative Hilbert's irreducibility theorem for inseparable polynomials.

THEOREM 4.3. *Let $f(\mathbf{T}, X) \in K[T_1, \ldots, T_n, X]$ be an irreducible polynomial. Let $H_f(d)$ be the number of tuples $\mathbf{t} \in M_d^n$ for which $f(\mathbf{t}, X)$ is irreducible. There exists a constant $\lambda = \lambda(\deg_{\mathbf{T},X} f)$ such that whenever $\lfloor d/2 \rfloor \geq \log_q \deg_u f + \log_q \lambda$ we have*

$$\frac{H_f(d)}{q^{dn}} = 1 + O_{\deg_{\mathbf{T},X} f}\left(dq^{-d/2+r}\right),$$

where $r \leq \deg_X(f)$ is the degree of the algebraic closure $\mathbb{F}_q^r$ of $\mathbb{F}_q$ in the splitting field of f over $K(\mathbf{T})$.

PROOF. If f is separable in X this follows from Theorem 3.6. Hence we assume that f is inseparable in X. By Corollary 2.3, we may assume w.l.o.g. that $f \in \mathbb{F}_q[u, \mathbf{T}, X]$ and that f is monic in X (moreover the condition on d is preserved with a possibly different value of λ).

Since f is inseparable, we may write $f(\mathbf{T}, X) = g(\mathbf{T}, X^{p^\alpha})$ for some g that is irreducible, separable in X and of degree $\deg_X g \leq \deg_X f$ and for some $\alpha > 0$. Let $\mathbb{F}_{q^r}$ be the algebraic closure of $\mathbb{F}_q$ inside a splitting field of g over $K(\mathbf{T})$.

Write $g(\mathbf{T}, X) = \sum_i g_i(\mathbf{T})X^i$, with $g_i \in \mathbb{F}_q[u, \mathbf{T}]$. Had all the g_i been p-th powers, we would have that f is reducible. Hence there exists g_i which is not a p-th power.

If $g_i(\mathbf{t})$ is not a p-th power and $g(\mathbf{t}, X)$ is irreducible, then by Lemma 4.1, $f(\mathbf{t}, X) = g(\mathbf{t}, X^{p^\alpha})$ is irreducible. Applying Theorem 3.6 and Lemma 4.2, gives that this happens for all but $O_{\deg_{\mathbf{T},X} f}(dq^{d(n-1/2)+r})$ of the $\mathbf{t}$-s (as long as $\lfloor d/2 \rfloor \geq \log_q \deg_u f + \log_q \lambda$ for a suitable $\lambda = \lambda(\deg_{\mathbf{T},X} f)$), hence the proof is complete. □

Acknowledgments

The authors thank Dan Haran for some illuminating discussions of p-th powers in characteristic p and Arno Fehm for helpful remarks that improved the presentation of the paper.

References

[1] S. D. Cohen, *The distribution of Galois groups and Hilbert's irreducibility theorem*, Proc. London Math. Soc. (3) **43** (1981), no. 2, 227–250, DOI 10.1112/plms/s3-43.2.227. MR628276

[2] Alexei Entin, *Monodromy of Hyperplane Sections of Curves and Decomposition Statistics over Finite Fields*, Int. Math. Res. Not. rnz120 (2019).

[3] Michael D. Fried and Moshe Jarden, *Field arithmetic*, 3rd ed., Ergebnisse der Mathematik und ihrer Grenzgebiete. 3. Folge. A Series of Modern Surveys in Mathematics [Results in Mathematics and Related Areas. 3rd Series. A Series of Modern Surveys in Mathematics], vol. 11, Springer-Verlag, Berlin, 2008. Revised by Jarden. MR2445111

[4] Wulf-Dieter Geyer, *Field theory*, Travaux mathématiques. Vol. XXII, Trav. Math., vol. 22, Fac. Sci. Technol. Commun. Univ. Luxemb., Luxembourg, 2013, pp. 5–177. MR3204847

[5] Robert M. Guralnick, *Cyclic quotients of transitive groups*, J. Algebra **234** (2000), no. 2, 507–532, DOI 10.1006/jabr.2000.8540. Special issue in honor of Helmut Wielandt. MR1800740

[6] Chih-Nung Hsu, *A large sieve inequality for rational function fields*, J. Number Theory **58** (1996), no. 2, 267–287, DOI 10.1006/jnth.1996.0077. MR1393616

[7] Michael Rosen, *Number theory in function fields*, Graduate Texts in Mathematics, vol. 210, Springer-Verlag, New York, 2002. MR1876657

[8] Jean-Pierre Serre, *Topics in Galois theory*, 2nd ed., Research Notes in Mathematics, vol. 1, A K Peters, Ltd., Wellesley, MA, 2008. With notes by Henri Darmon. MR2363329

[9] Jean-Pierre Serre, *Lectures on the Mordell-Weil theorem*, 3rd ed., Aspects of Mathematics, Friedr. Vieweg & Sohn, Braunschweig, 1997. Translated from the French and edited by Martin Brown from notes by Michel Waldschmidt; With a foreword by Brown and Serre. MR1757192

[10] Kôji Uchida, *Separably Hilbertian fields*, Kodai Math. J. **3** (1980), no. 1, 83–95. MR569535

Raymond and Beverly Sackler School of Mathematical Sciences, Tel Aviv University, Tel Aviv 69978, Israel

Email address: `barylior@tauex.tau.ac.il`

Raymond and Beverly Sackler School of Mathematical Sciences, Tel Aviv University, Tel Aviv 69978, Israel

Email address: `aentin@tauex.tau.ac.il`

Contemporary Mathematics
Volume **767**, 2021
https://doi.org/10.1090/conm/767/15403

A comparison between obstructions to local-global principles over semi-global fields

David Harbater, Julia Hartmann, Valentijn Karemaker, and Florian Pop

Abstract. We consider local-global principles for rational points on varieties, in particular torsors, over one-variable function fields over complete discretely valued fields. There are several notions of such principles, arising either from the valuation theory of the function field, or from the geometry of a regular model of the function field. Our results compare the corresponding obstructions, proving in particular that a local-global principle with respect to valuations implies a local-global principle with respect to a sufficiently fine regular model.

Introduction

Classical local-global principles originated in the study of algebraic structures over the rational numbers, and asserted that such a structure has a specified property over $\mathbb{Q}$ if and only if it has that property over $\mathbb{Q}_p$ for all p and also for $\mathbb{R}$. Later such results were generalized to arbitrary global fields. Typically a local-global principle can be rephrased as asserting that a variety Z has a rational point over a global field F if it has a rational point over each completion F_v of F. In many cases, Z is a homogeneous space under some algebraic group. When Z is a principal homogeneous space (i.e., a torsor) under an algebraic group G, the local-global principle can also be rephrased as asserting the vanishing of the Tate-Shafarevich set $\text{Ш}(F,G) := \ker(H^1(F,G) \to \prod_v H^1(F_v,G))$. More generally, the set $\text{Ш}(F,G)$ is the obstruction to a local-global principle holding.

In the classical case of global fields, local-global principles have been proven in a number of contexts; e.g., for quadratic forms (Hasse-Minkowski), Brauer groups (Albert-Brauer-Hasse-Noether), and torsors under connected rational linear algebraic groups ([**22**, Corollary 9.7], [**6**]). Motivated by that classical case, analogs of the above results have in recent years been proven in the case of *semi-global fields*; i.e., function fields of transcendence degree one over a complete discretely valued

2020 *Mathematics Subject Classification.* Primary 13F30, 14G05, 14H25; Secondary 14G27, 11E72.

Key words and phrases. Local-global principles, valuation theory, semi-global fields, completions, henselizations, torsors, linear algebraic groups.

The first and second author were supported on NSF grants DMS-1463733 and DMS-1805439; additional support was provided by a Simons Fellowship (JH).

The third author was supported by an Oberwolfach Leibniz Fellowship.

The fourth author was supported by NSF collaborative FRG grant DMS-1265290.

field K such as $\mathbb{Q}_p$. (For example, see [**7–9, 14, 16, 17, 20**].) In this situation, the geometry is richer, and there are various possible sets of overfields of F with respect to which local-global principles and the corresponding Tate-Shafarevich sets can be studied. The current manuscript aims to clarify the relationships among those obstruction sets, and among the corresponding local-global principles for rational points on smooth varieties.

In the global field case, the completions of F are taken with respect to isomorphism classes of non-trivial absolute values on F. For F of equal characteristic (the global function field case), this is equivalent to taking the completions of F with respect to the non-trivial valuations on F. All of these valuations have rank one, and they are all discrete. Moreover, they are in bijection with the closed points on the unique regular projective model C of F, which is a curve over the finite base field of F. Here the completion F_P corresponding to a point $P \in C$ is the fraction field of the complete local ring $\widehat{\mathscr{O}}_{C,P}$.

By contrast, the valuations on a semi-global field F are not all of rank one, and those of rank one are not all discrete. This leads to several choices for the set of completions, and therefore for the local-global principle being considered. More precisely, write T for the valuation ring of the complete discretely valued field K over which F is a function field. Let Ω_F be the set of all non-trivial T-valuations on F (i.e., those whose valuation ring contains T); let Ω_F^1 be the set of all non-trivial rank one T-valuations on F; and let Ω_F^{dvr} be the set of all discrete valuations on F (whose valuation ring automatically contains T, by [**17**, Corollary 7.2]). Thus $\Omega_F^{\mathrm{dvr}} \subseteq \Omega_F^1 \subseteq \Omega_F$. Each of these sets then gives rise to a set of completions, and therefore to a local-global principle. If one focuses on torsors over F under an algebraic group G, then one correspondingly obtains several versions of Ш, which will be denoted by $Ш_{\Omega_F}(F,G)$, $Ш_{\Omega_F^1}(F,G)$, and $Ш_{\mathrm{dvr}}(F,G)$.

In addition, there are many choices for a regular projective model $\mathscr{X}$ of F over the valuation ring $T = \mathscr{O}_K$ of the ground field K, each with a closed fiber $X \subset \mathscr{X}$; for each such model and point $P \in X$, the fraction field F_P of $\widehat{\mathscr{O}}_{\mathscr{X},P}$ will not in general be the completion of F with respect to a valuation. (In fact, it will be such if and only if P is the generic point of an irreducible component of X.) Taking the set of overfields $\{F_P\}_{P\in X}$ for a given model $\mathscr{X}$ leads to yet another choice of local-global principle. For a torsor over F under an algebraic group G, the corresponding obstruction will be denoted by $Ш_X(F,G)$.

In this paper, we prove the following (see Theorem 3.4):

THEOREM. *Let G be a linear algebraic group over a semi-global field F. Then for any regular projective model $\mathscr{X}$ of F with closed fiber X,*

$$Ш_X(F,G) \subseteq Ш_{\Omega_F}(F,G) \subseteq Ш_{\Omega_F^1}(F,G) \subseteq Ш_{\mathrm{dvr}}(F,G).$$

Moreover, taking the direct limit over all such models $\mathscr{X}$, we have

$$\varinjlim_{\mathscr{X}} Ш_X(F,G) = Ш_{\Omega_F}(F,G) = Ш_{\Omega_F^1}(F,G).$$

More generally, we prove the corresponding result concerning local-global principles for the existence of rational points on smooth varieties over a semi-global field; see Theorem 3.1.

The significance of these results is that one is usually interested in local-global principles with respect to valuations (of a suitable type), but these can be hard to study over semi-global fields. On the other hand, by using patching methods (as in [**14**] and [**17**]), one can understand the obstruction to local-global principles with respect to the points on the closed fiber of a regular projective model quite explicitly, prove that local-global principles hold under certain hypotheses, and thereby obtain results about field invariants. The results in this manuscript relate local-global principles with respect to points (or patches) to those with respect to rank one valuations (the natural notion in the context of Berkovich spaces; in particular see [**19**] for a Berkovich analog of the results in [**14**]). We should note, however, that it remains open whether local-global principles with respect to the set of *discrete* valuations are equivalent to the other types of local-global principles.

1. Completions and henselizations of semi-global fields

Let K be a complete discretely valued field with ring of integers $T = \mathscr{O}_K$ and residue field k. Let F be a *semi-global field* over K, i.e., a finitely generated field extension of transcendence degree one over K in which K is algebraically closed. (We do not assume that F/K is separable.) The aim of this section is to introduce certain overfields of F, defined via completions or henselizations, and to study their relationship. This discussion will be important when we study rational points and local-global principles in the later sections.

We will consider two types of completions in this paper, associated to prime ideals (or points) and to valuations that need not be discrete. In general, for any commutative ring A, let $\mathscr{I} = \{\mathfrak{a}_i\}_{i\in I}$ be a set of ideals of A such that for all $i', i'' \in I$ there exists $i \in I$ with $\mathfrak{a}_i \subset \mathfrak{a}_{i'}\mathfrak{a}_{i''}$. The $\mathscr{I}$*-completion* of A is the projective limit $\widehat{A}$ of the system of projections $A/\mathfrak{a}_{i'} \to A/\mathfrak{a}_i$, $\mathfrak{a}_{i'} \subset \mathfrak{a}_i$. Here the completion map $\hat{\imath} : A \to \widehat{A}$ has $\ker(\hat{\imath}) = \cap_i \mathfrak{a}_i$.

In particular, if $\mathfrak{p}$ is a prime ideal of A and we take $\mathfrak{a}_i = \mathfrak{p}^i$ for $i \geq 1$, then we get the usual $\mathfrak{p}$*-adic completion* of A. On the other hand, suppose we are given a field L and a valuation $w : L \to \Gamma \cup \{\infty\}$, with Γ an ordered abelian group (the *value group* of w), having valuation ring $\mathscr{O}_w$ and maximal ideal $\mathfrak{m}_w$. In this situation, consider a set $\mathscr{I} = \{\mathfrak{a}_i\}_{i\in I}$ of non-zero ideals of $\mathscr{O}_w$ as above with $\cap_i \mathfrak{a}_i = (0)$. (For example, we may take $\mathfrak{a}_i = \{a \in A \,|\, w(a) \geq i\}$ for all $i > 0$ in Γ.) The $\mathscr{I}$-completion $\widehat{\mathscr{O}}_w$ of $\mathscr{O}_w$ is independent of the choice of such a set $\mathscr{I}$; and $\widehat{\mathscr{O}}_w$ is a valuation ring of its fraction field L_w, with valuation ideal $\widehat{\mathfrak{m}}_w = \mathfrak{m}_w\widehat{\mathscr{O}}_w$. One calls $\widehat{\mathscr{O}}_w$ and L_w the *w-adic completions* of $\mathscr{O}_w$ and L, respectively. Note that $\mathscr{O}_w$ is a local ring with maximal ideal $\mathfrak{m}_w$; however, the w-adic completion of $\mathscr{O}_w$ is not in general the same as the $\mathfrak{m}_w$-adic completion of $\mathscr{O}_w$ (in fact, they are the same if and only if w is a discrete valuation; otherwise $\cap_i \mathfrak{m}_w^i$ is a non-zero prime ideal). See [**4**, Chapter VI] and [**10**] for more about valuations and their completions.

Throughout this paper, for a semi-global field F we will restrict attention to valuations that lie in Ω_F, the set of non-trivial T-valuations on F (i.e., those whose valuation ring contains T). By [**17**, Corollary 7.2], all discrete valuations lie in that set. For each valuation $v \in \Omega_F$, we let F_v denote the v-adic completion of F, as in the general discussion above. This gives a first collection of overfields of F.

We next define overfields that are obtained from the geometry of curves with function field F. A *normal model* of F is a normal integral T-scheme $\mathscr{X}$ with function field F that is flat and projective over T of relative dimension one. If in addition the scheme $\mathscr{X}$ is regular we say that it is a *regular model*. The *closed fiber* of $\mathscr{X}$ is $X := (\mathscr{X} \times_T k)^{\mathrm{red}}$.

Let $\mathscr{X}$ be a regular model of F with closed fiber X. The points of X consist of the closed points of $\mathscr{X}$ and the generic points of the irreducible components of X. For each point P in X, let $\mathscr{O}_{\mathscr{X},P}$ be the local ring of $\mathscr{X}$ at P with maximal ideal $\mathfrak{m}_P$, let $\widehat{\mathscr{O}}_{\mathscr{X},P}$ be its $\mathfrak{m}_P$-adic completion, and let F_P be the fraction field of $\widehat{\mathscr{O}}_{\mathscr{X},P}$. These give the second set of overfields of F.

An essential first step in understanding the relationship between the various local-global principles (or their obstructions) is to study the relationship between the overfields F_v and F_P of F introduced above. In order to do so, we need to recall a few basic facts about henselian local rings and henselization.

Recall that a local ring $\mathscr{O}$ with maximal ideal $\mathfrak{m}$ and residue field κ is *henselian* if Hensel's lemma holds in $\mathscr{O}$. That is, given a monic polynomial $f \in \mathscr{O}[x]$ with image $\bar{f} \in \kappa[x]$, any simple root $\bar{a} \in \kappa$ of $\bar{f}$ lifts to a simple root $a \in \mathscr{O}$ of f. A *henselization* of a local ring $\mathscr{O}$ with maximal ideal $\mathfrak{m}$ is a henselian local ring $\mathscr{O}^h$ with the following properties: $\mathscr{O}^h$ is a direct limit of étale $\mathscr{O}$-algebras, $\mathfrak{m}\mathscr{O}^h$ is the maximal ideal of $\mathscr{O}^h$, and $\mathscr{O}/\mathfrak{m} = \mathscr{O}^h/\mathfrak{m}\mathscr{O}^h$. A henselization exists for any local ring and it is unique up to isomorphism (see, for example, [**25**, Theorems 32.28 and 32.29]). If $\mathscr{O}$ is an integrally closed local domain with fraction field L, then the fraction field of $\mathscr{O}^h$ is correspondingly a direct limit of finite separable extensions of L; see also [**5**, p. 48]. In particular, for $\mathscr{X}$ and P as above, we may consider the henselization $\mathscr{O}^h_{\mathscr{X},P}$ of the local ring $\mathscr{O}_{\mathscr{X},P}$ of $\mathscr{X}$ at P. We may also consider the henselization $\mathscr{O}^h_w$ of a valuation ring $\mathscr{O}_w$ associated to a valuation w on some field L.

If w is a valuation on a field L, then w and L are said to be *henselian* if the valuation ring $\mathscr{O}_w$ is henselian. Equivalently, w is henselian if it has a unique extension to each algebraic field extension of L [**24**, Definition A.13]. Other equivalent formulations of the henselian property for valuation rings can be found in, e.g., [**24**, Theorem A.14].

In the above situation, the fraction field L^h_w of $\mathscr{O}^h_w$ is the *henselization* of L with respect to w. That is, it is the unique (up to unique isomorphism) valued field extension of L whose valuation is henselian and extends w, and which is universal with these properties.

The *rank* of a valuation is the Krull dimension of its valuation ring. A valuation is of rank one if and only if there is an order-preserving embedding of its value group into $\mathbb{R}$. In the situation of semi-global fields F, it follows from Abhyankar's inequality (see [**1**] or [**10**, Theorem 3.4.3]) that every valuation $v \in \Omega_F$ has either rank one or rank two. Moreover, the rank two valuations are of the form $v = v_2 \circ v_1$, where $v_1 \in \Omega^1_F$, where v_2 is a rank one valuation of the residue field of v_1, and where composition is in the sense of valuations; cf. [**23**, Chapter I, §4]. (See [**21**, Appendix A] for a classification of T-valuations on semi-global fields; although smoothness was assumed there, it was not essential.)

In studying henselizations and completions of semi-global fields, we distinguish between two cases, based on the rank of the valuation. For a rank one valuation w on an arbitrary field L, the completion L_w of L with respect to w is henselian (e.g., see [**10**, Proposition 1.2.2]), and the henselization of L with respect to w is the separable closure of L in its completion (this follows via [**25**, Theorem 32.19]). For valuations of rank two on a semi-global field, there is the following result.

LEMMA 1.1. *Let F be a semi-global field and let $v \in \Omega_F$ have rank two. As above, write $v = v_2 \circ v_1$ as a composition of a valuation v_1 of F of rank one, and a valuation v_2 of the residue field of v_1. Then*

(a) *the completions $F_v = F_{v_1}$ are equal; and*
(b) *the henselizations satisfy $F^h_{v_1} \subseteq F^h_v$.*

PROOF. Part (a) is proven in [**3**, Lemma 2.2.2 and Corollary 2.2.27]. By [**24**, Proposition A.31], there is a henselian valuation v_1' on F^h_v that extends v_1. Thus $F^h_{v_1}$ may be identified with the relative henselization of F in F^h_v, with respect to v_1 and v_1', proving (b). □

In the situation of the introduction, Proposition 1.4 below will provide an essential link between the fields F_v for $v \in \Omega_F$ and the fields F_P for $P \in X$. First we state two lemmas.

LEMMA 1.2. *Let $\mathscr{X}$ be a regular model of a semi-global field F, and let Q be a point of $\mathscr{X}$. If I is an ideal of $\widehat{\mathscr{O}}_{\mathscr{X},Q}$ such that $I \cap \mathscr{O}_{\mathscr{X},Q} = (0)$, then $I = (0)$.*

PROOF. The assertion is trivial if Q is the generic point. It is also clear if Q is a point of codimension one, since then $\mathscr{O}_{\mathscr{X},Q}$ is a discrete valuation ring, and contraction induces a bijection between the ideals of the completion $\widehat{\mathscr{O}}_{\mathscr{X},Q}$ and the ideals of $\mathscr{O}_{\mathscr{X},Q}$. So we now assume that Q has codimension two; i.e., Q is a closed point of $\mathscr{X}$, lying on the closed fiber X of $\mathscr{X}$.

First consider the case that $\mathscr{X} = \mathbb{P}^1_T$ and that Q is the origin on $X = \mathbb{P}^1_k$. Let g be an element of the ideal $I \subset \widehat{\mathscr{O}}_{\mathscr{X},Q}$. By the Weierstrass Preparation Theorem (e.g., see [**4**, Proposition VII.3.9.6]), there exist an element $f \in F$ and a unit $u \in \widehat{\mathscr{O}}^{\times}_{\mathscr{X},Q}$ such that $g = fu$. Thus $f = gu^{-1}$ lies in $\widehat{\mathscr{O}}_{\mathscr{X},Q} \cap F = \mathscr{O}_{\mathscr{X},Q}$ and in I. So $f \in I \cap \mathscr{O}_{\mathscr{X},Q}$ is equal to 0 by hypothesis. Hence $g = fu = 0$, concluding the proof of this case.

For a general $\mathscr{X}$, there exists a finite morphism $\mathscr{X} \to \mathscr{X}' := \mathbb{P}^1_T$ that takes Q to the origin $P \in \mathbb{P}^1_k \subset \mathscr{X}'$ (see [**13**, Proposition 6.6]). The inclusion $\mathscr{O}_{\mathscr{X}',P} \subseteq \mathscr{O}_{\mathscr{X},Q}$ induces a morphism $\widehat{\mathscr{O}}_{\mathscr{X}',P} \to \widehat{\mathscr{O}}_{\mathscr{X},Q}$, which is finite since $\mathscr{X} \to \mathscr{X}'$ is finite, and is injective because the $\mathfrak{m}_Q$-adic topology on $\mathscr{O}_{\mathscr{X},Q}$ restricts to the $\mathfrak{m}_P$-adic topology on $\mathscr{O}_{\mathscr{X}',P}$. The contraction of I to $\mathscr{O}_{\mathscr{X}',P} \subset \mathscr{O}_{\mathscr{X},Q}$ is trivial, hence so is the contraction to $\widehat{\mathscr{O}}_{\mathscr{X}',P}$ by the case of $\mathscr{X} = \mathbb{P}^1_T$ above. Since $\widehat{\mathscr{O}}_{\mathscr{X},Q}$ is finite over $\widehat{\mathscr{O}}_{\mathscr{X}',P}$, the ideal I itself is trivial. □

We note that as an alternative argument for a general $\mathscr{X}$, we could use [**15**, Proposition 3.4] to reduce to the case where the Weierstrass Preparation Theorem [**15**, Theorem 3.1(c)] applies, and then use that result to prove Lemma 1.2.

Recall that the *center* of a valuation w on the function field of a separated integral scheme $\mathscr{Z}$ is a point Q on $\mathscr{Z}$ such that $\mathscr{O}_w$ contains the local ring $\mathscr{O}_{\mathscr{Z},Q}$, and such that $\mathfrak{m}_{\mathscr{Z},Q} = \mathfrak{m}_v \cap \mathscr{O}_{\mathscr{Z},Q}$. If a center exists on $\mathscr{Z}$ then it is unique, and if w is non-trivial then the center is not the generic point of $\mathscr{Z}$.

The next lemma is well known to experts, but there does not seem to be a good reference for it in the literature. For the sake of completeness, we include a short proof. (This lemma remains true more generally for proper integral schemes by the valuative criterion for properness, but we do not need that stronger form here.)

LEMMA 1.3. *Let $\mathscr{X}$ be a projective integral scheme over a ring A, and let w be an A-valuation on the function field of $\mathscr{X}$ (i.e., the valuation ring of w contains A). Then w has a center on $\mathscr{X}$.*

PROOF. Write $\mathscr{X} = \operatorname{Proj} S$, where S is a graded A-algebra generated by finitely many elements $x_0, \ldots, x_n$ of degree one. Choose i such that $w(x_i/x_0) = \min_{j>0} w(x_j/x_0)$. Thus $w(x_j/x_i) \geq 0$ for all j. Hence $R := A[x_j/x_i]_{0\leq j\leq n} \subset \mathscr{O}_w$, with $\operatorname{Spec}(R) = D_{x_i}^+$, the open subset of $\mathscr{X}$ where $x_i \neq 0$. The prime ideal $\mathfrak{m}_w \cap R$ of R defines a point $Q \in D_{x_i}^+ \subset \mathscr{X}$ such that $\mathscr{O}_{\mathscr{X},Q} \subset \mathscr{O}_w$ and $\mathfrak{m}_{\mathscr{X},Q} = \mathfrak{m}_w \cap \mathscr{O}_Q$. That is, Q is the center of w on $\mathscr{X}$. □

PROPOSITION 1.4. *Let F be a semi-global field over a complete discretely valued field K, and let $\mathscr{X}$ be a regular model of F with closed fiber X. For every valuation $v \in \Omega_F$ there exists a point $P \in X$ (not necessarily closed) such that $F_P \subseteq F_v$.*

PROOF. First suppose that v is of rank one. By Lemma 1.3, v has a center Q on $\mathscr{X}$. Then $c := \min\{v(f) \mid f \in \mathfrak{m}_{\mathscr{X},Q}\} > 0$ since $\mathfrak{m}_{\mathscr{X},Q}$ is finitely generated; and

$$\mathfrak{m}^i_{\mathscr{X},Q} \subseteq \mathfrak{a}_i := \{a \in \mathscr{O}_v \,|\, v(a) \geq ic\}.$$

Since $v(F) \subseteq \mathbb{R}$, we have $\cap_i \mathfrak{a}_i = (0)$. Hence $\mathscr{O}_{\mathscr{X},Q} \hookrightarrow \mathscr{O}_v$ is a continuous map of topological rings, and there is an induced map $\widehat{\mathscr{O}}_{\mathscr{X},Q} \to \widehat{\mathscr{O}}_v$ between the respective completions. This map is injective, as can be seen by applying Lemma 1.2 to the kernel. Therefore it induces an inclusion of fraction fields $F_Q \hookrightarrow F_v$.

If Q lies on X, then $P = Q$ satisfies the required condition. On the other hand, if Q does not lie on X, then Q is a codimension one point of the generic fiber $\mathscr{X}_K$ of $\mathscr{X}$, and there is a discrete valuation v on F corresponding to Q. The closure of Q in $\mathscr{X}$ meets X at a closed point P, and the containment $\mathscr{O}_{\mathscr{X},P} \subset \mathscr{O}_{\mathscr{X},Q}$ induces an inclusion $\widehat{\mathscr{O}}_{\mathscr{X},P} \subset \widehat{\mathscr{O}}_{\mathscr{X},Q}$. Namely, this is clear if $\mathscr{X}$ is the projective line over the valuation ring of K and P is the origin on the closed fiber, and one can reduce to that case as in the proof of Lemma 1.2 above. Thus $F_P \subset F_Q = F_v$.

Next, suppose that v is of rank two, so that we may write $v = v_2 \circ v_1$ as in the discussion before Lemma 1.1, for some rank one valuation $v_1 \in \Omega^1_F$. By the previous case, F_{v_1} contains F_P for some point $P \in X$. But $F_v = F_{v_1}$ by Lemma 1.1(a), and so F_v contains F_P. □

2. Local points on F-varieties

In this section, we study varieties over a semi-global field F that have rational points over field extensions F_v (for $v \in \Omega_F$) or F_P (for P in the closed fiber X of a regular model $\mathscr{X}$). We begin by stating some consequences of the results in the previous section.

PROPOSITION 2.1. *Let Z be an F-variety. Then $Z(F_v) \neq \emptyset$ for* all *valuations $v \in \Omega_F$, if and only if $Z(F_v) \neq \emptyset$ for all valuations $v \in \Omega^1_F$.*

PROOF. The forward implication is immediate, so it suffices to prove the reverse implication. As above, a valuation in Ω_F that is not of rank one is a rank two valuation $v = v_2 \circ v_1$, with each v_i of rank one. By Lemma 1.1(a), the completion F_v at v is equal to the completion F_{v_1} at the rank one valuation v_1. Hence $Z(F_v) = Z(F_{v_1})$ is nonempty. □

PROPOSITION 2.2. *Let $\mathscr{X}$ be a regular model of F with closed fiber X. Let Z be an F-variety. If $Z(F_P) \neq \emptyset$ for all $P \in X$, then $Z(F_v) \neq \emptyset$ for all valuations $v \in \Omega_F$.*

PROOF. If v is a valuation in Ω_F, then F_v contains a field F_P for some $P \in X$ by Proposition 1.4. Hence $Z(F_P) \subseteq Z(F_v)$, and so $Z(F_v) \neq \emptyset$. □

The next proposition allows us to deduce the existence of F_v^h-points on a variety from F_v-points. This will be important in proving the main assertion in this section, Theorem 2.5.

PROPOSITION 2.3. *Let Z be a smooth F-variety.*

(a) *Let $v \in \Omega_F^1$. If $Z(F_v) \neq \emptyset$, then $Z(F_v^h) \neq \emptyset$.*

(b) *Let $\mathscr{X}$ be a regular model of F and let P be a closed point on its closed fiber. If $Z(F_P) \neq \emptyset$, then $Z(\mathrm{Frac}(\mathscr{O}^h_{\mathscr{X},P})) \neq \emptyset$.*

PROOF. Part (a) of the proposition follows from [**11**, Proposition 3.5.2], in the case of a smooth variety. (Equivalently, it follows from the generalized Implicit Function Theorem (see [**12**, Theorem 9.2]), after choosing an étale morphism from an affine open subset to some $\mathbb{A}^d_F$, using smoothness.)

For part (b), recall that $F_P = \mathrm{Frac}(\widehat{\mathscr{O}}_{\mathscr{X},P})$, where $\widehat{\mathscr{O}}_{\mathscr{X},P}$ is the $\mathfrak{m}_P$-adic completion of $\mathscr{O}_{\mathscr{X},P}$. Let $U = \mathrm{Spec}(B)$ be an affine open neighborhood of P in $\mathscr{X}$. Then B is a T-algebra of finite type, with a maximal ideal $\mathfrak{m}'_P$ corresponding to P, and with fraction field F. The henselization of the localization of B at $\mathfrak{m}'_P$ is $\mathscr{O}^h_{\mathscr{X},P}$.

After replacing Z by an affine open subset that contains an F_P-point, we may assume that Z is an affine F-variety, say $\mathrm{Spec}(F[t_1, t_2, \ldots, t_k]/(f_1, f_2, \ldots, f_n))$. Since $F_P = \widehat{\mathscr{O}}_{\mathscr{X},P} \otimes_{\mathscr{O}_{\mathscr{X},P}} F$ by Lemma 1.2, after clearing denominators we may assume that the coefficients of the polynomials f_i lie in B; that the B-variety $W := \mathrm{Spec}(B[t_1, t_2, \ldots, t_k]/(f_1, f_2, \ldots, f_n))$ has generic fiber Z; and that W has an $\widehat{\mathscr{O}}_{\mathscr{X},P}$-point whose general fiber is the given F_P-point of Z. By the existence of this $\widehat{\mathscr{O}}_{\mathscr{X},P}$-point, there is a solution in $\widehat{\mathscr{O}}_{\mathscr{X},P}$ to $f_1 = \ldots = f_n = 0$. Since the complete discrete valuation ring T is excellent, the Artin Approximation Theorem (see [**2**, Theorem 1.10]) applies; and so there is also a solution in $\mathscr{O}^h_{\mathscr{X},P}$ to $f_1 = \ldots = f_n = 0$. That is, W has an $\mathscr{O}^h_{\mathscr{X},P}$-point, so Z has a $\mathrm{Frac}(\mathscr{O}^h_{\mathscr{X},P})$-point. □

The next lemma is the key ingredient to the main theorem of this section.

LEMMA 2.4. *Consider an infinite sequence of regular models of F*

$$\mathscr{X} = \mathscr{X}_0 \leftarrow \mathscr{X}_1 \leftarrow \mathscr{X}_2 \leftarrow \ldots$$

and non-empty finite sets $\mathscr{P}_i$ of closed points on the respective closed fibers X_i of the models $\mathscr{X}_i$, where each $\mathscr{X}_i$ is obtained by blowing up $\mathscr{X}_{i-1}$ at the ideal defined by the set $\mathscr{P}_{i-1}$. Let $P_0, P_1, P_2, \ldots$ be an infinite sequence of points $P_i \in \mathscr{P}_i$ such that P_{i+1} maps to P_i for all $i \geq 0$, and consider the direct limit $\mathscr{O} := \varinjlim_i \mathscr{O}_{\mathscr{X}_i,P_i}$. Then $\mathscr{O}$ is a valuation ring on F, with valuation $v \in \Omega_F$ (i.e., $\mathscr{O} = \mathscr{O}_v$). Moreover,

$$\mathscr{O}^h_v = \varinjlim_i \mathscr{O}^h_{\mathscr{X}_i,P_i}. \tag{1}$$

In particular, $F^h_v = \varinjlim_i \mathrm{Frac}(\mathscr{O}^h_{\mathscr{X}_i,P_i}) \subseteq \varinjlim_i F_{P_i}$.

PROOF. We first prove that $\mathcal{O}$ is a valuation ring of F; equivalently, for any $f \in F^\times$, either $f \in \mathcal{O}$ or $f^{-1} \in \mathcal{O}$. For this, it suffices to show that for any fixed $f \in F^\times$, there exists an index i such that either $f \in \mathcal{O}_{\mathscr{X}_i,P_i}$ or $f^{-1} \in \mathcal{O}_{\mathscr{X}_i,P_i}$. This condition does in fact hold, by [**18**, Theorem 26.2], which states that there does not exist an infinite sequence of points $P_i \in X_i \subseteq \mathscr{X}_i$ as in the hypothesis of this lemma, for which each P_i is a point of indeterminacy for f. So $\mathcal{O}$ is a valuation ring of F. Its valuation v lies in Ω_F, because each local ring $\mathcal{O}_{\mathscr{X},P_i}$ contains T by definition, and hence so does $\mathcal{O}$.

To prove (1), it suffices to show that $R := \varinjlim_i \mathcal{O}^h_{\mathscr{X}_i,P_i}$ satisfies the properties that characterize the henselization of $\mathcal{O}$ with respect to v. This ring is henselian because it is a direct limit of henselian rings. For each $(\mathscr{X}_i, P_i)$, the ring $\mathcal{O}^h_{\mathscr{X}_i,P_i}$ is a direct limit of étale $\mathcal{O}_{\mathscr{X}_i,P_i}$-algebras $A_{\mathscr{X}_i,P_i,j}$, and so R is the direct limit of the étale $\mathcal{O}_v$-algebras $\mathcal{O}_v A_{\mathscr{X}_i,P_i,j}$, as i,j vary. Since $\mathfrak{m}\mathcal{O}^h_{\mathscr{X}_i,P_i}$ is the unique maximal ideal of the henselization $\mathcal{O}^h_{\mathscr{X}_i,P_i}$ of $\mathcal{O}_{\mathscr{X}_i,P_i}$ for all i, it follows that $\mathfrak{m}R$ is the unique maximal ideal of R. Similarly, since $\mathcal{O}^h_{\mathscr{X}_i,P_i}/\mathfrak{m}\mathcal{O}^h_{\mathscr{X}_i,P_i} = \mathcal{O}/\mathfrak{m}$ for all i, it follows that $R/\mathfrak{m}R = \mathcal{O}/\mathfrak{m}$. So (1) holds. The last statement follows since construction of fraction fields commutes with direct limits and respects inclusions, and since $\mathcal{O}_{\mathscr{X}_i,P_i} \subseteq \mathcal{O}^h_{\mathscr{X}_i,P_i} \subseteq \widehat{\mathcal{O}}_{\mathscr{X}_i,P_i}$ for all i. □

As above, F is a semi-global field over a complete discretely valued field K.

THEOREM 2.5. *Let Z be a smooth F-variety that has an F_v-point for all $v \in \Omega^1_F$. Then there exists a regular model $\mathscr{X}$ of F such that $Z(F_P) \neq \emptyset$ for all P in the closed fiber X of $\mathscr{X}$.*

PROOF. Fix a regular model $\mathscr{X}_0$ of F, and let X_0 denote its closed fiber. Let Y_0 be an irreducible component of X_0. Then the generic point η of Y_0 corresponds to a valuation v of rank one (centered on η). So by assumption, $Z(F_v) = Z(F_\eta)$ is nonempty. Let t be a uniformizer of K. For a nonempty affine open subset $U \subseteq Y_0$ that does not meet any other irreducible component of X_0, let F_U denote the fraction field of the t-adic completion of the ring of rational functions on $\mathscr{X}_0$ that are regular along U (see [**14**, Notation 3.3]). By [**17**, Proposition 5.8], there exists such an open subset U for which $Z(F_U) \neq \emptyset$. (This could also be deduced from Proposition 2.3(a) above.) If $P \in U$, then $F_U \subseteq F_P$ (loc. cit.), and thus $Z(F_P) \neq \emptyset$ for such P. Since $Y_0 \setminus U$ is finite, there are at most finitely many points $P \in Y_0$ for which the set $Z(F_P)$ is empty. Ranging over the finitely many components of X_0, we obtain a finite (possibly empty) set $\mathscr{P}_0 \subseteq X_0$ consisting of exactly those points $P \in X_0$ for which $Z(F_P) = \emptyset$.

Let $\mathscr{X}_1$ be the blowup of $\mathscr{X}_0$ at all points of $\mathscr{P}_0$ and let X_1 be its closed fiber. By the same argument as above, there exists a finite (possibly empty) set $\mathscr{P}_1 \subseteq X_1$ such that $Z(F_P) = \emptyset$ exactly for $P \in \mathscr{P}_1$. Let $\mathscr{X}_2$ be the blowup of $\mathscr{X}_1$ at all points of $\mathscr{P}_1$, etc. This process yields a chain of models $\mathscr{X}_0 \leftarrow \mathscr{X}_1 \leftarrow \mathscr{X}_2 \leftarrow \dots$ and corresponding sets $\mathscr{P}_0, \mathscr{P}_1, \mathscr{P}_2, \dots$ of closed points on their respective closed fibers.

We claim that the disjoint union $\mathscr{P} := \bigcup_{i\geq 0} \mathscr{P}_i$ is finite. This claim immediately implies that the chain of models $\mathscr{X}_0 \leftarrow \mathscr{X}_1 \leftarrow \mathscr{X}_2 \leftarrow \dots$ terminates in some model $\mathscr{X}_M =: \mathscr{X}$ for which $\mathscr{P}_M = \emptyset$. That is, the model $\mathscr{X}$ satisfies $Z(F_P) \neq \emptyset$ for all points P in its closed fiber X, as required.

It remains to prove the claim. Suppose to the contrary that $\mathscr{P}$ is infinite. For $j \geq i$, we say that a point Q on X_j *lies over* a point P on X_i if Q maps to P under the sequence of blowups $\mathscr{X}_i \leftarrow \cdots \leftarrow \mathscr{X}_j$. Note that each point of $\mathscr{P}_1$ lies over some point of $\mathscr{P}_0$, and similarly each point of $\mathscr{P}_{i+1}$ lies over some point of $\mathscr{P}_i$, for all $i \geq 1$. Since $\mathscr{P}$ is infinite and $\mathscr{P}_0$ is finite, there exists a point $P_0 \in \mathscr{P}_0$ such that infinitely many points in $\mathscr{P}$ lie over P_0. Inductively, if $P_i \in \mathscr{P}_i$ is a point such that infinitely many points in $\mathscr{P}$ lie over P_i, there is a point $P_{i+1} \in \mathscr{P}_{i+1}$ lying over P_i, such that infinitely many points in $\mathscr{P}$ lie over P_{i+1}. This defines an infinite sequence of points $P_1, P_2, \ldots$, since $\mathscr{P}$ is infinite. Here $\mathscr{O}_{\mathscr{X}_i,P_i} \subseteq \mathscr{O}_{\mathscr{X}_{i+1},P_{i+1}}$, $\widehat{\mathscr{O}}_{\mathscr{X}_i,P_i} \subseteq \widehat{\mathscr{O}}_{\mathscr{X}_{i+1},P_{i+1}}$, and $F_{P_i} \subseteq F_{P_{i+1}}$.

Consider the direct limit $\mathscr{O} := \varinjlim_i \mathscr{O}_{\mathscr{X}_i,P_i}$. By Lemma 2.4, the ring $\mathscr{O}$ is a valuation ring with respect to some valuation v on F. If v has rank one, then $Z(F_v) \neq \emptyset$ by assumption, and consequently $Z(F_v^h) \neq \emptyset$ by Proposition 2.3(a). If v has rank two, then Lemma 1.1 implies that there exists a rank one valuation v_1 of v for which $F_v = F_{v_1}$ and $F_{v_1}^h \subseteq F_v^h$, so again $Z(F_v) = Z(F_{v_1}) \neq \emptyset$, and hence $\emptyset \neq Z(F_{v_1}^h) \subseteq Z(F_v^h)$ by Proposition 2.3(a). By Lemma 2.4, $F_v^h \subseteq \varinjlim_i F_{P_i}$, and thus

$$\emptyset \neq Z(F_v^h) \subseteq Z(\varinjlim_i F_{P_i}) = Z(\cup_i F_{P_i}) = \cup_i Z(F_{P_i}) = \emptyset,$$

where the equalities hold because $F_{P_i} \subseteq F_{P_{i+1}}$. This contradiction proves the claim. □

REMARK 2.6. *(a)* The model given by Theorem 2.5 also satisfies the a priori stronger assertion that $Z(\mathrm{Frac}(\mathscr{O}^h_{\mathscr{X},P})) \neq \emptyset$ for all points P in the closed fiber X of $\mathscr{X}$, by Proposition 2.3(b).

(b) In Theorem 2.5, instead of assuming that Z is smooth, we could assume the existence of a smooth F_v-point for every v, since we could apply the theorem to the smooth locus of Z.

3. Local-global principles for rational points

In this section, we use the results of Section 2 to deduce results on local-global principles. Subsection 3.1 treats local-global principles for varieties, while Subsection 3.2 concerns local-global principles for torsors.

As before, F is a semi-global field over a complete discretely valued field K with ring of integers T. Let Ω_F denote the set of valuations on F whose valuation ring contains T, write Ω_F^1 for the subset of valuations of rank one whose valuation ring contains T, and write Ω_F^{dvr} for the subset of discrete valuations. For a regular model $\mathscr{X}$ of F with closed fiber X, let $\Omega_{\mathscr{X}}$ denote the set of points P of X.

3.1. Local-global principles for F-varieties. The results in Section 2 immediately give the following theorem:

THEOREM 3.1. *Let Z be a smooth variety over a semi-global field F. Then the following are equivalent:*

(a) There is a regular model $\mathscr{X}$ of F such that $Z(F_P) \neq \emptyset$ for all points P in its closed fiber X.

(b) $Z(F_v) \neq \emptyset$ for all valuations $v \in \Omega_F^1$.

(c) $Z(F_v) \neq \emptyset$ for all valuations $v \in \Omega_F$.

PROOF. The assertion (a) implies (c) by Proposition 2.2. Trivially (c) implies (b). The fact that (b) implies (a) holds by Theorem 2.5. □

We say that a class $\mathscr{C}$ of F-varieties *satisfies a local-global principle* with respect to a set $\{F_i\}_i$ of overfields of F if every Z in $\mathscr{C}$ has the property that if $Z(F_i) \neq \emptyset$ for all F_i then $Z(F) \neq \emptyset$. In this language, Theorem 3.1 immediately yields the following corollary.

COROLLARY 3.2. *Let $\mathscr{C}$ be a class of smooth varieties over a semi-global field F. Then $\mathscr{C}$ satisfies a local-global principle with respect to $\{F_v \mid v \in \Omega_F\}$ (or equivalently, with respect to $\{F_v \mid v \in \Omega_F^1\}$) if and only if $\mathscr{C}$ satisfies a local-global principle with respect to $\{F_P \mid P \in X\}$ for every regular model $\mathscr{X}$ of F with closed fiber X.*

EXAMPLE 3.3. Let G be a rational connected linear algebraic group over the semi-global field F, and let $\mathscr{C}$ be the class of transitive homogeneous G-spaces Z over F (i.e., $G(E)$ acts transitively on $Z(E)$ for every overfield E of F). Then for every regular model $\mathscr{X}$ of F, say with closed fiber X, the class $\mathscr{C}$ satisfies a local-global principle with respect to $\{F_P \mid P \in X\}$ (see [**17**, Theorem 9.1]). As a consequence of Corollary 3.2, it then follows that the class $\mathscr{C}$ also satisfies a local-global principle with respect to the set $\{F_v \mid v \in \Omega_F^1\}$.

3.2. Local-global principles for torsors. Let G be a linear algebraic group (i.e., a smooth affine group scheme of finite type) over F. We now define and compare several obstruction sets to the existence of global F-points on torsors under G.

Recall that each G-torsor over F is represented by a class in the pointed set $H^1(F, G)$, and that this class is trivial if and only if the torsor has an F-point. As discussed in the introduction, we have the following obstruction sets to local-global principles:

$$\text{Ш}_{\Omega_F}(F, G) = \ker\big(H^1(F, G) \to \prod_{v \in \Omega_F} H^1(F_v, G)\big),$$

$$\text{Ш}_{\Omega_F^1}(F, G) = \ker\big(H^1(F, G) \to \prod_{v \in \Omega_F^1} H^1(F_v, G)\big),$$

$$\text{Ш}_{\mathrm{dvr}}(F, G) = \ker\big(H^1(F, G) \to \prod_{v \in \Omega_F^{\mathrm{dvr}}} H^1(F_v, G)\big).$$

Here the *kernel* of a map of pointed sets is by definition the preimage of the trivial element.

Since $\Omega_F^{\mathrm{dvr}} \subseteq \Omega_F^1 \subseteq \Omega_F$, the above obstruction sets are related by containments as well:

$$\text{Ш}_{\Omega_F}(F, G) \subseteq \text{Ш}_{\Omega_F^1}(F, G) \subseteq \text{Ш}_{\mathrm{dvr}}(F, G).$$

Finally, for any regular model $\mathscr{X}$ of F with closed fiber X, we let

$$\text{Ш}_X(F, G) = \ker\big(H^1(F, G) \to \prod_{P \in X} H^1(F_P, G)\big).$$

By the previous subsection, we obtain:

THEOREM 3.4. *Let G be a linear algebraic group over a semi-global field F. Then*

$$\text{Ш}_X(F, G) \subseteq \text{Ш}_{\Omega_F}(F, G)$$

for any regular model $\mathscr{X}$ of F. Moreover, taking the direct limit over all such models $\mathscr{X}$, we have

$$\varinjlim_{\mathscr{X}} Ш_X(F,G) = Ш_{\Omega_F}(F,G) = Ш_{\Omega^1_F}(F,G).$$

Proof. If Z is a G-torsor over F, then Z is smooth (because G is). Proposition 2.2 shows that $Ш_X(F,G) \subseteq Ш_{\Omega_F}(F,G)$ for any regular model $\mathscr{X}$ of F, and so $\varinjlim_{\mathscr{X}} Ш_X(F,G) \subseteq Ш_{\Omega^1_F}(F,G)$. The asserted equalities follow by applying Theorem 3.1 to Z. □

It is an interesting open problem to understand the relationship between $Ш_{\Omega^1_F}(F,G)$ and $Ш_{\mathrm{dvr}}(F,G)$; in particular, whether the inclusion is an equality.

References

[1] Shreeram Abhyankar, *On the valuations centered in a local domain*, Amer. J. Math. **78** (1956), 321–348, DOI 10.2307/2372519. MR82477

[2] M. Artin, *Algebraic approximation of structures over complete local rings*, Inst. Hautes Études Sci. Publ. Math. **36** (1969), 23–58. MR268188

[3] Matthias Aschenbrenner, Lou van den Dries, and Joris van der Hoeven, *Asymptotic differential algebra and model theory of transseries*, Annals of Mathematics Studies, vol. 195, Princeton University Press, Princeton, NJ, 2017. MR3585498

[4] Nicolas Bourbaki, *Commutative Algebra*, Hermann and Addison-Wesley, 1972.

[5] Siegfried Bosch, Werner Lütkebohmert, and Michel Raynaud, *Néron models*, Ergebnisse der Mathematik und ihrer Grenzgebiete (3) [Results in Mathematics and Related Areas (3)], vol. 21, Springer-Verlag, Berlin, 1990. MR1045822

[6] V. I. Chernousov, *The Hasse principle for groups of type E_8* (Russian), Dokl. Akad. Nauk SSSR **306** (1989), no. 5, 1059–1063; English transl., Soviet Math. Dokl. **39** (1989), no. 3, 592–596. MR1014762

[7] J.-L. Colliot-Thélène, M. Ojanguren, and R. Parimala, *Quadratic forms over fraction fields of two-dimensional Henselian rings and Brauer groups of related schemes*, Algebra, arithmetic and geometry, Part I, II (Mumbai, 2000), Tata Inst. Fund. Res. Stud. Math., vol. 16, Tata Inst. Fund. Res., Bombay, 2002, pp. 185–217. MR1940669

[8] Jean-Louis Colliot-Thélène, Raman Parimala, and Venapally Suresh, *Patching and local-global principles for homogeneous spaces over function fields of p-adic curves*, Comment. Math. Helv. **87** (2012), no. 4, 1011–1033, DOI 10.4171/CMH/276. MR2984579

[9] J.-L. Colliot-Thélène, R. Parimala, and V. Suresh, *Lois de réciprocité supérieures et points rationnels* (French, with English and French summaries), Trans. Amer. Math. Soc. **368** (2016), no. 6, 4219–4255, DOI 10.1090/tran/6519. MR3453370

[10] Antonio J. Engler and Alexander Prestel, *Valued fields*, Springer Monographs in Mathematics, Springer-Verlag, Berlin, 2005. MR2183496

[11] Ofer Gabber, Philippe Gille, and Laurent Moret-Bailly, *Fibrés principaux sur les corps valués henséliens* (French, with English and French summaries), Algebr. Geom. **1** (2014), no. 5, 573–612, DOI 10.14231/AG-2014-025. MR3296806

[12] B. Green, F. Pop, and P. Roquette, *On Rumely's local-global principle*, Jahresber. Deutsch. Math.-Verein. **97** (1995), no. 2, 43–74. MR1341772

[13] David Harbater and Julia Hartmann, *Patching over fields*, Israel J. Math. **176** (2010), 61–107, DOI 10.1007/s11856-010-0021-1. MR2653187

[14] David Harbater, Julia Hartmann, and Daniel Krashen, *Applications of patching to quadratic forms and central simple algebras*, Invent. Math. **178** (2009), no. 2, 231–263, DOI 10.1007/s00222-009-0195-5. MR2545681

[15] David Harbater, Julia Hartmann, and Daniel Krashen, *Weierstrass preparation and algebraic invariants*, Math. Ann. **356** (2013), no. 4, 1405–1424, DOI 10.1007/s00208-012-0888-8. MR3072806

[16] David Harbater, Julia Hartmann, and Daniel Krashen, *Local-global principles for Galois cohomology*, Comment. Math. Helv. **89** (2014), no. 1, 215–253, DOI 10.4171/CMH/317. MR3177913

[17] David Harbater, Julia Hartmann, and Daniel Krashen, *Local-global principles for torsors over arithmetic curves*, Amer. J. Math. **137** (2015), no. 6, 1559–1612, DOI 10.1353/ajm.2015.0039. MR3432268

[18] Joseph Lipman, *Rational singularities, with applications to algebraic surfaces and unique factorization*, Inst. Hautes Études Sci. Publ. Math. **36** (1969), 195–279. MR276239

[19] Vlerë Mehmeti, *Patching over Berkovich curves and quadratic forms*, Compos. Math. **155** (2019), no. 12, 2399–2438, DOI 10.1112/s0010437x19007632. MR4030183

[20] R. Parimala, R. Preeti, and V. Suresh, *Local-global principle for reduced norms over function fields of p-adic curves*, Compos. Math. **154** (2018), no. 2, 410–458, DOI 10.1112/S0010437X17007618. MR3732207

[21] Florian Pop and Jakob Stix, *Arithmetic in the fundamental group of a p-adic curve. On the p-adic section conjecture for curves*, J. Reine Angew. Math. **725** (2017), 1–40, DOI 10.1515/crelle-2014-0077. MR3630116

[22] J.-J. Sansuc, *Groupe de Brauer et arithmétique des groupes algébriques linéaires sur un corps de nombres* (French), J. Reine Angew. Math. **327** (1981), 12–80, DOI 10.1515/crll.1981.327.12. MR631309

[23] O. F. G. Schilling, *The Theory of Valuations*, Mathematical Surveys, No. 4, American Mathematical Society, New York, N. Y., 1950. MR0043776

[24] Jean-Pierre Tignol and Adrian R. Wadsworth, *Value functions on simple algebras, and associated graded rings*, Springer Monographs in Mathematics, Springer, Cham, 2015. MR3328410

[25] Seth Warner, *Topological fields*, North-Holland Mathematics Studies, vol. 157, North-Holland Publishing Co., Amsterdam, 1989. Notas de Matemática [Mathematical Notes], 126. MR1002951

Department of Mathematics, University of Pennsylvania, Philadelphia, Pennsylvania 19104-6395

Email address: harbater@math.upenn.edu

Department of Mathematics, University of Pennsylvania, Philadelphia, Pennsylvania 19104-6395

Email address: hartmann@math.upenn.edu

Mathematical Institute, Utrecht University, P.O. Box 80010, 3508 TA Utrecht, the Netherlands

Email address: V.Z.Karemaker@uu.nl

Department of Mathematics, University of Pennsylvania, Philadelphia, Pennsylvania 19104-6395

Email address: pop@math.upenn.edu

Contemporary Mathematics
Volume **767**, 2021
https://doi.org/10.1090/conm/767/15404

The generic monodromy of Drinfeld modular varieties in special characteristic

Gebhard Böckle and Florian Breuer

Dedicated to Gerhard Frey on the occasion of his 75th *birthday.*

Abstract. By combining theorems of Drinfeld and Strauch, we show that the monodromy representation on the special fiber of a Drinfeld modular variety, with level not divisible by the characteristic, is surjective. We illustrate this result in the special case of Drinfeld $\mathbb{F}_q[t]$-modules in level t, and apply this to show that the Kronecker factors of a Drinfeld modular polynomial in rank r are irreducible.

1. Statement of the main result

Throughout this paper we fix a global function field F of characteristic $p > 0$ with exact field of constants the finite field $\mathbb{F}_q$ of cardinality q. We fix a place ∞ of F, and let A denote the ring of elements of F which are regular away from ∞. This is a Dedekind domain with finite class group $\mathrm{Cl}(A)$ and unit group $A^\times = \mathbb{F}_q^\times$.

Let $I \subset A$ denote a proper non-zero ideal and n_I the order of I in the ideal class group $\mathrm{Cl}(A)$ of A. Let $g_I \in A$ be a generator of I^{n_I}; it is unique up to multiplication by $\mathbb{F}_q^\times$. Hence $A[1/I] := A[1/g_I]$ is independent of the choice of g_I. We also write $\mathrm{Cl}_I(A)$ for the I-class group of A, i.e., the group of fractional ideals of A of support prime to I modulo its subgroup of principal fractional ideals that possess a generator which is congruent to 1 modulo I. One has a short exact sequence $0 \longrightarrow (A/I)^\times/\mathbb{F}_q^\times \longrightarrow \mathrm{Cl}_I(A) \longrightarrow \mathrm{Cl}(A) \longrightarrow 0$.

Let $r \geq 1$ be a positive integer and consider the functor $\mathcal{M}_I^r$ from $A[1/I]$-schemes S to **Sets**, which to any such S assigns the set of isomorphism classes of tuples $(\mathcal{L}, \phi, \alpha)$, where $\mathcal{L}$ is a line bundle on S, where ϕ (together with $\mathcal{L}$) is a Drinfeld A-module $\phi\colon A \longrightarrow \mathrm{End}_{\mathbb{F}_q\text{-gp sch.}/S}(\mathcal{L})$ of rank r, and where α denotes a level I-structure on $(\mathcal{L}, \phi)$, subject to the condition that the characteristic $\partial\phi\colon A \longrightarrow \mathrm{End}_S(\mathrm{Lie}\,\mathcal{L}) = \mathcal{O}_S(S)$ coincides with the structure morphism $S \longrightarrow \mathrm{Spec}\,A[1/I]$ composed with the open immersion $\mathrm{Spec}\,A[1/I] \longrightarrow \mathrm{Spec}\,A$. By [**4**, Prop. 5.3 and Cor. to 5.4], the functor $\mathcal{M}_I^r$ is representable by a smooth finite type morphism $\mathfrak{M}_I^r \longrightarrow \mathrm{Spec}\,A[1/I]$ of relative dimension $r-1$. The universal Drinfeld module on $\mathfrak{M}_I^r$ we denote by

$$\phi_I^r\colon A \longrightarrow \mathrm{End}_{\mathbb{F}_q\text{-gp sch.}/\mathfrak{M}_I^r}(\mathcal{L}_I^r).$$

2020 *Mathematics Subject Classification.* Primary 11G09, 11F80, 14L05.

Let now $\mathfrak{p} \subset A$ denote a maximal ideal that is prime to I. We write $\kappa_\mathfrak{p}$ for its residue field, $A_\mathfrak{p}$ for the completion of A at $\mathfrak{p}$, and let $\overline{\kappa}_\mathfrak{p}$ be an algebraic closure of $\kappa_\mathfrak{p}$.

DEFINITION 1. We call $\mathfrak{M}^r_{I,\mathfrak{p}} := \mathfrak{M}^r_I \times_{\operatorname{Spec} A[1/I]} \operatorname{Spec} \kappa_\mathfrak{p}$ the *special fiber of* $\mathfrak{M}^r_I$ *at* $\mathfrak{p}$.

Let $\mathfrak{M}^r_{I,\overline{\mathfrak{p}}}$ be the base change $\mathfrak{M}^r_{I,\mathfrak{p}} \times_{\kappa_\mathfrak{p}} \overline{\kappa}_\mathfrak{p}$ and let $\phi^r_{I,\overline{\mathfrak{p}}}$ be the corresponding universal Drinfeld module . The scheme $\mathfrak{M}^r_{I,\overline{\mathfrak{p}}}$ is regular. Its connected components can be naturally labelled by $\mathrm{Cl}_I(A)$: By [**5**, proof of Cor. 4.6] the connected components of $\mathfrak{M}^r_{I,\overline{\mathfrak{p}}}$ are in bijection with those of $\mathfrak{M}^1_{I,\overline{\mathfrak{p}}} \cong \operatorname{Spec}\left(R_I \otimes_{A[1/I]} \overline{\kappa}_\mathfrak{p}\right)$, where by [**4**, Thm. 1, §7] R_I is the integral closure of A in the class field of F associated to $\mathrm{Cl}_I(A)$. Now class field theory gives the desired labeling.

For each ideal class $\mathfrak{c}$ in $\mathrm{Cl}_I(A)$, denote by $\eta_\mathfrak{c} = \operatorname{Spec} \kappa_{\eta_\mathfrak{c}}$ the generic point of the corresponding component, and let $\overline{\eta}_\mathfrak{c} = \operatorname{Spec} \overline{\kappa}_{\eta_\mathfrak{c}}$ be a geometric point above $\kappa_{\eta_\mathfrak{c}}$. Observe that $\kappa_{\eta_\mathfrak{c}}$ contains $\overline{\kappa}_\mathfrak{p}$.

Let $\phi^r_{\eta_\mathfrak{c}}$ denote the pullback of $\phi^r_{I,\overline{\mathfrak{p}}}$ to $\eta_\mathfrak{c}$. By [**4**, Prop. 5.5], it is a Drinfeld A-module of characteristic $\mathfrak{p}$ and height 1, i.e., $\phi^r_{\eta_\mathfrak{c}}$ is ordinary. This means that for any $n \geq 1$ the group of $\mathfrak{p}^n$-torsion points $\phi^r_{\eta_\mathfrak{c}}[\mathfrak{p}^n](\overline{\kappa}_{\eta_\mathfrak{c}})$ is a free $A/\mathfrak{p}^n$-module of rank $r-1$.

Let $g_\mathfrak{p}$ be a generator of the principal ideal $\mathfrak{p}^{n_\mathfrak{p}}$, so that $\phi^r_{\eta_\mathfrak{c}}[\mathfrak{p}^{nn_\mathfrak{p}}](\overline{\kappa}_{\eta_\mathfrak{c}})$ is the set of roots of $\phi^r_{\eta_\mathfrak{c},g^n_\mathfrak{p}}(X)$. We define the $\mathfrak{p}$-adic Tate module of $\phi^r_{\eta_\mathfrak{c}}$ as

$$\operatorname{Ta}_\mathfrak{p} \phi^r_{\eta_\mathfrak{c}} = \varprojlim_n \phi^r_{\eta_\mathfrak{c}}[\mathfrak{p}^{nn_\mathfrak{p}}](\overline{\kappa}_{\eta_\mathfrak{c}}),$$

where multiplication by $g_\mathfrak{p}$ defines the transition map in the inverse system. The limit is independent of the choice of $g_\mathfrak{p}$. By ordinariness of $\phi^r_{\eta_\mathfrak{c}}$ it is free of rank $r-1$ over $A_\mathfrak{p}$.

Observe that $\phi^r_{\eta_\mathfrak{c},g^n_\mathfrak{p}}(X) = h_n \circ (X \mapsto X^{q^{n \deg \mathfrak{p}}})$ for some unique $\mathbb{F}_q$-linear polynomial $h_n \in \kappa_{\eta_\mathfrak{c}}[X]$ with non-vanishing linear term. The étale quotient $\phi^r_{\eta_\mathfrak{c}}[\mathfrak{p}^{nn_\mathfrak{p}}]^{\text{ét}}$ of the finite flat A-module scheme $\phi^r_{\eta_\mathfrak{c}}[\mathfrak{p}^{nn_\mathfrak{p}}]$ is $\operatorname{Spec} \kappa_{\eta_\mathfrak{c}}[X]/(h_n(X))$. The group schemes $\phi^r_{\eta_\mathfrak{c}}[\mathfrak{p}^{nn_\mathfrak{p}}]^{\text{ét}}$ also form an inverse system, and one has $\phi^r_{\eta_\mathfrak{c}}[\mathfrak{p}^{nn_\mathfrak{p}}](\overline{\kappa}_{\eta_\mathfrak{c}}) \cong \phi^r_{\eta_\mathfrak{c}}[\mathfrak{p}^{nn_\mathfrak{p}}]^{\text{ét}}(\kappa^{\text{sep}}_{\eta_\mathfrak{c}})$ as finite A-modules. Because the polynomials h_n are defined over $\kappa_{\eta_\mathfrak{c}}$, the absolute Galois group $G_{\kappa_{\eta_\mathfrak{c}}} = \operatorname{Gal}(\kappa^{\text{sep}}_{\eta_\mathfrak{c}}/\kappa_{\eta_\mathfrak{c}})$ acts on $\operatorname{Ta}_\mathfrak{p} \phi^r_{\eta_\mathfrak{c}}$ and by the very construction of $\operatorname{Ta}_\mathfrak{p} \phi^r_{\eta_\mathfrak{c}}$, the group $G_{\kappa_{\eta_\mathfrak{c}}}$ acts $A_\mathfrak{p}$-linearly. This yields a continuous group homomorphism

$$\rho_{\mathfrak{p},\eta_\mathfrak{c}} : G_{\kappa_{\eta_\mathfrak{c}}} \longrightarrow \operatorname{Aut}_{A_\mathfrak{p}}(\operatorname{Ta}_\mathfrak{p} \phi^r_{\eta_\mathfrak{c}}) \cong \mathrm{GL}_{r-1}(A_\mathfrak{p}).$$

Our main result is the following:

THEOREM 1. *The map* $\rho_{\mathfrak{p},\eta_\mathfrak{c}}$ *is surjective.*

The proof is a simple consequence of the results [**4**, **9**] by Drinfeld and Strauch, which seems not to have been recorded in the literature. In fact, combining the work of Drinfeld and Strauch, it even follows that the image under $\rho_{\mathfrak{p},\eta_\mathfrak{c}}$ of a decomposition group of $G_{\kappa_{\eta_\mathfrak{c}}}$ at a supersingular point of $\mathfrak{M}^r_{I,\overline{\mathfrak{p}}}$ in the component of $\eta_\mathfrak{c}$ is already surjective; cf. Remark 1.

One might wonder about refinements of Theorem 1. For any point x of $\mathfrak{M}^r_{I,\overline{\mathfrak{p}}}$ denote by $\rho_{\mathfrak{p},x} \colon G_x \longrightarrow \operatorname{Aut}_\mathfrak{p}(\operatorname{Ta}_\mathfrak{p} \phi^r_x)$ the action of the absolute Galois group $G_x = \operatorname{Gal}(\kappa^{\text{sep}}_x/\kappa_x)$ of the residue field at x on the corresponding Tate module $\operatorname{Ta}_\mathfrak{p} \phi^r_x$ of rank r_x with $0 \leq r_x \leq r-1$.

QUESTION 1. Suppose that $\mathrm{End}_{\overline{\kappa}_x}(\phi_x^r) = A$. Is $\rho_{\mathfrak{p},x}(G_x)$ open in $\mathrm{Aut}_{\mathfrak{p}}(\mathrm{Ta}_{\mathfrak{p}}\,\phi_x^r) \cong \mathrm{GL}_{r_x}(A_{\mathfrak{p}})$?

This appears to be a natural analog of the results [**3**] of Devic and Pink on adelic openness for Drinfeld modules in special characteristic. They consider the Galois action of a Drinfeld A-module ϕ of rank r and characteristic $\mathfrak{p} \neq 0$ over a finitely generated field. If $\mathrm{End}_{\overline{\kappa}_x}(\phi_x) = A$, their results imply that the associated adelic Galois representation of G_x away from $\mathfrak{p}$ and ∞ has open image in $\mathrm{SL}_r(\prod_{v\neq\mathfrak{p}} A_v)$. They also give a complete answer with no condition on $\mathrm{End}_{\overline{\kappa}_x}(\phi_x)$. This leads to.

QUESTION 2. Describe for any point x of $\mathfrak{M}^d_{I,\overline{\mathfrak{p}}}$ the Zariski closure $\mathcal{G}_x$ of $\rho_{\mathfrak{p},x}(G_x)$ in $\mathrm{Aut}_{\mathfrak{p}}(\mathrm{Ta}_{\mathfrak{p}}\,\phi_x^r) \cong \mathrm{GL}_{r_x}(A_{\mathfrak{p}})$. Is $\rho_{\mathfrak{p},x}(G_x)$ an open subgroup in $\mathcal{G}_x(A_{\mathfrak{p}})$?

We end this introduction with a quick survey of the content of the individual sections. Section 2 recalls the relevant work of Drinfeld on formal $\mathcal{O}$-modules and $\mathcal{O}$-divisible groups from [**4**]. Section 3 recalls the main theorem of Strauch, so that in Section 4 we can combine the two and deduce the proof of Theorem 1. Section 5 illustrates the main result in the special case of $A = \mathbb{F}_q[t]$ and level t, where the moduli scheme can be described explicitly. In Section 6 we shall answer in Proposition 2 a question raised in [**2**] related to the reduction of modular polynomials of level $\mathfrak{p}$ in the case $A = \mathbb{F}_q[t]$. We shall prove that certain special polynomials which are the natural building blocks of the mod $\mathfrak{p}$ reduction of modular polynomials are irreducible as asked in [**2**, Question 4.5].

2. Formal $\mathcal{O}$-modules, $\mathcal{O}$-divisible groups and deformations of Drinfeld modules

Let K be a non-archimedean local field with ring of integers $\mathcal{O}$ and finite residue field k. The normalized valuation on K is v_K, its uniformizer ϖ_K and the cardinality of k will be q_K. Let $\breve{K}$ be the completion of the maximal unramified extension of K and write $\breve{\mathcal{O}}$ for its ring of integers. The residue field $\breve{k}$ of $\breve{\mathcal{O}}$ is an algebraic closure of k. Denote by $\mathrm{CNL}_{\breve{\mathcal{O}}}$ the category of complete noetherian local $\breve{\mathcal{O}}$-algebras C with residue field $\breve{k}$, and with morphisms being $\breve{\mathcal{O}}$-algebra homomorphisms $f\colon C \longrightarrow C'$ such that $f(\mathfrak{m}_C) \subset \mathfrak{m}_{C'}$, where for $C \in \mathrm{CNL}_{\breve{\mathcal{O}}}$ we denote by $\mathfrak{m}_C$ its maximal ideal.

Let B be a ring. The power series ring over B in indeterminates $x_1, .., x_n$ will be $B[[x_1, .., x_n]]$.

DEFINITION 2 ([**4**, § 1]). A *formal group*[1] over B is a series Φ in $B[[x, y]]$ such that $\Phi(x, y) = \Phi(y, x)$, $\Phi(x, 0) = x$ and $\Phi(\Phi(x, y), z) = \Phi(x, \Phi(y, z))$.

A *homomorphism* from a formal group Φ to a formal group Ψ over B is a series $\beta \in xB[[x]]$ such that $\Psi(\beta(x), \beta(y)) = \beta(\Phi(x, y))$. Composition of homomorphisms is composition of formal power series; it is well-defined because the series have zero constant term.

The endomorphism ring of a formal group Φ is denoted $\mathrm{End}(\Phi)$. It comes with a natural homomorphism $D\colon \mathrm{End}(\Phi) \longrightarrow B$, given by differentiation at zero $\beta \mapsto D\beta = \left(\frac{\mathrm{d}}{\mathrm{d}x}\beta\right)(0)$.

Suppose that B is an $\mathcal{O}$-algebra via a map α. A *formal $\mathcal{O}$-module over B* is a pair $X = (\Phi, [\cdot]_X)$ where Φ is a formal group over B and $[\cdot]_X$ is a homomorphism

[1] More correctly we should add the attributes one-dimensional and commutative; but we shall not deal with any other kind of formal group; and so for the sake of brevity we suppress them.

$\mathcal{O} \longrightarrow \mathrm{End}(\Phi)$ such that $D \circ [\cdot]_X = \alpha$. Morphisms of formal $\mathcal{O}$-modules are defined in the obvious way, see also [**4**, §1] and [**10**, §1.3].

If $\alpha(\varpi_K) = 0$, then $[\varpi_K]_X \in xB[[x]]$ can be written as the composition $\gamma \circ (x \mapsto x^{q_K^h})$ for some unique $\gamma \in xB[[x]]$ with linear term $D\gamma \neq 0$ and a unique $h \geq 1$. One calls h the *height* of the formal $\mathcal{O}$-module $X = (\Phi, [\cdot]_X)$.

EXAMPLE 1 ([**4**, Rem. after Prop. 2.2], [**8**, § 4]). Let $s\colon A \longrightarrow \mathcal{O}$ be a ring homomorphism with $s(\mathfrak{p})\mathcal{O} = \varpi_K \mathcal{O}$, and let C be in $\mathrm{CNL}_{\breve{\mathcal{O}}}$. This induces an A-algebra structure on C, which we denote by γ. Let $\phi\colon A \longrightarrow C\{\tau\}, a \mapsto \phi_a$ be a Drinfeld A-module in standard form of rank r and characteristic γ; cf. [**4**, Rem. after Prop. 5.2]. Let $\Phi(x,y) = x + y$ be the additive formal group law. Then $\mathrm{End}(\Phi) = C\{\{\tau\}\}$, the subring, under addition and composition, of $xC[[x]]$ of power series in the monomials x^{q^i}, $i \geq 0$, with coefficients in C. It can be shown that ϕ extends uniquely to a continuous ring homomorphism $\widehat{\phi}\colon A_\mathfrak{p} \longrightarrow \mathrm{End}(\Phi), a \mapsto \widehat{\phi}_a$. This uses that elements in $\mathfrak{p}$ map to topologically nilpotent elements in C under γ. This defines the structure of a formal $A_\mathfrak{p}$-module $\widehat{\phi}_\mathfrak{p} = (\Phi, \widehat{\phi})$ on Φ. Moreover the height of the formal $A_\mathfrak{p}$-module $\widehat{\phi}_\mathfrak{p} \pmod{\mathfrak{m}_C}$ agrees with the height of the Drinfeld A-module $\phi \pmod{\mathfrak{m}_C}$.

Let $\breve{k}$ be an $\mathcal{O}$-algebra via reduction, i.e., via the canonical maps $\mathcal{O} \longrightarrow \breve{\mathcal{O}} \longrightarrow \breve{k}$. Let $\overline{X}$ be a formal $\mathcal{O}$-module over $\breve{k}$ of finite height $h > 0$. A *deformation of* $\overline{X}$ *to* $C \in \mathrm{CNL}_{\breve{\mathcal{O}}}$ is a formal $\mathcal{O}$-module X_C over C whose reduction modulo $\mathfrak{m}_C$ is equal to $\overline{X}$. Two deformations X_C and X'_C to C are isomorphic if there exists an isomorphism of formal $\mathcal{O}$-modules over C that reduces to the identity modulo $\mathfrak{m}_C$. Since h is finite, by [**4**, Prop. 4.1] there is at most one such isomorphism.

THEOREM 2 ([**4**, Prop. 4.2]). *The functor* $\mathrm{CNL}_{\breve{\mathcal{O}}} \longrightarrow$ **Sets** *that associates to* $C \in \mathrm{CNL}_{\breve{\mathcal{O}}}$ *the set of deformations of* $\overline{X}$ *to* C *up to isomorphism is representable by a ring* $R_{\overline{X}}$ *in* $\mathrm{CNL}_{\breve{\mathcal{O}}}$. *The universal ring* $R_{\overline{X}}$ *is a power series ring over* $\breve{\mathcal{O}}$ *in* $h-1$ *indeterminates.*

DEFINITION 3. The universal formal group over $R_{\overline{X}}$ is denoted by $X_{\overline{X}}$.

To recall the notion of $\mathcal{O}$-divisible module (again of dimension 1), we need some preparations. We fix a ring B in $\mathrm{CNL}_{\breve{\mathcal{O}}}$. Following [**10**], for R any ring, we define an *R-module scheme* over B to be a pair $(\mathcal{G}, \phi)$, where $\mathcal{G}$ is a commutative group scheme over B and $\phi\colon R \longrightarrow \mathrm{End}(\mathcal{G})$ is a ring homomorphism. A *map* $(\mathcal{G}, \phi) \longrightarrow (\mathcal{G}', \phi')$ *of R-module schemes* is a map $\mathcal{G} \longrightarrow \mathcal{G}'$ of group schemes over B that is equivariant for the R-action.

If $\mathcal{G}$ is finite flat over B, then one can define the étale and connected parts $\mathcal{G}^{\text{ét}}$, $\mathcal{G}^{\text{loc}}$ of $\mathcal{G}$, and one has a short exact sequence $0 \longrightarrow \mathcal{G}^{\text{loc}} \longrightarrow \mathcal{G} \longrightarrow \mathcal{G}^{\text{ét}} \longrightarrow 0$; see [**11**, 1.4]. Because any endomorphism of $\mathcal{G}$ preserves $\mathcal{G}^{\text{loc}}$, if $\mathcal{G}$ carries an R-action, then the short exact sequence is one of R-module schemes. For the following, we assume that K has positive characteristic. Then the field k is canonically a subring of $\mathcal{O}$.

DEFINITION 4 ([**4**, § 4], [**10**, § 1]). Let r be in $\mathbb{N}$. An *$\mathcal{O}$-divisible module of rank r over B* is an inductive system $\mathcal{F} = (\mathcal{F}_n, i_n)_{n \in \mathbb{N}}$ such that for all $n \in \mathbb{N}$ the following hold:

(1) $\mathcal{F}_n$ is a finite flat group scheme over B that carries an $\mathcal{O}$-module structure.

(2) There is a closed immersion $\mathcal{F}_n \hookrightarrow \mathbb{G}_{a,B}$ of k-module schemes.[2]
(3) The order of $\mathcal{F}_n$ over B is q_K^{rn},
(4) The following sequence of $\mathcal{O}$-module schemes over B is exact

$$0 \longrightarrow \mathcal{F}_n \xrightarrow{i_n} \mathcal{F}_{n+1} \xrightarrow{\varpi_K^n} \mathcal{F}_{n+1}.$$

A *morphism of $\mathcal{O}$-divisible modules over B* is a morphism of inductive systems of $\mathcal{O}$-module schemes.

Given an $\mathcal{O}$-divisible module $\mathcal{F} = (\mathcal{F}_n)_{n\geq 1}$, the connected and étale parts $\mathcal{F}^{\mathrm{loc}} = (\mathcal{F}_n^{\mathrm{loc}})_{n\geq 1}$ and $\mathcal{F}^{\mathrm{\acute{e}t}} = (\mathcal{F}_n^{\mathrm{\acute{e}t}})_{n\geq 1}$ form $\mathcal{O}$-divisible modules as well and one has a degree-wise short exact sequence of $\mathcal{O}$-divisible modules $0 \longrightarrow \mathcal{F}^{\mathrm{loc}} \longrightarrow \mathcal{F} \longrightarrow \mathcal{F}^{\mathrm{\acute{e}t}} \longrightarrow 0$. If $\mathcal{F}^{\mathrm{\acute{e}t}} = 0$, we call $\mathcal{F}$ *local*.

Concerning $\mathcal{F}^{\mathrm{\acute{e}t}}$ note that since $\breve{k}$ is algebraically closed, one has an isomorphism of $\mathcal{O}$-module schemes between $\mathcal{F}_n^{\mathrm{\acute{e}t}}$ mod $\mathfrak{m}_B$ and the constant $\mathcal{O}$-module scheme $\underline{\mathcal{O}^s/\varpi_K^n\mathcal{O}^s}$ for s the rank of $\mathcal{F}^{\mathrm{\acute{e}t}}$. Hensel lifting shows that the same isomorphism holds over B. Drinfeld writes $\mathcal{F}^{\mathrm{\acute{e}t}} = \underline{K^s/\mathcal{O}^s}$.

To analyze $\mathcal{F}^{\mathrm{loc}}$, we present in the following paragraphs, up to and including Proposition 1, some results that are implicitly stated in [**4**, § 4] and are straightforward to deduce from [**10**, § 1]. Suppose that $\mathcal{F}^{\mathrm{loc}}$ is non-trivial. Then $\varinjlim \mathcal{F}_n^{\mathrm{loc}} \cong \operatorname{Spf} B[[x]]$, and this isomorphism is one of formal k-module schemes, if we identify $\operatorname{Spf} B[[x]]$ with the formal completion of $\mathbb{G}_{a,B}$ at the zero section. The action of $\mathcal{O}$ on $\mathcal{F}^{\mathrm{loc}}$ induces an $\mathcal{O}$-action on the formal additive group over B. The resulting formal $\mathcal{O}$-module will be denoted by $X_{\mathcal{F}}$.

Conversely, let $X = (\Phi, [\cdot]_X)$ be a formal $\mathcal{O}$-module over B whose reduction to $\breve{k}$ has finite height h. Then a local divisible $\mathcal{O}$-module is defined as follows: For $n \in \mathbb{N}$, write $[\varpi_K^n]_X = H_n u_n$ uniquely with $H_n \in B[x]$ monic of degree q_K^{hn} and $H_n \pmod{\mathfrak{m}_B} = x^{q_K^{hn}}$, and $u_n \in B[[x]]$ a unit. Then $X[\varpi_K^n] := \operatorname{Spec} B[x]/(H_n) = \operatorname{Spec} B[[x]]/([\varpi_K^n]_X)$ is a finite flat scheme over B and one can verify for all $n \geq 1$ that

(1) The formal $\mathcal{O}$-module structure $[\cdot]_X$ defines an $\mathcal{O}$-action on $X[\varpi_K^n]$, and in such a way that the closed immersion $X[\varpi_K^n] \hookrightarrow \mathbb{G}_{a,B}$ is one of k-module schemes.
(2) One has a short exact sequence $0 \longrightarrow X[\varpi_K^n] \longrightarrow X[\varpi_K^{n+1}] \xrightarrow{\varpi_K} X[\varpi_K^{n+1}]$ of $\mathcal{O}$-module schemes.

The resulting $\mathcal{O}$-divisible local group is denoted by $\mathcal{F}_X$.

PROPOSITION 1. *The constructions $X \mapsto \mathcal{F}_X$ and $\mathcal{F} \mapsto X_{\mathcal{F}}$ define mutual inverses between the set of local divisible $\mathcal{O}$-modules $\mathcal{F}$ of rank h and the set of formal $\mathcal{O}$-modules $X = (\Phi, [\cdot]_X)$ such that $x \pmod{\mathfrak{m}_B}$ has height h.*

EXAMPLE 2 ([**4**, before Prop. 5.4]). Let C, ϕ, $\mathcal{O}$, $\widehat{\phi}$ be as in Example 1, and let $g_{\mathfrak{p}}$ be a generator of the ideal $\mathfrak{p}^{n_{\mathfrak{p}}} \subset A$. Let r be the rank of ϕ and h its height. One verifies the following:

(1) For $n \geq 0$, the scheme $\phi[\mathfrak{p}^{nn_{\mathfrak{p}}}] := C[x]/(\phi_{g_{\mathfrak{p}}}^n(x))$ is finite flat over C and possess an $A_{\mathfrak{p}}$-module structure via ϕ.

[2] We restrict to $\mathcal{O}$-modules of dimension 1 and therefore suppress the dimension in the definition.

(2) The sequence $(\phi[\mathfrak{p}^n])_n$ with $\phi[\mathfrak{p}^n] \hookrightarrow \phi[\mathfrak{p}^{n+1}]$ given by inclusion defines a divisible $A_\mathfrak{p}$-module $\phi[\mathfrak{p}^\infty]$ over C of height r.
(3) One has an isomorphism $\mathcal{F}^{\mathrm{loc}} \cong (\widehat{\phi}_\mathfrak{p}[\mathfrak{p}^n])_{n\geq 1}$.
(4) The rank h of $\mathcal{F}^{\mathrm{loc}}$ is the height of $\phi \pmod{\mathfrak{m}_C}$ over $\breve{k}$, and one has $\mathcal{F}^{\text{ét}} \cong (F_\mathfrak{p}/A_\mathfrak{p})^{r-h}$.

Let $\mathcal{F}_{\breve{k}}$ be an $\mathcal{O}$-divisible module of rank r over $\breve{k}$. There is an obvious notion of a deformation $\mathcal{F}_{\breve{k}}$ to $\mathcal{O}$-divisible modules over rings C in $\mathrm{CNL}_{\breve{\mathcal{O}}}$ and this defines a functor $\mathrm{CNL}_{\breve{\mathcal{O}}} \longrightarrow$ **Sets**.

THEOREM 3 ([**4**, Prop. 4.5]). *Suppose that $\mathcal{F}^{\mathrm{loc}}_{\breve{k}}$ has height $h > 0$. Then the functor $\mathrm{CNL}_{\breve{\mathcal{O}}} \longrightarrow$ **Sets** that associates to $C \in \mathrm{CNL}_{\breve{\mathcal{O}}}$ the set of deformations of $\mathcal{F}_{\breve{k}}$ to C up to isomorphism is representable by some ring $R_{\mathcal{F}_{\breve{k}}}$ in $\mathrm{CNL}_{\breve{\mathcal{O}}}$. The universal ring $R_{\mathcal{F}_{\breve{k}}}$ is a power series ring over $\breve{\mathcal{O}}$ in $r-1$ indeterminates.*

From here on, we let $\mathcal{O} := A_\mathfrak{p}$ with $\mathfrak{p}$ a closed point of $\operatorname{Spec} A$ as in the introduction. We let $\phi_0 \colon A \longrightarrow \breve{k}\{\tau\}$ be a Drinfeld-module of rank r whose characteristic is given by $A \longrightarrow A_\mathfrak{p} = \mathcal{O} \longrightarrow \breve{\mathcal{O}} \longrightarrow \breve{k}$, for our chosen $\breve{\mathcal{O}}$. Let $I \subset A$ be a proper non-zero ideal with $I + \mathfrak{p} = A$. Choosing a level I-structure for ϕ_0, which can be done over $\breve{k}$, defines a point of $\mathfrak{M}^r_{I,\mathfrak{p}}(\breve{k})$ which we denote by x. Then x defines a Drinfeld A-module ϕ_x that is isomorphic to ϕ_0 (over $\breve{k}$) together with a level I-structure. A *deformation of ϕ_0 to $C \in \mathrm{CNL}_{\breve{\mathcal{O}}}$* is a Drinfeld A-module $\phi \colon A \longrightarrow C\{\tau\}$, in standard form, up to isomorphism, which reduces modulo $\mathfrak{m}_C$ to ϕ_0. By Hensel's Lemma, the level I-structure on ϕ_0 extends uniquely to a level I-structure of ϕ over C. Hence one can identify deformations of ϕ_0 with morphisms $\operatorname{Spec} C \longrightarrow \mathfrak{M}^r_I$ that when composed with $\operatorname{Spec} \breve{k} \longrightarrow \operatorname{Spec} C$ yield x. The following is Drinfeld's analog of the Serre-Tate theorem for Drinfeld A-modules.

THEOREM 4 ([**4**, 5.C, in part. Prop. 5.4]). *The following holds*

(1) *The functor $\mathrm{CNL}_{\breve{\mathcal{O}}} \longrightarrow$ **Sets** which associates to $C \in \mathrm{CNL}_{\breve{\mathcal{O}}}$ the set of deformations of ϕ_0 to C is representable by the completion of the stalk of $\mathcal{O}_{\mathfrak{M}^r_I \otimes_{A[1/I]} \breve{\mathcal{O}}}$ at x; in particular, this completion is independent of the choice of I.*
(2) *The natural transformation from deformations of ϕ_0 to deformations of the $\mathcal{O}$-divisible group $\phi_0[\mathfrak{p}^\infty]$ defined in Example 2, is an isomorphism. I.e., there is a natural isomorphism of $\breve{\mathcal{O}}$-algebras*

$$R_{\phi_0[\mathfrak{p}^\infty]} \longrightarrow \widehat{\mathcal{O}}_{\mathfrak{M}^r_I \otimes A[1/I] \breve{\mathcal{O}}, x}.$$

3. The result of Strauch

Let $K, \mathcal{O}, k, \breve{K}, \breve{\mathcal{O}}, \breve{k}$ and $\mathrm{CNL}_{\breve{\mathcal{O}}}$ be as in the previous section. Let $\overline{X}$ be a formal group over k of height h and let $R_{\overline{X}}$ and $X_{\overline{X}}$ be as in Theorem 2. The following is from [**9**, § 1,2]. First one may choose an identification $R_{\overline{X}} \cong \breve{\mathcal{O}}[[u_1, \dots, u_{h-1}]]$ such that the multiplication by ϖ_K on $X_{\overline{X}}$ is given by a power series $[\varpi_K]_{X_{\overline{X}}} \in R_{\overline{X}}[[x]]$ with the property that for all $i = 0, \dots, h$ one has

$$[\varpi_K]_{X_{\overline{X}}} \equiv u_i x^{q_K^i} \pmod{(u_0, \dots, u_{i-1}, x^{q_K^i+1})}, \tag{1}$$

with the conventions $u_0 = \varpi_K$ and $u_h = 1$.

For $m \in \{0, \ldots, h-1\}$ put

$$R_m := \breve{\mathcal{O}}[[u_1, \ldots, u_{h-1}]]/(u_0, \ldots, u_m)$$

with the convention that $R_0 = R_{X_{\overline{X}}}$. Then the closed reduced subscheme of $\operatorname{Spec} R_0$ where the height of the connected component of $X_{\overline{X}}[\varpi_K^\infty]$ is at least m is equal to $\operatorname{Spec} R_m$, and the open part of $\operatorname{Spec} R_m$ where the height of the connected component is equal to m is

$$U_m := \operatorname{Spec} R_m \setminus V(u_m).$$

Let κ_m be the field of fractions of R_m and put $\eta_m = \operatorname{Spec} \kappa_m$. Let $\overline{\kappa}_m$ be an algebraic closure of κ_m and put $\overline{\eta}_m = \operatorname{Spec} \overline{\kappa}_m$. Fix a positive integer n. Denote by

$$\mathrm{Ta}_{X_{\overline{X}}, \eta_m} := \varprojlim_n X_{\overline{X}}[\varpi_K^n]_{\eta_m}(\overline{\kappa}_m)$$

the Tate-module of $X_{\overline{X}}$ at η_m. It is a free $\mathcal{O}$-module of rank $h - m$. The absolute Galois group $\pi_1(\eta_m, \overline{\eta}_m)$ of κ_m acts $\mathcal{O}$-linearly on it. We denote the resulting representation by

$$\rho_{X_{\overline{X}}, m} \colon \pi_1(\eta_m, \overline{\eta}_m) \longrightarrow \mathrm{Aut}_{\mathcal{O}}(\mathrm{Ta}_{X_{\overline{X}}, \eta_m}) \cong \mathrm{GL}_{h-m}(\mathcal{O}).$$

It clearly factors via $\pi_1(U_m, \overline{\eta}_m)$. Then [**9**, Thm. 2.1], asserts:

THEOREM 5. *For any $m \in \{0, \ldots, h-1\}$ the homomorphism $\rho_{X_{\overline{X}}, m}$ is surjective.*

4. Proof of Theorem 1

Let the notation be as in Section 1. Set in addition $\mathcal{O} = A_{\mathfrak{p}}$, $K = \operatorname{Frac} \mathcal{O}$, $k = A/\mathfrak{p}$ and take $\breve{K}, \breve{\mathcal{O}}, \breve{k}$ as in Section 2. Let $\xi_{\mathfrak{c}} \in \mathfrak{M}^r_{I,\mathfrak{p}}(\breve{k})$ be a supersingular point in the component of $\mathfrak{M}^r_I$ labelled by $\mathfrak{c}$.[3] Consider the following canonical morphisms of schemes

$$\operatorname{Spec} \widehat{\mathcal{O}}_{\mathfrak{M}^r_{I,\overline{\mathfrak{p}}}, \xi_{\mathfrak{c}}} \longrightarrow \operatorname{Spec} \mathcal{O}_{\mathfrak{M}^r_{I,\overline{\mathfrak{p}}}, \xi_{\mathfrak{c}}} \stackrel{\iota}{\longrightarrow} \mathfrak{M}^r_{I,\overline{\mathfrak{p}}}, \qquad (\text{composite } \widehat{\iota})$$

with $\widehat{\mathcal{O}}_{\mathfrak{M}^r_{I,\overline{\mathfrak{p}}}, \xi_{\mathfrak{c}}}$ the completion of the local ring $\mathcal{O}_{\mathfrak{M}^r_{I,\overline{\mathfrak{p}}}, \xi_{\mathfrak{c}}}$. Denote by $\widehat{\eta}_{\mathfrak{c}}$ the generic point of $\operatorname{Spec} \widehat{\mathcal{O}}_{\mathfrak{M}^r_{I,\overline{\mathfrak{p}}}, \xi_{\mathfrak{c}}}$ and choose a minimal geometric point $\overline{\widehat{\eta}}_{\mathfrak{c}}$ over $\widehat{\eta}_{\mathfrak{c}}$ together with a map $\overline{\widehat{\eta}}_{\mathfrak{c}} \longrightarrow \overline{\eta}_{\mathfrak{c}}$. Let $\mathfrak{M}^{r,\mathrm{ord}}_{I,\overline{\mathfrak{p}}} \subset \mathfrak{M}^r_{I,\overline{\mathfrak{p}}}$ be the locus of ordinary Drinfeld A-modules. It is an open subscheme since its complement is defined by the vanishing of the coefficient of $(\phi^r_{I,\overline{\mathfrak{p}}})_{g_{\mathfrak{p}}} \in M^r_{I,\overline{\mathfrak{p}}}[\tau]$ in degree $\deg g_{\mathfrak{p}}$, where $M^r_{I,\overline{\mathfrak{p}}}$ is the coordinate ring of the affine scheme $\mathfrak{M}^r_{I,\overline{\mathfrak{p}}}$, and hence its complement is closed in $\mathfrak{M}^r_{I,\overline{\mathfrak{p}}}$. We obtain a corresponding diagram of fundamental groups with continuous group homomorphisms

$$\text{(2)} \qquad \begin{array}{ccccc} \pi_1(\widehat{\iota}^{-1}(\mathfrak{M}^{r,\mathrm{ord}}_{I,\overline{\mathfrak{p}}}), \overline{\widehat{\eta}}_{\mathfrak{c}}) & \longrightarrow & \pi_1(\iota^{-1}(\mathfrak{M}^{r,\mathrm{ord}}_{I,\overline{\mathfrak{p}}}), \overline{\eta}_{\mathfrak{c}}) & \longrightarrow & \pi_1(\mathfrak{M}^{r,\mathrm{ord}}_{I,\overline{\mathfrak{p}}}, \overline{\eta}_{\mathfrak{c}}) \\ \uparrow & & \uparrow & \nearrow & \\ \pi_1(\widehat{\eta}_{\mathfrak{c}}, \overline{\widehat{\eta}}_{\mathfrak{c}}) & \longrightarrow & \pi_1(\eta_{\mathfrak{c}}, \overline{\eta}_{\mathfrak{c}}). & & \end{array}$$

[3]Supersingular points exist; and via the action of Hecke correspondences, which preserves the supersingular locus, they can be seen to lie in every component.

Over $\mathfrak{M}^{r,\mathrm{ord}}_{I,\overline{\mathfrak{p}}}$ the Tate-module

$$\mathrm{Ta}_{\mathfrak{p}}\,\phi^r_{\eta_{\mathfrak{c}}} = \varprojlim_n \phi^r_{\eta_{\mathfrak{c}}}[\mathfrak{p}^{nn_{\mathfrak{p}}}](\overline{\kappa}_{\eta_{\mathfrak{c}}}),$$

is free over $\mathcal{O}$ of rank $r-1$, and we have continuous homomorphisms

$$\pi_1(\widehat{\eta}_{\mathfrak{c}}, \overline{\widehat{\eta}}_{\mathfrak{c}}) \longrightarrow \pi_1(\eta_{\mathfrak{c}}, \overline{\eta}_{\mathfrak{c}}) \longrightarrow \pi_1(\mathfrak{M}^{r,\mathrm{ord}}_{I,\overline{\mathfrak{p}}}, \overline{\eta}_{\mathfrak{c}}) \longrightarrow \mathrm{Aut}_{\mathcal{O}}(\mathrm{Ta}_{\mathfrak{p}}\,\phi^r_{\eta_{\mathfrak{c}}}). \tag{3}$$

By Theorem 4, the ring $\mathrm{Spec}\,\widehat{\mathcal{O}}_{\mathfrak{M}^r_{I,\overline{\mathfrak{p}}},\xi_{\mathfrak{c}}}$ is naturally identified with the special fiber of the universal deformation ring of the $\mathcal{O}$-divisible module $\phi^r_{\xi_{\mathfrak{c}}}[\mathfrak{p}^\infty]$. Because $\xi_{\mathfrak{c}}$ is supersingular, it is a local $\mathcal{O}$-divisible module of rank r, and thus by Proposition 1 it arises from a formal $\mathcal{O}$-module of height r. Now by Theorem 5 of Strauch, the composition of the maps in (3) is surjective.

We have thus proved the following result.

THEOREM 6. *The monodromy representation* $\pi_1(\widehat{\eta}_{\mathfrak{c}}, \overline{\widehat{\eta}}_{\mathfrak{c}}) \longrightarrow \mathrm{Aut}_{\mathcal{O}}(\mathrm{Ta}_{\mathfrak{p}}\,\phi^r_{\eta_{\mathfrak{c}}})$ *is surjective.*

Hence the map $\pi_1(\eta_{\mathfrak{c}}, \overline{\eta}_{\mathfrak{c}}) \longrightarrow \mathrm{Aut}_{\mathcal{O}}(\mathrm{Ta}_{\mathfrak{p}}\,\phi^r_{\eta_{\mathfrak{c}}})$ is surjective, as well. This completes the proof of Theorem 1.

REMARK 1. One can think of the image of $\pi_1(\widehat{\eta}_{\mathfrak{c}}, \overline{\widehat{\eta}}_{\mathfrak{c}})$ in $\pi_1(\mathfrak{M}^{r,\mathrm{ord}}_{I,\overline{\mathfrak{p}}}, \overline{\eta}_{\mathfrak{c}})$ as the decomposition group at $\xi_{\mathfrak{c}}$. From this viewpoint, Theorem 5 says that already the image of this decomposition group surjects onto $\mathrm{Aut}_{\mathcal{O}}(\mathrm{Ta}_{\mathfrak{p}}\,\phi^r_{\eta_{\mathfrak{c}}}) \cong \mathrm{GL}_{r-1}(\mathcal{O})$. According to the same theorem, the decomposition groups at points of height $m<r$ map onto a natural subgroup of $\mathrm{Aut}_{\mathcal{O}}(\mathrm{Ta}_{\mathfrak{p}}\,\phi^r_{\eta_{\mathfrak{c}}})$ isomorphic to $\mathrm{GL}_{m-1}(\mathcal{O})$.

5. An example

For the remainder of this article we specialize to the case $A = \mathbb{F}_q[t]$ and level $I = tA$. We also set $B = A[\frac{1}{t}] = \mathbb{F}_q[t, \frac{1}{t}]$, we let $\mathfrak{p} \in A$ be a non-zero prime (monic irreducible polynomial) and suppose $\mathfrak{p} \neq t$ (otherwise, just replace t by $t+1$), and we write $|\mathfrak{p}| = q^{\deg(\mathfrak{p})}$. Let $\kappa_{\mathfrak{p}} = A/\mathfrak{p}$ with algebraic closure $\bar{\kappa}_{\mathfrak{p}}$. As a preparation for Section 6, in the present section we will work out an explicit example of the main result.

We start by recalling Pink's explicit description of $\mathfrak{M}^r_t$ [**6, 7**], see also [**1**, Theorem 2] for more details: Let V be an $\mathbb{F}_q$-vector space of dimension $r \geq 1$ and write $V' = V \smallsetminus \{0\}$. Denote by $S_V = \mathrm{Sym}_B(V)$ the symmetric algebra of V over B and by K_V the fraction field of S_V. Denote by $RS_{V,0} = B[\frac{v}{v'} \mid v, v \in V']$ the subalgebra of K_V generated over B by quotients of non-zero elements of V. Then the base-change of $\mathfrak{M}^r_t$ to $\mathrm{Spec}\,B$ is given by

$$\mathfrak{M}^r_{t,B} = \mathrm{Spec}\,RS_{V,0},$$

which has geometrically irreducible fibres. Furthermore, for any fixed $v_1 \in V'$, the universal Drinfeld module $\phi = \phi^r_\eta$ on $\mathfrak{M}^r_{t,B}$ is determined by

$$t \mapsto \phi_t(X) = tX \prod_{v \in V'} \left(1 - \frac{v_1}{v}X\right) \in RS_{V,0}[X]$$

with level structure

$$V \xrightarrow{\sim} \phi[t]; \quad v \mapsto \frac{v}{v_1}.$$

The base change of the moduli scheme $\mathfrak{M}^r_{t,B}$ to $\overline{\kappa}_\mathfrak{p}$ is $\mathfrak{M}^r_{t,\overline{\mathfrak{p}}} = \operatorname{Spec}\left(RS_{V,0}\otimes_B\overline{\kappa}_\mathfrak{p}[X]\right)$, with universal Drinfeld module the reduction of ϕ modulo $\mathfrak{p}$, which we denote $\overline{\phi}$. We have

$$\overline{\phi}_{\mathfrak{p}^n}(X) = \overline{\phi}^{\text{ét}}_{\mathfrak{p}^n}(X^{|\mathfrak{p}|^n}),$$

where $\overline{\phi}^{\text{ét}}_{\mathfrak{p}^n}(X) \in RS_{V,0}\otimes_B \overline{\kappa}_\mathfrak{p}[X]$ is a separable $\mathbb{F}_q$-linear polynomial of degree $|\mathfrak{p}|^{n(r-1)}$. The outer terms in the local-étale decomposition

$$0 \longrightarrow \overline{\phi}[\mathfrak{p}^n]^{\text{loc}} \longrightarrow \overline{\phi}[\mathfrak{p}^n] \longrightarrow \overline{\phi}[\mathfrak{p}^n]^{\text{ét}} \longrightarrow 0$$

are given by

$$\overline{\phi}[\mathfrak{p}^n]^{\text{loc}} = \operatorname{Spec}\left(RS_{V,0}\otimes_B \overline{\kappa}_\mathfrak{p}[X]/\langle X^{|\mathfrak{p}|^n}\rangle\right)$$

and

$$\overline{\phi}[\mathfrak{p}^n]^{\text{ét}} = \operatorname{Spec}\left(RS_{V,0}\otimes_B \overline{\kappa}_\mathfrak{p}[X]/\langle \overline{\phi}^{\text{ét}}_{\mathfrak{p}^n}(X)\rangle\right).$$

Denote by $\overline{\kappa}_\eta$ the fraction field of $RS_{V,0}\otimes_B \overline{\kappa}_\mathfrak{p}$, which is the function field of $\mathfrak{M}^r_{t,\overline{\mathfrak{p}}}$ over $\overline{\kappa}_\mathfrak{p}$, and by $\overline{\kappa}_\eta(\overline{\phi}[\mathfrak{p}^n]^{\text{ét}})$ the splitting field of $\overline{\phi}^{\text{ét}}_{\mathfrak{p}^n}(X)$ over $\overline{\kappa}_\eta$.

Now Theorem 1 says the following: For every positive integer n, we have

$$\operatorname{Gal}\left(\overline{\kappa}_\eta(\overline{\phi}[\mathfrak{p}^n]^{\text{ét}})/\overline{\kappa}_\eta\right) \cong \mathrm{GL}_{r-1}(A/\mathfrak{p}^n). \tag{4}$$

6. An application

In this last section, we consider a variant of the above example and answer a question posed in [**2**].

Suppose $g_1, g_2, \dots, g_{r-1}$ are algebraically independent over $\mathbb{F}_q(t)$ and set $L = \mathbb{F}_q(t, g_1, \dots, g_{r-1})$, a rational function field of transcendence degree r over $\mathbb{F}_q$.

We define the Drinfeld module $\psi : A \longrightarrow L\{\tau\}$ by

$$t \mapsto \psi_t(X) = tX + g_1X^q + \cdots + g_{r-1}X^{q^{r-1}} + X^{q^r} \in L[X].$$

It is shown in [**1**, Thm. 6] that, for every non-zero proper ideal $\mathfrak{n} \subset A$,

$$\operatorname{Gal}(L(\psi[\mathfrak{n}])/L) \simeq \mathrm{GL}_r(A/\mathfrak{n}).$$

Our goal is to prove a similar result in special characteristic.

Denote by $L_t = L(\psi[t])$ the splitting field of $\psi_t(X)$ over L, and set

$$RS_t = B\big[v, \frac{1}{v} \mid 0 \neq v \in \psi[t]\big] \subset L_t,$$

the subalgebra of L_t generated over B by v and $\frac{1}{v}$ for $0 \neq v \in \psi[t]$. It is a graded ring if we set $\deg(v) = 1$ for all $0 \neq v \in \psi[t]$. We have $\psi_t(X) \in RS_t[X]$.

Fix a non-zero t-torsion point $0 \neq v_1 \in \psi[t]$, and consider the isomorphic Drinfeld module $\phi = v_1^{-1}\psi v_1$ over L_t. We denote by $L_{t,0} = L(\phi[t]) \subset L_t$ the splitting field of $\phi_t(X)$ over L, and set

$$RS_{t,0} = B\big[\frac{v}{v'} \mid v, v' \in \psi[t],\ v' \neq 0\big] \subset L_{t,0}.$$

This is the degree zero component of RS_t.

We have $\phi_t(X) = tX\prod_{0\neq v\in\psi[t]}\left(1 - \frac{v_1}{v}X\right) \in RS_{t,0}[X]$.

By [**1**, Thm. 5] and its proof, there is an isomorphism

$$\theta : \operatorname{Spec}(RS_{t,0}) \xrightarrow{\sim} \mathfrak{M}^r_{t,B}, \tag{5}$$

and ϕ is the pullback via θ of the universal Drinfeld module described in Section 5.

Now consider the reduced Drinfeld modules $\overline{\psi}$ and $\overline{\phi}$ over $RS_t \otimes_B \kappa_{\mathfrak{p}}$ and $RS_{t,0} \otimes_B \kappa_{\mathfrak{p}}$, respectively.

Again, for every positive integer n, we have $\overline{\psi}_{\mathfrak{p}^n}(X) = \overline{\psi}^{\text{ét}}_{\mathfrak{p}^n}(X^{|\mathfrak{p}|^n})$, where $\overline{\psi}^{\text{ét}}_{\mathfrak{p}^n}(X) \in RS_t \otimes_B \overline{\kappa}_{\mathfrak{p}}[X]$ is a separable $\mathbb{F}_q$-linear polynomial of degree $|\mathfrak{p}|^{n(r-1)}$, and

$$\overline{\psi}[\mathfrak{p}^n]^{\text{ét}} = \operatorname{Spec}\left(RS_t \otimes_B \overline{\kappa}_{\mathfrak{p}}[X]/\langle \overline{\psi}^{\text{ét}}_{\mathfrak{p}^n}(X)\rangle\right).$$

Analoguous statements hold for $\overline{\phi}$.

We define the A-field $\ell = \overline{\kappa}_{\mathfrak{p}}(g_1, g_2, \ldots, g_{r-1})$, equipped with the homomorphism $A \longrightarrow A/\mathfrak{p} \subset \overline{\kappa}_{\mathfrak{p}} \subset \ell$. The Drinfeld module $\overline{\psi}$ is defined over ℓ. Denote by $\ell(\overline{\psi}[\mathfrak{p}^n]^{\text{ét}})$ the splitting field of $\overline{\psi}^{\text{ét}}_{\mathfrak{p}^n}(X)$ over ℓ. We have

THEOREM 7. $\operatorname{Gal}\left(\ell(\overline{\psi}[\mathfrak{p}^n]^{\text{ét}})/\ell\right) \cong \mathrm{GL}_{r-1}(A/\mathfrak{p}^n)$.

PROOF. We define the following fields.

$$\begin{aligned} \ell_{t,0} &= \overline{\kappa}_{\mathfrak{p}}(\overline{\phi}[t]) = \operatorname{Frac}\left(RS_{t,0} \otimes_B \overline{\kappa}_{\mathfrak{p}}\right), \\ \ell_t &= \ell(\overline{\psi}[t]) = \ell_{t,0}(v_1) = \operatorname{Frac}\left(RS_t \otimes_B \overline{\kappa}_{\mathfrak{p}}\right), \end{aligned}$$

and consider the following field extensions.

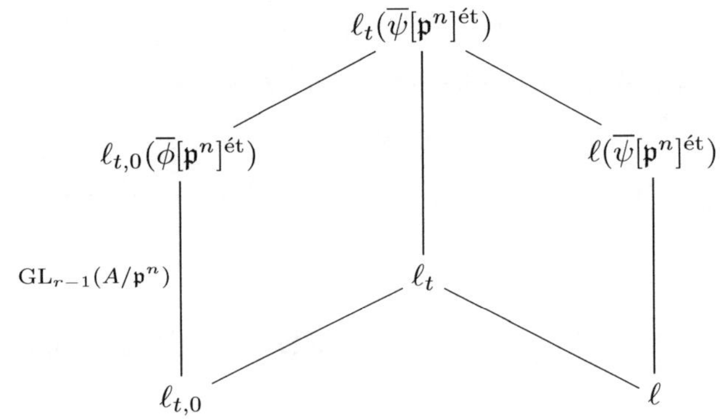

The field $\ell_{t,0}$ is (isomorphic via θ to) the function field of $\mathfrak{M}^r_{t,\overline{\mathfrak{p}}} = \mathfrak{M}^r_{t,B} \times_{\operatorname{Spec} B} \operatorname{Spec} \overline{\kappa}_{\mathfrak{p}}$ over $\overline{\kappa}_{\mathfrak{p}}$. It follows from (4) that $\operatorname{Gal}\left(\ell_{t,0}(\overline{\phi}[\mathfrak{p}^n]^{\text{ét}})/\ell_{t,0}\right) \cong \mathrm{GL}_{r-1}(A/\mathfrak{p}^n)$, and our goal is to show that the other two vertical extensions have this same Galois group.

We write

$$\overline{\phi}_t(X) = \bar{t}X + c_1 X^q + \cdots + c_r X^{q^r} \in RS_{t,0} \otimes_B \overline{\kappa}_{\mathfrak{p}}[X],$$

where $\bar{t}$ denotes the image of t in $A \longrightarrow A/\mathfrak{p} \subset \overline{\kappa}_{\mathfrak{p}}$, $c_i = v_1^{q^i-1} g_i$ for $i = 1, \ldots, r-1$ and $c_r = v_1^{q^r-1}$. Because 1 is a t-torsion point, we have the algebraic relation $0 = \overline{\phi}_t(1) = \bar{t} + c_1 + c_2 + \cdots + c_r$. Observe also that $\overline{\kappa}_{\mathfrak{p}} \subset \ell_{t,0}$ contains roots of unity of all orders prime to p. It follows that $\ell_t = \ell_{t,0}(v_1) = \ell_{t,0}(\sqrt[q^r-1]{c_r})$ is a Kummer extension of $\ell_{t,0}$.

Let $\xi \in \mathfrak{M}^r_{t,\overline{\mathfrak{p}}}(\overline{\kappa}_{\mathfrak{p}})$ correspond to a supersingular Drinfeld module

$$\phi^{\xi}_t(X) = \bar{t}X + s_1 X^q + \cdots + s_r X^{q^r} \in \overline{\kappa}_{\mathfrak{p}}[X].$$

Then the completion $\widehat{\ell_{t,0}}$ of $\ell_{t,0}$ at ξ contains the ring of formal power series

$$\overline{\kappa}_{\mathfrak{p}}[[c_2 - s_2, \cdots, c_r - s_r]].$$

This ring, in turn, contains

$$v_1 = \sqrt[q^r-1]{c_r} = \sqrt[q^r-1]{(c_r - s_r) + s_r} = \sqrt[q^r-1]{s_r} \sum_{i=0}^{\infty} \binom{\frac{1}{q^r-1}}{i} \left(\frac{c_r - s_r}{s_r}\right)^i$$

since $s_r \in \overline{\kappa}_{\mathfrak{p}}^{\times}$ and $q^r - 1$ is not divisible by the characteristic p.

This implies that $\overline{\psi}$ and $\overline{\phi}$ are isomorphic over $\widehat{\ell_{t,0}}$. Also, $\widehat{\ell_{t,0}}$ contains ℓ_t and $\widehat{\ell_{t,0}}(\overline{\phi}[\mathfrak{p}^n]^{\text{ét}}) = \widehat{\ell_{t,0}}(\overline{\psi}[\mathfrak{p}^n]^{\text{ét}})$.

Theorem 6 implies that the Galois representation

$$\mathrm{Gal}\left(\widehat{\ell_{t,0}}^{\,\mathrm{sep}}/\widehat{\ell_{t,0}}\right) \longrightarrow \mathrm{Ta}_{\mathfrak{p}}\, \overline{\phi}$$

is surjective. Since $\overline{\phi}$ and $\overline{\psi}$ are isomorphic over $\widehat{\ell_{t,0}}$, the same holds for $\mathrm{Ta}_{\mathfrak{p}}\, \overline{\psi}$, and in particular the Galois representation

$$\mathrm{Gal}\left(\ell_t^{\mathrm{sep}}/\ell_t\right) \longrightarrow \mathrm{Ta}_{\mathfrak{p}}\, \overline{\psi}$$

is surjective. This implies that

$$\mathrm{Gal}\left(\ell_t(\overline{\psi}[\mathfrak{p}^n]^{\text{ét}})/\ell_t\right) \cong \mathrm{GL}_{r-1}(A/\mathfrak{p}^n).$$

Finally, we have

$$[\ell(\overline{\psi}[\mathfrak{p}^n]^{\text{ét}}) : \ell] \geq [\ell_t(\overline{\psi}[\mathfrak{p}^n]^{\text{ét}}) : \ell_t] = \#\,\mathrm{GL}_{r-1}(A/\mathfrak{p}^n).$$

Since $\mathrm{Gal}\left(\ell(\overline{\psi}[\mathfrak{p}^n]^{\text{ét}})/\ell\right)$ is isomorphic to a subgroup of $\mathrm{GL}_{r-1}(A/\mathfrak{p}^n)$, it must be isomorphic to the whole group. This completes the proof of Theorem 7. □

Finally, we address [**2**, Question 4.5]. For this, we must first recall the construction of Drinfeld modular polynomials from [**2**].

Denote by C the subring of $A[g_1, \ldots, g_{r-1}] \subset L$ generated by monomials of the form $a g_1^{e_1} \cdots g_{r-1}^{e_{r-1}}$ satisfying $a \in A$ and $\sum_{k=1}^{r-1} e_k(q^k - 1) \equiv 0 \bmod q^r - 1$. Then the elements of C are the isomorphism invariants of rank r Drinfeld A-modules, i.e. $\mathrm{Spec}\, C$ is the coarse moduli scheme of Drinfeld modules of rank r and no level structure, see [**2**, Prop. 1.1].

Let $1 \leq s \leq r-1$. An isogeny $f : \psi \longrightarrow \psi^{(f)}$ is said to have *type* $(A/\mathfrak{p})^s$ if $\ker f(\overline{L}) \cong (A/\mathfrak{p})^s$, and such an isogeny is called *special* if $\ker f$ contains $U_0 := \ker\left(\psi[\mathfrak{p}](\overline{L}) \longrightarrow \overline{\psi}[\mathfrak{p}](\overline{\ell})\right)$. Because ψ has ordinary reduction $\overline{\psi}$, and so $U_0 \cong A/\mathfrak{p}$, f is special if and only if its reduction is inseparable.

To each invariant $J \in C$ we associate the *Drinfeld modular polynomial of type* $(A/\mathfrak{p})^s$, defined by

$$\Phi_{J,(A/\mathfrak{p})^s}(X) = \prod_{f : \psi \longrightarrow \psi^{(f)} \text{ of type } (A/\mathfrak{p})^s} (X - J(\psi^{(f)})) \in C[X].$$

This is irreducible over L if its roots in $\overline{L}$ are distinct (there always exist $J \in C$ for which the roots are distinct).

Modulo $\mathfrak{p}$, we have the Kronecker congruence relation [**2**, Thm. 4.4]:

$$\Phi_{J,(A/\mathfrak{p})^s}(X) \equiv \Phi^{\mathrm{spec}}_{J,(A/\mathfrak{p})^s}(X) \cdot \left(\Phi^{\mathrm{spec}}_{J,(A/\mathfrak{p})^{s+1}}(X^{|\mathfrak{p}|})\right)^{|\mathfrak{p}|^{s-1}} \bmod \mathfrak{p}, \tag{6}$$

where

$$\Phi^{\mathrm{spec}}_{J,(A/\mathfrak{p})^s}(X) := \prod_{f : \psi \longrightarrow \psi^{(f)} \text{ special of type } (A/\mathfrak{p})^s} \left(X - (J(\psi^{(f)}) \bmod \mathfrak{p})\right) \in \ell[X].$$

We answer [**2**, Question 4.5] in the affirmative, as follows.

PROPOSITION 2. *Suppose $J \in C$ is such that the roots of $\Phi^{\mathrm{spec}}_{J,(A/\mathfrak{p})^s}(X)$ in $\overline{\ell}$ are distinct. Then $\Phi^{\mathrm{spec}}_{J,(A/\mathfrak{p})^s}(X) \in \ell[X]$ is irreducible.*

PROOF. If $s = 1$, then $\Phi^{\mathrm{spec}}_{J,(A/\mathfrak{p})}(X) = X - J^{|\mathfrak{p}|}$ by [**2**, Example 5.1], and we are done.

Now suppose that $s > 1$. Let R be the integral closure of $A[g_1, g_2, \dots, g_{r-1}]$ in $L(\psi[\mathfrak{p}])$, then $\psi[\mathfrak{p}] \subset R$. Let $f : \phi \longrightarrow \phi^{(f)}$ be a special isogeny of type $(A/\mathfrak{p})^s$. Then $f(X) \in R[X]$ is an $\mathbb{F}_q$-linear polynomial, and $f(X) \equiv f^{\text{ét}}(X^{|\mathfrak{p}|}) \bmod \mathfrak{p}$, where $f^{\text{ét}}(X) \in R \otimes_A \kappa_{\mathfrak{p}}[X]$ is separable and $\ker f^{\text{ét}}$ is an A-submodule of $\overline{\psi}[\mathfrak{p}]^{\text{ét}}$ isomorphic to $(A/\mathfrak{p})^{s-1}$.

By Theorem 7 $\mathrm{Gal}(\ell^{\mathrm{sep}}/\ell)$ acts transitively on the set of such submodules of $\overline{\psi}[\mathfrak{p}]^{\text{ét}}$, and thus also on the set of Drinfeld modules $\overline{\psi}^{(f)}$. Because J maps different $\overline{\psi}^{(f)}$ to different elements of $\overline{\ell}$, the group $\mathrm{Gal}(\ell^{\mathrm{sep}}/\ell)$, in turn, acts transitively on the roots of $\Phi^{\mathrm{spec}}_{J,(A/\mathfrak{p})^s}(X)$. □

REMARK 2. Equation (6) thus describes the decomposition of the Hecke correspondence associated to $(A/\mathfrak{p})^s$-isogenies on $\mathfrak{M}^r_{I,\overline{\mathfrak{p}}}$ into irreducible components with multiplicities.

Acknowledgements

The authors are grateful to Judith Ludwig for helpful discussions and comments, and also thank the referee for a very careful reading of the manuscript. The second author would like to thank the University of Heidelberg for its hospitality and the Alexander-von-Humboldt Foundation for financial support.

References

[1] Florian Breuer, *Explicit Drinfeld moduli schemes and Abhyankar's generalized iteration conjecture*, J. Number Theory **160** (2016), 432–450, DOI 10.1016/j.jnt.2015.08.021. MR3425215

[2] Florian Breuer and Hans-Georg Rück, *Drinfeld modular polynomials in higher rank II: Kronecker congruences*, J. Number Theory **165** (2016), 1–14, DOI 10.1016/j.jnt.2016.01.001. MR3479213

[3] Anna Devic and Richard Pink, *Adelic openness for Drinfeld modules in special characteristic*, J. Number Theory **132** (2012), no. 7, 1583–1625, DOI 10.1016/j.jnt.2012.01.015. MR2903172

[4] V. G. Drinfel′ d, *Elliptic modules* (Russian), Mat. Sb. (N.S.) **94(136)** (1974), 594–627, 656. MR0384707

[5] Mihran Papikian, *The number of rational points on Drinfeld modular varieties over finite fields*, Int. Math. Res. Not., posted on 2006, Art. ID 94356, 36, DOI 10.1155/IMRN/2006/94356. MR2276350

[6] Richard Pink, *Compactification of Drinfeld modular varieties and Drinfeld modular forms of arbitrary rank*, Manuscripta Math. **140** (2013), no. 3-4, 333–361, DOI 10.1007/s00229-012-0544-3. MR3019130

[7] Richard Pink and Simon Schieder, *Compactification of a Drinfeld period domain over a finite field*, J. Algebraic Geom. **23** (2014), no. 2, 201–243, DOI 10.1090/S1056-3911-2013-00605-0. MR3166390

[8] Michael Rosen, *Formal Drinfeld modules*, J. Number Theory **103** (2003), no. 2, 234–256, DOI 10.1016/S0022-314X(03)00111-2. MR2020270

[9] Matthias Strauch, *Galois actions on torsion points of one-dimensional formal modules*, J. Number Theory **130** (2010), no. 3, 528–533, DOI 10.1016/j.jnt.2009.10.004. MR2584836

[10] Yuichiro Taguchi, *Semi-simplicity of the Galois representations attached to Drinfel′ d modules over fields of "infinite characteristics"*, J. Number Theory **44** (1993), no. 3, 292–314, DOI 10.1006/jnth.1993.1055. MR1233291

[11] J. T. Tate, *p-divisible groups*, Proc. Conf. Local Fields (Driebergen, 1966), Springer, Berlin, 1967, pp. 158–183. MR0231827

COMPUTATIONAL ARITHMETIC GEOMETRY, IWR, UNIVERSITY OF HEIDELBERG, HEIDELBERG, GERMANY

Email address: gebhard.boeckle@iwr.uni-heidelberg.de

SCHOOL OF MATHEMATICAL AND PHYSICAL SCIENCES, UNIVERSITY OF NEWCASTLE, NEWCASTLE, AUSTRALIA

Email address: florian.breuer@newcastle.edu.au

Contemporary Mathematics
Volume **767**, 2021
https://doi.org/10.1090/conm/767/15405

Symmetric tensor products of octonion algebras

Aharon Razon

To Gerhard Frey on the occasion of his 75th birthday

ABSTRACT. Let A be a finite dimensional central simple algebra over a field F of characteristic zero. Consider a positive integer n. The symmetric group $\mathfrak{S}_n$ acts naturally on $A^{\otimes n}$. Let $\operatorname{Sym}^n A$ be the ring of symmetric tensors of $A^{\otimes n}$, i.e. the fixed ring of $A^{\otimes n}$ under the action of $\mathfrak{S}_n$. If A is associative, there is a group homomorphism $\mathfrak{S}_n \to (A^{\otimes n})^\times$ that can be extended to an F-algebra homomorphism $F[\mathfrak{S}_n] \to A^{\otimes n}$. As a consequence, $\operatorname{Sym}^n A$ can be presented as a direct sum of central simple algebras such that the central idempotents of this semisimple algebra are the nonzero images of the central idempotents of $F[\mathfrak{S}_n]$, corresponding to a subset of all partitions of n. In this work we generalize this result to a finite dimensional central simple alternative algebra A. The only strictly non-associative finite dimensional central simple alternative algebras are the octonion algebras. Our result states that the fixed ring of the tensor product of n copies of an octonion algebra, under the action of $\mathfrak{S}_n$, can be presented as a direct sum of central simple (non-associative) algebras. Moreover, we give an explicit description of this presentation by its central idempotents, which correspond to all partitions of n of length 2.

Introduction

Let A be a finite dimensional central simple alternative (see §2) algebra over a field F of characteristic 0. Consider a positive integer n and let $A^{\otimes n}$ be the tensor product of n copies of A over F. The symmetric group $\mathfrak{S}_n$ acts on $A^{\otimes n}$ by $\sigma(a_1 \otimes \cdots \otimes a_n) = a_{\sigma^{-1}(1)} \otimes \cdots \otimes a_{\sigma^{-1}(n)}$ for $\sigma \in \mathfrak{S}_n$ and $a_1, \dots, a_n \in A$. Let

$$\operatorname{Sym}^n A = \{r \in A^{\otimes n} \mid \sigma(r) = r \text{ for all } \sigma \in \mathfrak{S}_n\}.$$

Our main theorem (Theorem 5.6) is:

THEOREM A. *$\operatorname{Sym}^n A$ is a direct sum of central simple algebras over F.*

In §1 we deal with the case that A is associative. In this case Theorem A is well known (e.g. [**Sal99**, Chap. 5]). The case $A = \operatorname{End}(V)$, where V is a finite dimensional complex vector space, is a consequence of Weyl's construction: The symmetric group $\mathfrak{S}_n$ acts on $V^{\otimes n} = V \otimes \cdots \otimes V$ by permuting the factors and it commutes with the action of $\operatorname{GL}(V)$. Any partition λ of n gives rise to an irreducible

---

2020 *Mathematics Subject Classification.* Primary 17A75; Secondary 16W10.
Key words and phrases. Symmetric tensor products, octonion algebra, central simple algebras.

representation V_λ of $\mathfrak{S}_n$ of dimension m_λ. Let $\mathbb{S}_\lambda V$ be its image on $V^{\otimes n}$. Then each $\mathbb{S}_\lambda V$ is an irreducible representation of $\mathrm{GL}(V)$ and

$$V^{\otimes n} \cong \bigoplus_\lambda \mathbb{S}_\lambda V^{\otimes m_\lambda}$$

[**FuH04**, Thm. 6.3] or [**Wey46**, Chapter IV, especially Thm. (4.4. A), p. 127].

The proof of Theorem A in the associative case relies on the fact that there exists a group homomorphism $\theta\colon \mathfrak{S}_n \to (A^{\otimes n})^\times$ such that $\sigma(a) = \theta(\sigma)\cdot a\cdot\theta(\sigma)^{-1}$ for $\sigma \in \mathfrak{S}_n$ and $a \in A^{\otimes n}$ (Proposition 1.2). We extend θ to a homomorphism of F-algebras $\theta\colon F[\mathfrak{S}_n] \to A^{\otimes n}$. Then $\mathrm{Sym}^n A$ is the centralizer of $\theta(F[\mathfrak{S}_n])$ in $A^{\otimes n}$. Moreover, any central element of $F[\mathfrak{S}_n]$ is mapped under θ to a central element of $\mathrm{Sym}^n A$ (Lemma 1.3). The semisimple ring $F[\mathfrak{S}_n]$ can be represented as a direct sum of simple rings:

$$F[\mathfrak{S}_n] = \bigoplus_{\lambda\in\Lambda} F[\mathfrak{S}_n]c_\lambda\,,$$

where c_λ are the central idempotents of $F[\mathfrak{S}_n]$ for λ in the set Λ of all partitions of n. Then

$$\mathrm{Sym}^n A = \bigoplus_{\lambda\in\Lambda} \mathrm{Sym}^n A\cdot\theta(c_\lambda)\,.$$

We prove that if λ is a partition of n such that $\theta(c_\lambda)\neq 0$, then $\mathrm{Sym}^n A\cdot\theta(c_\lambda)$ is a central simple algebra over F (Theorem 1.5). As a corollary we get Theorem A in the case that A is associative (Corollary 1.6).

Suppose now that A is a strictly non-associative finite dimensional central simple alternative algebra over F. By a result of M. Zorn [**Sch43**], A is an octonion algebra $\mathcal{C}$, also called a Cayley algebra or a Cayley-Dickson algebra. After two sections of preparations, we construct in §4 a central idempotent e_λ in the algebra $\mathrm{Sym}^n\mathcal{C}$, for each partition λ of n of length 2, and prove in Corollary 4.7 that $\mathrm{Sym}^n\mathcal{C}$ is a direct sum of $\mathrm{Sym}^n\mathcal{C}\cdot e_\lambda$ over all such partitions. Finally, in §5, we prove that each $\mathrm{Sym}^n\mathcal{C}\cdot e_\lambda$ is a central simple algebra.

The algebra of real octonions $\mathbb{O}$, i.e. the octonion algebra over the real numbers, plays a significant role in the classification and construction of other mathematical entities. For example, the exceptional Lie group G_2 is the automorphism group of $\mathbb{O}$, and the other exceptional Lie groups F_4, E_6, E_7 and E_8 can be understood as the isometries of certain projective planes defined using the real octonions (see §4 in the survey paper by John C. Baez [**Ba02**]). It also relates to Clifford algebras and spinors, Bott periodicity, projective and Lorentzian geometry, and Jordan algebras [**Ba02**, §2 and §3]. The representation theory of some Lie algebras can be used to prove Theorem 5.4 in the case that $\mathcal{C}$ is an octonion algebra over an algebraic closed field of characteristic 0 (see Remark 4.9). In physics, the algebra of real octonions has applications in quantum logic, special relativity and supersymmetry [**Ba02**]. For example, in the 1970s, attempts were made to understand quarks by way of an octonionic Hilbert space ([**GG73**] and [**GG74**]). We hope that this work on the decomposition of symmetric tensors of the octonions as a direct sum of central simple algebras will contribute to deeper understanding of the role of the octonion algebra both in mathematics and physics.

The author is indebted to the anonymous referee for the remarks that make some of the calculations of this paper a little bit less involved.

1. The associative case

Let A be a finite dimensional central simple associative algebra over a field F and let $\mathrm{tr}_A\colon A \to F$ be its reduced trace.

Fix a positive integer n. Denote the tensor product over F of n copies of A by $A^{\otimes n}$. The symmetric group $\mathfrak{S}_n$ acts naturally on $A^{\otimes n}$. The action is defined using linearity by

$$\sigma(a_1 \otimes \cdots \otimes a_n) = a_{\sigma^{-1}(1)} \otimes \cdots \otimes a_{\sigma^{-1}(n)}$$

for $\sigma \in \mathfrak{S}_n$ and $a_1, \ldots, a_n \in A$.

REMARK 1.1. [**Row88**, Prop. 7.2.24]. There is a unique element $\mathfrak{g} = \sum x_i \otimes y_i \in A \otimes A$, called the **Goldman's element**, such that $\mathrm{tr}_A(a) = \sum x_i a y_i$ for all a in A. It satisfies $\mathfrak{g}^2 = 1 \otimes 1$ and $\mathfrak{g}(a \otimes b) = (b \otimes a)\mathfrak{g}$ for all $a, b \in A$. That is, the element $\mathfrak{g}$ plays the role of the transposition $(1\,2)$ in $\mathfrak{S}_2$ that acts on $A \otimes A$.

Applying Remark 1.1 to higher tensor products, we get Goldman like elements for $n \geq 2$:

PROPOSITION 1.2 ([**Row88**, Cor. 7.2.25] or [**Sal99**, Thm. 5.2].). *There exists a group homomorphism θ from $\mathfrak{S}_n$ into the multiplicative group of $A^{\otimes n}$ such that*

$$\sigma(a) = \theta(\sigma) \cdot a \cdot \theta(\sigma)^{-1} \quad \text{for } \sigma \in \mathfrak{S}_n \text{ and } a \in A^{\otimes n}. \tag{1.1}$$

We extend θ to a homomorphism of the group algebra $F[\mathfrak{S}_n]$ into $A^{\otimes n}$. Let $\mathrm{Sym}^n A$ be the ring of symmetric tensors of $A^{\otimes n}$, i.e. the fixed ring of $A^{\otimes n}$ under the action of $\mathfrak{S}_n$.

LEMMA 1.3. *Any central element of $F[\mathfrak{S}_n]$ is mapped under θ to a central element of* $\mathrm{Sym}^n A$.

PROOF. Let c be in the center of $F[\mathfrak{S}_n]$. Then, for each $\sigma \in \mathfrak{S}_n$, $\sigma c = c\sigma$. Hence, by (1.1),

$$\sigma(\theta(c)) = \theta(\sigma) \cdot \theta(c) \cdot \theta(\sigma)^{-1} = \theta(\sigma c \sigma^{-1}) = \theta(c)$$

for each $\sigma \in \mathfrak{S}_n$. Thus, $\theta(c) \in \mathrm{Sym}^n A$.

Now write $c = \sum_{\sigma \in \mathfrak{S}_n} a_\sigma \sigma$ with $a_\sigma \in F$. Then, by (1.1) again,

$$\theta(c) \cdot r = \sum_{\sigma \in \mathfrak{S}_n} a_\sigma \theta(\sigma) \cdot r = \sum_{\sigma \in \mathfrak{S}_n} a_\sigma \sigma(r) \cdot \theta(\sigma) = \sum_{\sigma \in \mathfrak{S}_n} a_\sigma r \cdot \theta(\sigma) = r \cdot \theta(c),$$

for each $r \in \mathrm{Sym}^n A$. Hence $\theta(c)$ is in the center of $\mathrm{Sym}^n A$. □

Assume that the characteristic of F is zero or of prime degree exceeding n. Then, the group algebra $F[\mathfrak{S}_n]$ is a direct sum of central simple algebras over F. In fact, it is isomorphic to a direct sum of matrix algebras over F [**Jac89**, Thm. 5.9]:

$$F[\mathfrak{S}_n] \cong M_{m_1}(F) \oplus M_{m_2}(F) \oplus \cdots \oplus M_{m_{p(n)}}(F),$$

where $p(n)$ is the number of conjugacy classes of $\mathfrak{S}_n$ (which is the number of partitions of n). Hence, $F[\mathfrak{S}_n] = \bigoplus_{i=1}^{p(n)} F[\mathfrak{S}_n]c_i$, where $c_i \in F[\mathfrak{S}_n]$ is a central idempotent, $i = 1, \ldots, p(n)$, and the center of $F[\mathfrak{S}_n]$ is $\bigoplus_{i=1}^{p(n)} Fc_i$. Thus,

$$1 = c_1 + \cdots + c_{p(n)}, \quad \text{and} \tag{1.2a}$$

$$c_i^2 = c_i, \quad \text{and} \quad c_i c_j = 0 \ \text{ if } i \neq j, \quad \text{for } i, j = 1, \ldots, p(n). \tag{1.2b}$$

It follows from (1.2) that

$$1^{\otimes n} = \theta(c_1) + \cdots + \theta(c_{p(n)}), \quad \text{and} \tag{1.3a}$$

$$\theta(c_i)^2 = \theta(c_i), \qquad \text{and } \theta(c_i)\theta(c_j) = 0 \quad \text{if } i \neq j, \qquad \text{for } i, j = 1, \ldots, p(n). \tag{1.3b}$$

Therefore, $A^{\otimes n} = \bigoplus_{i=1}^{p(n)} A^{\otimes n} \cdot \theta(c_i)$. Since $\theta(c_i)$ is in the center of $\operatorname{Sym}^n A$ (Lemma 1.3), we get

COROLLARY 1.4. $\operatorname{Sym}^n A \cdot \theta(c_i)$ *is an ideal of* $\operatorname{Sym}^n A$, $i = 1, \ldots, p(n)$, *and*

$$\operatorname{Sym}^n A = \operatorname{Sym}^n A \cdot \theta(c_1) \oplus \cdots \oplus \operatorname{Sym}^n A \cdot \theta(c_{p(n)}).$$

Ingredients of the proof below of the following theorem can be found in [**Sal99**, Chap. 5, p. 34]. We give its proof here for the convenience of the reader.

THEOREM 1.5. *Suppose* $\theta(c_j) \neq 0$. *Then* $\operatorname{Sym}^n A \cdot \theta(c_j)$ *is a central simple algebra of degree* $\frac{\operatorname{tr}_{A^{\otimes n}}(\theta(c_j))}{m_j}$ *over* F.

PROOF. Let $e = \theta(c_j)$ and $B = \theta(F[\mathfrak{S}_n])$. Then, since $F[\mathfrak{S}_n]c_j \cong M_{m_j}(F)$ and $\theta(c_j) \neq 0$, $Be = \theta(F[\mathfrak{S}_n]c_j) \cong F[\mathfrak{S}_n]c_j$ is a central simple algebra over F with unit e. Since e is an idempotent of the central simple algebra $A^{\otimes n}$, $eA^{\otimes n}e$ is also a central simple algebra over F (with unit e). Let $C = C_{eA^{\otimes n}e}(Be)$ be the centralizer of Be in $eA^{\otimes n}e$. By the double centralizer theorem [**Row88**, Thm. 7.1.9], C is simple and the center of C is the center of Be which is Fe. Moreover, $\deg(C) = \frac{\deg(eA^{\otimes n}e)}{\deg(Be)} = \frac{\operatorname{tr}_{A^{\otimes n}}(e)}{m_j}$.

Let $c = eae \in C$ with $a \in A^{\otimes n}$. Since e is a central idempotent of B,

$$cb = eaeeb = (eae)(be) = (be)(eae) = beae = bc$$

for each $b \in B$. By (1.1), $\operatorname{Sym}^n A = C_{A^{\otimes n}}(B)$ is the centralizer of B in $A^{\otimes n}$. Hence, since e is in the center of $\operatorname{Sym}^n A$, $C = \operatorname{Sym}^n A \cap (eA^{\otimes n}e) = \operatorname{Sym}^n A \cdot e$. By the preceding paragraph, $\operatorname{Sym}^n A \cdot e$ is a central simple algebra of degree $\frac{\operatorname{tr}_{A^{\otimes n}}(e)}{m_j}$ over F with unit e. □

By Corollary 1.4 and Theorem 1.5 we get:

COROLLARY 1.6. $\operatorname{Sym}^n A$ *is a direct sum of central simple algebras over* F.

2. Octonion algebras

Let F be a field of characteristic $\neq 2$. A **composition algebra** (A, N) over F is a unital non-associative algebra A over F with a non-degenerate quadratic form N such that $N(xy) = N(x)N(y)$ for all $x, y \in A$. It is **alternative**, i.e. $x(xy) = (xx)y$ and $(yx)x = y(xx)$ for all $x, y \in A$, and has an involution $x \mapsto \bar{x}$ such that $\bar{x}x = N(x)1$ for all $x \in A$ [**Jac85**, Thm. 7.5]. Define $\operatorname{tr}(x) = N(x+1) - N(x) - N(1)$ for each $x \in A$. Then, $\operatorname{tr}(x)1 = x + \bar{x}$ and

$$x^2 - \operatorname{tr}(x)x + N(x)1 = 0 \tag{2.1}$$

for each $x \in A$ [**Jac85**, §7.6]. In particular, $N(1) = 1$ and $\operatorname{tr}(1) = 2 \neq 0$. Also, an element $x \in A$ is invertible if and only if $N(x) \neq 0$. We call N and tr the **norm** and **trace**, respectively, of A. Artin's theorem states that in an alternative algebra the subalgebra generated by any two elements is associative [**Sch66**, p. 29]. Denote the multiplicatively closed set of all invertible elements in A by $A^\times$. It follows from

(2.1) that $\operatorname{tr}(xy) = \operatorname{tr}(yx)$ for $x, y \in A^\times$. Hence, since tr is F-linear and $A^\times$ spans A over F, $\operatorname{tr}(xy) = \operatorname{tr}(yx)$ for all $x, y \in A$.

Let $\mathcal{C}$ be an octonion algebra over a field F of characteristic $\neq 2, 3$. By definition, $\mathcal{C}$ is an 8-dimensional composition algebra over F. Let N and tr be the norm and trace, respectively, of $\mathcal{C}$. Recall that $\mathcal{C}$ has a basis $e_0 = 1, e_1, e_2, e_3, e_4, e_5, e_6, e_7$ with multiplication table

$\cdot$	e_0	e_1	e_2	e_3	e_4	e_5	e_6	e_7
e_0	e_0	e_1	e_2	e_3	e_4	e_5	e_6	e_7
e_1	e_1	αe_0	e_3	αe_2	e_5	αe_4	$-e_7$	$-\alpha e_6$
e_2	e_2	$-e_3$	βe_0	$-\beta e_1$	e_6	e_7	βe_4	βe_5
e_3	e_3	$-\alpha e_2$	βe_1	$-\alpha\beta e_0$	e_7	αe_6	$-\beta e_5$	$-\alpha\beta e_4$
e_4	e_4	$-e_5$	$-e_6$	$-e_7$	γe_0	$-\gamma e_1$	$-\gamma e_2$	$-\gamma e_3$
e_5	e_5	$-\alpha e_4$	$-e_7$	$-\alpha e_6$	γe_1	$-\alpha\gamma e_0$	γe_3	$\alpha\gamma e_2$
e_6	e_6	e_7	$-\beta e_4$	βe_5	γe_2	$-\gamma e_3$	$-\beta\gamma e_0$	$-\beta\gamma e_1$
e_7	e_7	αe_6	$-\beta e_5$	$\alpha\beta e_4$	γe_3	$-\alpha\gamma e_2$	$\beta\gamma e_1$	$\alpha\beta\gamma e_0$

where $\alpha, \beta, \gamma \in \mathcal{C}^\times$ and with involution defined by $\bar{e}_0 = e_0$, $\bar{e}_i = -e_i$, $i = 1, 2, 3, 4, 5, 6, 7$ [**Jac85**, §7.6]. Set $\mathcal{I} = \{0, 1, 2, 3, 4, 5, 6, 7\}$ and $\mathcal{I}^\times = \mathcal{I} \setminus \{0\} = \{1, 2, 3, 4, 5, 6, 7\}$. In particular, $\operatorname{tr}(e_i) = e_i + \bar{e}_i = 0$ for $i \in \mathcal{I}^\times$. Therefore $\operatorname{tr}(e_i e_j) = 0$ for $i \neq j$ in $\mathcal{I}$. By (2.1),

$$e_i^{-1} = -\frac{1}{N(e_i)} e_i \quad \text{for } i \in \mathcal{I}^\times . \tag{2.2}$$

Hence

$$\operatorname{tr}(e_i^{-1} e_j) = 2\delta_{ij} \quad \text{for } i, j \in \mathcal{I} . \tag{2.3}$$

In particular, tr is **non-degenerate**, i.e. if $\operatorname{tr}(xy) = 0$ for all $y \in \mathcal{C}$, then $x = 0$. By (2.1) (or by the multiplication table),

$$e_j e_i = N(e_i) N(e_j) (e_i e_j)^{-1} = -e_i e_j \quad \text{for } i \neq j \in \mathcal{I}^\times . \tag{2.4}$$

Denote $\beta_i = 2\delta_{i0} - 1$ for $i \in \mathcal{I}$, so $\beta_0 = 1$ and $\beta_i = -1$ for $i \in \mathcal{I}^\times$. Then, by (2.4),

$$e_j e_i = (2\delta_{ij} - \beta_i \beta_j) \cdot e_i e_j \quad \text{for } i, j \in \mathcal{I} . \tag{2.5}$$

Also, by the multiplication table, for each $i, j, k \in \mathcal{I}$, either $(e_i e_j) e_k = e_i (e_j e_k)$ or $(e_i e_j) e_k = -e_i (e_j e_k)$ and in the latter case $\operatorname{tr}((e_i e_j) e_k) = \operatorname{tr}(e_i (e_j e_k)) = 0$. Hence $\operatorname{tr}((e_i e_j) e_k) = \operatorname{tr}(e_i (e_j e_k))$ for all $i, j, k \in \mathcal{I}$. Thus, by linearity,

$$\operatorname{tr}((xy)z) = \operatorname{tr}(x(yz)) \quad \text{for all } x, y, z \in \mathcal{C} .^1 \tag{2.6}$$

Let $x = \sum_{j \in \mathcal{I}} \alpha_j e_j \in \mathcal{C}$ with $\alpha_j \in F$. Then, $\operatorname{tr}(x) = \sum_{j \in \mathcal{I}} \alpha_j \operatorname{tr}(e_j) = 2\alpha_0$. Hence, since $\sum_{i \in \mathcal{I}} \beta_i = -6$, we get by (2.5),

$$\begin{aligned} \sum_{i \in \mathcal{I}} e_i^{-1} x e_i &= \sum_{i,j \in \mathcal{I}} \alpha_j e_i^{-1} e_j e_i = \sum_{j \in \mathcal{I}} \sum_{i \in \mathcal{I}} (2\delta_{ij} - \beta_i \beta_j) \alpha_j e_j = \sum_{j \in \mathcal{I}} (2 + 6\beta_j) \alpha_j e_j \\ &= \sum_{j \in \mathcal{I}} (12\delta_{j0} - 4) \alpha_j e_j = 12\alpha_0 1 - 4x = 6 \operatorname{tr}(x) 1 - 4x . \end{aligned}$$

[1]Warning: $\mathcal{C}$ is not associative under tr, e.g. $\operatorname{tr}((e_1 e_2)(e_4 e_7)) = 2\alpha\beta\gamma = -\operatorname{tr}(e_1 (e_2 e_4) e_7)$.

(Note that by Artin's theorem, the algebra $F[e_i, x]$ is associative.) Hence

$$\mathrm{tr}(x)1 = \frac{1}{6}(4x + \sum_{i\in\mathcal{I}} e_i^{-1}xe_i) = \sum_{i\in\mathcal{I}}(\frac{2}{3}\delta_{i0} + \frac{1}{6}) \cdot e_i^{-1}xe_i \qquad \text{for } x \in \mathcal{C}\,. \tag{2.7}$$

As $\mathrm{tr}(xy)$ defines a non-degenerate bilinear form of $\mathcal{C}$ over F, any basis $(f_i)_{i\in\mathcal{I}}$ of $\mathcal{C}$ has a dual basis $(f_i^{\#})_{i\in\mathcal{I}}$ with respect to tr so that

$$\mathrm{tr}(f_i^{\#} f_j) = 2\delta_{ij} \qquad \text{for } i, j \in \mathcal{I}\,.$$

In particular, $(e_i^{-1})_{i\in\mathcal{I}}$ is the dual basis of $(e_i)_{i\in\mathcal{I}}$. Let $C = (c_{ij})_{i,j\in\mathcal{I}}, D = (d_{ij})_{i,j\in\mathcal{I}}$ be matrices in $\mathrm{GL}(8, F)$ with $f_j = \sum_{l\in\mathcal{I}} c_{jl}e_l$ and $f_i^{\#} = \sum_{k\in\mathcal{I}} d_{ik}e_k^{-1}$ for $i, j \in \mathcal{I}$. They satisfy

$$2\delta_{ij} = \mathrm{tr}(f_i^{\#} f_j) = \sum_{k,l\in\mathcal{I}} d_{ik}c_{jl}\,\mathrm{tr}(e_k^{-1}e_l) = 2\sum_{k\in\mathcal{I}} d_{ik}c_{jk}\,. \tag{2.8}$$

By (2.8) $DC^t = \mathbf{1}$, so $CD^t = \mathbf{1}$, hence $D^tC = \mathbf{1}$. In other words, $\sum_{i\in\mathcal{I}} d_{ik}c_{il} = \delta_{kl}$. Hence, for each $x \in \mathcal{C}$,

$$\sum_{i\in\mathcal{I}}(f_i^{\#}x)f_i = \sum_{i,k,l\in\mathcal{I}} d_{ik}c_{il}(e_k^{-1}x)e_l = \sum_{k,l\in\mathcal{I}} (\sum_{i\in\mathcal{I}} d_{ik}c_{il})(e_k^{-1}x)e_l = \sum_{k\in\mathcal{I}} e_k^{-1}xe_k\,.$$

Similarly, $\sum_{i\in\mathcal{I}} f_i^{\#}(xf_i) = \sum_{k\in\mathcal{I}} e_k^{-1}xe_k$.

In the same way, the element $\varepsilon = \sum_{i\in\mathcal{I}} f_i^{\#} \otimes f_i = \sum_{i\in\mathcal{I}} e_i^{-1} \otimes e_i$ in $\mathcal{C}\otimes\mathcal{C}$ does not depend on the basis $(f_i)_{i\in\mathcal{I}}$ of $\mathcal{C}$. Note that, by (2.2), ε belongs to the ring of symmetric tensors $\mathrm{Sym}^2\,\mathcal{C}$ in $\mathcal{C}\otimes\mathcal{C}$.

LEMMA 2.1. (a) *$\varepsilon(1\otimes x) = (x\otimes 1)\varepsilon$ and $\varepsilon(x\otimes 1) = (1\otimes x)\varepsilon$ for all $x \in \mathcal{C}$.*
(b) *ε commutes with each element of* $\mathrm{Sym}^2\,\mathcal{C}$.

PROOF. Let $x \in \mathcal{C}^\times$. We claim that $(x^{-1}f_i^{\#})_{i\in\mathcal{I}}$ is the dual basis of $(f_ix)_{i\in\mathcal{I}}$. Indeed, for each $i, j \in \mathcal{I}$ we have by (2.6),

$$\begin{aligned}\mathrm{tr}((x^{-1}f_i^{\#})(f_jx)) &= \mathrm{tr}(x^{-1}f_i^{\#}(f_jx)) = \mathrm{tr}(f_i^{\#}(f_jx)x^{-1}) \\ &= \mathrm{tr}(f_i^{\#}(f_jxx^{-1})) = \mathrm{tr}(f_i^{\#} f_j) = 2\delta_{ij}\,.\end{aligned}$$

Hence

$$\varepsilon(1\otimes x) = \sum_{i\in\mathcal{I}} f_i^{\#} \otimes f_ix = \sum_{i\in\mathcal{I}} xx^{-1}f_i^{\#} \otimes f_ix = (x\otimes 1)\sum_{i\in\mathcal{I}} x^{-1}f_i^{\#} \otimes f_ix = (x\otimes 1)\varepsilon\,.$$

Since $\mathcal{C}^\times$ spans $\mathcal{C}$ over F, it follows that $\varepsilon(1\otimes x) = (x\otimes 1)\varepsilon$ for all $x \in \mathcal{C}$. Similarly, $\varepsilon(x\otimes 1) = (1\otimes x)\varepsilon$ for all $x \in \mathcal{C}$. This proves (a).

By (a),

$$\varepsilon(x\otimes x) = \varepsilon(x\otimes 1)(1\otimes x) = (1\otimes x)\varepsilon(1\otimes x) = (1\otimes x)(x\otimes 1)\varepsilon = (x\otimes x)\varepsilon\,.$$

(Note that by Artin's theorem there is no issue with associativity.)

Since $\{x\otimes x \mid x \in \mathcal{C}\}$ spans $\mathrm{Sym}^2\,\mathcal{C}$ (Lemma 3.1), we get that ε commutes with each element of $\mathrm{Sym}^2\,\mathcal{C}$. This proves (b). $\square$

We denote

$$\triangle = \frac{1}{4}1\otimes 1 - \frac{1}{8}\varepsilon = \frac{1}{8}\sum_{i\in\mathcal{I}} \beta_ie_i^{-1} \otimes e_i\,. \tag{2.9}$$

LEMMA 2.2. *$\triangle(1\otimes x+x\otimes 1)=\operatorname{tr}(x)\triangle$ and $\triangle(x\otimes x)=N(x)\triangle$ for all $x\in\mathcal{C}$. In particular, $\triangle$ lies in the* **center** *of $\operatorname{Sym}^2\mathcal{C}$, i.e. it commutes and associates with all elements of $\operatorname{Sym}^2\mathcal{C}$, and $\operatorname{Sym}^2\mathcal{C}\cdot\triangle=F\cdot\triangle$.*

PROOF. Let $x\in\mathcal{C}$.

CLAIM. $\triangle(1\otimes x+x\otimes 1)=\operatorname{tr}(x)\triangle$.

Indeed, it suffices to show that

$$\triangle(1\otimes e_j+e_j\otimes 1)=2\delta_{j0}\triangle \text{ for each } j\in\mathcal{I}.$$

By Lemma 2.1,

$$\begin{aligned}\triangle(e_j\otimes 1)&=(\tfrac{1}{4}1\otimes 1-\tfrac{1}{8}\varepsilon)(e_j\otimes 1)=\tfrac{1}{4}(e_j\otimes 1)-\tfrac{1}{8}(1\otimes e_j)\varepsilon\\&=(1\otimes e_j)\triangle+\tfrac{1}{4}(e_j\otimes 1-1\otimes e_j).\end{aligned}\tag{2.10}$$

By (2.5) and since $\beta_j=2\delta_{j0}-1$,

$$\begin{aligned}(1\otimes e_j)\triangle&=\frac{1}{8}\sum_{i\in\mathcal{I}}\beta_i e_i^{-1}\otimes e_je_i=\frac{1}{8}\sum_{i\in\mathcal{I}}\beta_i(2\delta_{ij}-\beta_i\beta_j)e_i^{-1}\otimes e_ie_j\\&=\beta_j\big(\tfrac{1}{4}e_j\otimes 1-\tfrac{1}{8}\varepsilon(1\otimes e_j)\big)=\beta_j\big(\triangle(1\otimes e_j)+\tfrac{1}{4}(e_j\otimes 1-1\otimes e_j)\big)\\&=2\delta_{j0}\triangle-\triangle(1\otimes e_j)-\tfrac{1}{4}(e_j\otimes 1-1\otimes e_j).\end{aligned}$$

Putting the last equation in (2.10) proves the claim.

By the claim and by (2.1),

$$\begin{aligned}\triangle(x\otimes x)&=\triangle(x\otimes 1)(1\otimes x)=\triangle(\operatorname{tr}(x)1\otimes 1-1\otimes x)(1\otimes x)\\&=\triangle\big(1\otimes(\operatorname{tr}(x)x-x^2)\big)=N(x)\triangle,\end{aligned}$$

as desired.

By Lemma 2.1(b), $\triangle$ commutes with each element of $\operatorname{Sym}^2\mathcal{C}$. To show that $\triangle$ associates with each element of $\operatorname{Sym}^2\mathcal{C}$, it suffices to show (by Lemma 3.1) for each $x,y\in\mathcal{C}$ that

$$\begin{aligned}&(\triangle(x\otimes x))(y\otimes y)=\triangle((x\otimes x)(y\otimes y)),\\&((x\otimes x)\triangle)(y\otimes y)=(x\otimes x)(\triangle(y\otimes y)),\text{ and}\\&((x\otimes x)(y\otimes y))\triangle=(x\otimes x)((y\otimes y)\triangle).\end{aligned}$$

Indeed, all of these equalities hold since $N(xy)=N(x)N(y)$. □

3. Algebras of symmetric tensors

Fix a positive integer n. Let A be an algebra over a field F of characteristic 0 or of positive characteristic exceeding n. Let $A^{\otimes n}$ be the tensor product over F of n copies of A and define the action of the symmetric group $\mathfrak{S}_n$ on $A^{\otimes n}$ by $\sigma(a_1\otimes\cdots\otimes a_n)=a_{\sigma^{-1}(1)}\otimes\cdots\otimes a_{\sigma^{-1}(n)}$ for $\sigma\in\mathfrak{S}_n$ and $a_1,\dots,a_n\in A$. Let

$$a^\vee:=\frac{1}{n!}\sum_{\sigma\in S_n}\sigma(a)$$

for $a\in A^{\otimes n}$. Then $\{a^\vee\mid a\in A^{\otimes n}\}$ is the fixed ring

$$\operatorname{Sym}^n A=\{a\in A^{\otimes n}\mid \sigma(a)=a \text{ for all } \sigma\in\mathfrak{S}_n\}$$

of $A^{\otimes n}$ under the action of $\mathfrak{S}_n$. Note that

$$(\sigma(a))^\vee = a^\vee \text{ for } a \in A^{\otimes n} \text{ and } \sigma \in \mathfrak{S}_n\,, \tag{3.1a}$$

$$(a \otimes b)^\vee = (a \otimes b^\vee)^\vee \text{ for } a \in A^{\otimes k}, b \in A^{\otimes(n-k)}, \text{ and} \tag{3.1b}$$

$$(ar)^\vee = a^\vee r \text{ and } (ra)^\vee = ra^\vee \text{ for } a \in A^{\otimes n}, r \in \operatorname{Sym}^n A\,. \tag{3.1c}$$

It is known that $\{a^{\otimes n} \mid a \in A\}$ spans $\operatorname{Sym}^n A$. For the convenience of the reader we give its proof here with some addition. An algebra is said to be **power-associative** if the subalgebra generated by any element is associative. If A is power-associative, we define the **degree** of A to be the minimal number (possibly infinite) d such that for each $x \in A$ there exists a polynomial $f \in F[X]$ of degree $\le d$ with $f(x) = 0$. For example, by (2.1), the degree of an octonion algebra is 2.

LEMMA 3.1. (a) *Let $a_1, \dots, a_n \in A$. Then*

$$(a_1 \otimes \cdots \otimes a_n)^\vee = \frac{1}{n!} \sum_{J \subseteq \{1,\dots,n\}} (-1)^{n-|J|} \Big(\sum_{j\in J} a_j\Big)^{\otimes n}.$$

(b) *Suppose that A is power-associative and has a unity* 1 *and finite degree d such that $|F| > n(d-1)+1$. Then the set $\{a^{\otimes n} \mid a \in A^\times\}$ spans* $\operatorname{Sym}^n A$.

PROOF. (a) Let $I = \{1, \dots, n\}$ and $B = \big\{(i_1, \dots, i_n) \in I^n \mid \{i_1, \dots, i_n\} = I\big\}$. Then $B = I^n \smallsetminus \bigcup_{j=1}^n B_j$ with $B_j = (I \smallsetminus \{j\})^n$. For $(\mathbf{i}) = (i_1, \dots, i_n) \in I^n$, we denote $a_{\mathbf{i}} = a_{i_1} \otimes \cdots \otimes a_{i_n}$. Note that for each $J \subseteq I$

$$\Big(\sum_{i\in I \smallsetminus J} a_i\Big)^{\otimes n} = \sum_{(i_1,\dots,i_n)\in(I\smallsetminus J)^n} a_{i_1} \otimes \cdots \otimes a_{i_n} = \sum_{\mathbf{i}\in\bigcap_{j\in J} B_j} a_{\mathbf{i}}\,.$$

Note also that $B_1 \cap \cdots \cap B_n = \emptyset$. Hence,

$$\begin{aligned}
\sum_{\sigma\in\mathfrak{S}_n} \sigma(a_1 \otimes \cdots \otimes a_n) &= \sum_{(i_1,\dots,i_n)\in B} a_{i_1} \otimes \cdots \otimes a_{i_n} = \sum_{\mathbf{i}\in I^n} a_{\mathbf{i}} - \sum_{\mathbf{i}\in\bigcup_{j=1}^n B_j} a_{\mathbf{i}} \\
&= \sum_{\mathbf{i}\in I^n} a_{\mathbf{i}} - \Big(\sum_{j=1}^n \sum_{\mathbf{i}\in B_j} a_{\mathbf{i}} - \sum_{j<k} \sum_{\mathbf{i}\in B_j\cap B_k} a_{\mathbf{i}} + \cdots + (-1)^n \sum_{\mathbf{i}\in B_1\cap\cdots\cap B_n} a_{\mathbf{i}}\Big) \\
&= \Big(\sum_{i\in I} a_i\Big)^{\otimes n} - \Big(\sum_{j=1}^n \Big(\sum_{i\in I\smallsetminus\{j\}} a_i\Big)^{\otimes n} - \sum_{j<k}\Big(\sum_{i\in I\smallsetminus\{j,k\}} a_i\Big)^{\otimes n} + \cdots + (-1)^{n-1} \sum_{i=1}^n a_i^{\otimes n}\Big) \\
&= \sum_{J\subseteq I} (-1)^{n-|J|} \Big(\sum_{j\in J} a_j\Big)^{\otimes n},
\end{aligned}$$

as claimed.

(b) By part (a), it suffices to show for a non-invertible element $b \in A$ that $b^{\otimes n}$ is a linear combination of elements of the form $a^{\otimes n}$ with $a \in A^\times$. Let $f \in F[X]$ be a nonzero polynomial of degree $\le d$ such that $f(b) = 0$. Since b is not invertible, $f(0) = 0$. Let $\alpha_1, \dots, \alpha_s$ be all the nonzero roots of f in F. Then $s \le d-1$. Since $|F^\times| > n(d-1)$, there is $\alpha \in F^\times \smallsetminus \{m\alpha_i \mid i = 1, \dots, s,\ m = 1, \dots, n\}$. Let $\beta = -\frac{1}{\alpha}$. Then, $f(-\frac{1}{m\beta}) = f(\frac{\alpha}{m}) \ne 0$, $m = 1, \dots, n$. Let $c = \beta b$ and $g_m(X) = f(\frac{X-1}{m\beta})$, $m = 1, \dots, n$. Then $g_m(1 + mc) = 0$ and $g_m(0) = f(-\frac{1}{m\beta}) \ne 0$, $m = 1, \dots, n$.

Hence $1 + mc \in A^\times$, $m = 0, 1, \ldots, n$. We claim that $c^{\otimes n} = \beta^n b^{\otimes n}$ is a linear combination of $(1 + mc)^{\otimes n}$, $m = 0, 1, \ldots, n$. Indeed,

$$(1+mc)^{\otimes n} = \sum_{r=0}^{n} \binom{n}{r} \left(1^{\otimes(n-r)} \otimes (mc)^{\otimes r}\right)^\vee = \sum_{r=0}^{n} m^r \cdot \left[\binom{n}{r} (1^{\otimes(n-r)} \otimes c^{\otimes r})^\vee\right],$$

$m = 0, 1, \ldots, n$. Therefore, since $\det\left((m^r)_{m,r=0}^n\right) = \prod_{0 \le i < j \le n}(j - i) \ne 0$, each $\binom{n}{r}(1^{\otimes(n-r)} \otimes c^{\otimes r})^\vee$ and, in particular, $c^{\otimes n}$ (for $r = n$) is a linear combination of $(1 + mc)^{\otimes n}$, $m = 0, 1, \ldots, n$, as claimed. □

We return to the notation of §2. We assume here that F is a field of characteristic 0 or of positive characteristic exceeding $n + 1$. We extend $\mathrm{tr}\colon \mathcal{C} \to F$ to $\mathrm{tr}_{\mathcal{C}^{\otimes n}} \colon \mathcal{C}^{\otimes n} \to F$ using linearity by

$$\mathrm{tr}_{\mathcal{C}^{\otimes n}}(a_1 \otimes \cdots \otimes a_n) = \mathrm{tr}(a_1) \cdots \mathrm{tr}(a_n) \quad \text{for } a_1, \ldots, a_n \in \mathcal{C}.$$

Then, $\mathrm{tr}_{\mathcal{C}^{\otimes n}}$ is a **non-degenerate trace map** on $\mathcal{C}^{\otimes n}$, i.e. $\mathrm{tr}_{\mathcal{C}^{\otimes n}}(ab) = \mathrm{tr}_{\mathcal{C}^{\otimes n}}(ba)$ for all $a, b \in \mathcal{C}^{\otimes n}$ and if $\mathrm{tr}_{\mathcal{C}^{\otimes n}}(ab) = 0$ for all $b \in \mathcal{C}^{\otimes n}$, then $a = \mathbf{0}$. Also

(3.2) $$\mathrm{tr}_{\mathcal{C}^{\otimes n}}(a^\vee) = \mathrm{tr}_{\mathcal{C}^{\otimes n}}(a) \quad \text{for all } a \in \mathcal{C}^{\otimes n}$$

and

(3.3) $$\mathrm{tr}_{\mathcal{C}^{\otimes n}}(a \otimes b) = \mathrm{tr}_{\mathcal{C}^{\otimes k}}(a)\, \mathrm{tr}_{\mathcal{C}^{\otimes (n-k)}}(b) \quad \text{for all } a \in \mathcal{C}^{\otimes k} \text{ and } b \in \mathcal{C}^{\otimes(n-k)}.$$

Note that $\mathrm{tr}_{\mathcal{C}^{\otimes 2}}(\triangle) = \mathrm{tr}_{\mathcal{C}^{\otimes 2}}(\frac{1}{8}\sum_{i \in \mathcal{I}} \beta_i e_i^{-1} \otimes e_i) = \frac{1}{8}\mathrm{tr}(1)^2 = \frac{1}{2} \ne 0$.

For each $\mathbf{i} = (i_1, \ldots, i_n) \in \mathcal{I}^n$ we denote

$$e_{\mathbf{i}} = e_{i_1} \otimes \cdots \otimes e_{i_n} \quad \text{and} \quad e_{\mathbf{i}}^{-1} = e_{i_1}^{-1} \otimes \cdots \otimes e_{i_n}^{-1}.$$

Then both $(e_{\mathbf{i}})_{\mathbf{i} \in \mathcal{I}^n}$ and $(e_{\mathbf{i}}^{-1})_{\mathbf{i} \in \mathcal{I}^n}$ are bases of $\mathcal{C}^{\otimes n}$ over F. If $a = \sum_{\mathbf{i} \in \mathcal{I}^n} \alpha_{\mathbf{i}} e_{\mathbf{i}}^{-1}$ with $\alpha_{\mathbf{i}} \in F$, then $\mathrm{tr}_{\mathcal{C}^{\otimes n}}(a e_{\mathbf{j}}) = \alpha_{\mathbf{j}}$ for each $\mathbf{j} \in \mathcal{I}^n$. Hence, $a = \mathbf{0}$ if and only if $\mathrm{tr}(a e_{\mathbf{i}}) = 0$ for each $\mathbf{i} \in \mathcal{I}^n$. Moreover,

$$\left\{ \left(e_0^{\otimes k_0} \otimes \cdots \otimes e_7^{\otimes k_7}\right)^\vee \mid k_0 + \cdots + k_7 = n \right\}$$

forms a basis of $\mathrm{Sym}^n\, \mathcal{C}$. Hence

(3.4) $$\dim_F \mathrm{Sym}^n\, \mathcal{C} = \binom{n+7}{n}.$$

LEMMA 3.2. *Suppose $n > 2$ and let $r \in \mathrm{Sym}^{n-2}\, \mathcal{C}$. Then, $(\triangle \otimes r)^\vee = \mathbf{0}$ in $\mathrm{Sym}^n\, \mathcal{C}$ if and only if $r = \mathbf{0}$.*

PROOF. Since the implication $r = \mathbf{0} \Rightarrow (\triangle \otimes r)^\vee = \mathbf{0}$ is trivial, it suffices to show the other implication. Indeed, by (3.1), (3.2), (3.3), Lemma 3.1(b), and

Lemma 2.2,

$$\begin{aligned}
(\triangle\otimes r)^\vee = \mathbf{0} &\Rightarrow \operatorname{tr}_{\mathcal{C}^{\otimes n}}\left((\triangle\otimes r)^\vee a^{\otimes n}\right) = 0 && \text{for all } a\in\mathcal{C}^\times\\
&\Rightarrow \operatorname{tr}_{\mathcal{C}^{\otimes n}}\left((N(a)\triangle\otimes r a^{\otimes(n-2)})^\vee\right) = 0 && \text{for all } a\in\mathcal{C}^\times\\
&\Rightarrow \operatorname{tr}_{\mathcal{C}^{\otimes n}}(\triangle\otimes r a^{\otimes(n-2)}) = 0 && \text{for all } a\in\mathcal{C}^\times\\
&\Rightarrow \operatorname{tr}_{\mathcal{C}^{\otimes 2}}(\triangle)\operatorname{tr}_{\mathcal{C}^{\otimes(n-2)}}(r a^{\otimes(n-2)}) = 0 && \text{for all } a\in\mathcal{C}^\times\\
&\Rightarrow \operatorname{tr}_{\mathcal{C}^{\otimes(n-2)}}(r r') = 0 && \text{for all } r'\in\operatorname{Sym}^{n-2}\mathcal{C}\\
&\Rightarrow \operatorname{tr}_{\mathcal{C}^{\otimes(n-2)}}(r e_{\mathbf{i}}^\vee) = 0 && \text{for all } \mathbf{i}\in\mathcal{I}^{n-2}\\
&\Rightarrow \operatorname{tr}_{\mathcal{C}^{\otimes(n-2)}}\left((r e_{\mathbf{i}})^\vee\right) = 0 && \text{for all } \mathbf{i}\in\mathcal{I}^{n-2}\\
&\Rightarrow \operatorname{tr}_{\mathcal{C}^{\otimes(n-2)}}(r e_{\mathbf{i}}) = 0 && \text{for all } \mathbf{i}\in\mathcal{I}^{n-2}\\
&\Rightarrow r = \mathbf{0},
\end{aligned}$$

as desired. □

4. Central idempotents

Let $n \geq 2$ be an integer. Assume that F is a field of characteristic 0 or of positive characteristic exceeding n. Denote

$$\triangle^{(n)} = (\triangle\otimes 1^{\otimes(n-2)})^\vee$$

in $\operatorname{Sym}^n\mathcal{C}$.

LEMMA 4.1. *$\triangle^{(n)}$ lies in the center of $\operatorname{Sym}^n\mathcal{C}$. In particular, $F[\triangle^{(n)}]$ is a commutative and associative F-algebra.*

PROOF. By Lemma 2.2, for all $a, b\in\mathcal{C}$,

$$\triangle^{(n)}a^{\otimes n} = N(a)(\triangle\otimes a^{\otimes(n-2)})^\vee = a^{\otimes n}\triangle^{(n)}$$

and

$$\begin{aligned}
(\triangle^{(n)}a^{\otimes n})b^{\otimes n} &= N(a)(\triangle\otimes a^{\otimes(n-2)})^\vee b^{\otimes n} = N(a)N(b)(\triangle\otimes a^{\otimes(n-2)}b^{\otimes(n-2)})^\vee\\
&= N(ab)(\triangle\otimes(ab)^{\otimes(n-2)})^\vee = \triangle^{(n)}(ab)^{\otimes n} = \triangle^{(n)}(a^{\otimes n}b^{\otimes n}).
\end{aligned}$$

Similarly, $(a^{\otimes n}\triangle^{(n)})b^{\otimes n} = a^{\otimes n}(\triangle^{(n)}b^{\otimes n})$ and $(a^{\otimes n}b^{\otimes n})\triangle^{(n)} = a^{\otimes n}(b^{\otimes n}\triangle^{(n)})$.

It follows from Lemma 3.1 that $\triangle^{(n)}$ commutes and associates with all elements of $\operatorname{Sym}^n\mathcal{C}$. □

LEMMA 4.2. *$\triangle^2 = \triangle$, $(\triangle\otimes 1)\triangle^{(3)} = \frac{5}{12}\triangle\otimes 1$, and for $n\geq 4$,*

$$(\triangle\otimes 1^{\otimes(n-2)})\cdot\triangle^{(n)} = \frac{n+2}{2n(n-1)}\triangle\otimes 1^{\otimes(n-2)} + \frac{(n-2)(n-3)}{n(n-1)}\triangle\otimes\triangle^{(n-2)}.$$

PROOF. We can write $\triangle$ in the form

$$\triangle = \frac{1}{8}\sum_{i=0}^{7}\frac{1}{N(e_i)}e_i\otimes e_i.$$

Then, since $\triangle(x\otimes x) = N(x)\triangle$ for each $x\in\mathcal{C}$ (Lemma 2.2), it follows that $\triangle^2 = \triangle$.

Note that

$$\triangle^{(3)} = \frac{1}{6}\sum_{\sigma\in\mathfrak{S}_3}\sigma(\triangle\otimes 1) = \frac{1}{6}\cdot\frac{1}{8}\cdot 2\sum_{i=0}^{7}\frac{1}{N(e_i)}(e_i\otimes e_i\otimes 1 + e_i\otimes 1\otimes e_i + 1\otimes e_i\otimes e_i).$$

Therefore, since $\triangle(1\otimes x + x\otimes 1) = \mathrm{tr}(x)\triangle$ for each $x \in \mathcal{C}$ (Lemma 2.2) and since $\mathrm{tr}(e_i) = 2\delta_{i0}$,

$$(\triangle\otimes 1)\triangle^{(3)} = \frac{1}{24}\sum_{i=0}^{7}\left(\triangle\otimes 1 + \frac{\mathrm{tr}(e_i)}{N(e_i)}\triangle\otimes e_i\right) = \frac{1}{24}(8\cdot\triangle\otimes 1 + 2\cdot\triangle\otimes 1) = \frac{5}{12}\triangle\otimes 1 .$$

Suppose now that $n \geq 4$. We can write $\triangle^{(n)}$ in the form

$$\triangle^{(n)} = \frac{1}{n!}\sum_{\sigma\in\mathfrak{S}_n}\sigma(\triangle\otimes 1^{\otimes(n-2)}) = \frac{1}{8}\sum_{i=0}^{7}\frac{1}{N(e_i)}\cdot\frac{1}{n!}\sum_{\sigma\in\mathfrak{S}_n}\sigma(e_i\otimes e_i\otimes 1^{\otimes(n-2)}) .$$

For each $i < j$ between 1 and n, let $\mathfrak{T}_{ij} = \{\sigma \in \mathfrak{S}_n \mid \sigma\{1,2\} = \{i,j\}\}$ and note that $|\mathfrak{T}_{ij}| = 2(n-2)!$. Also, let $\mathfrak{T}_{3..n} = \{\sigma \in \mathfrak{S}_n \mid \sigma\{1,2\} \subseteq \{3,\dots,n\}\}$. Then,

$$\mathfrak{S}_n = \mathfrak{T}_{12} \uplus \biguplus_{j=3}^{n}\mathfrak{T}_{1j} \uplus \biguplus_{j=3}^{n}\mathfrak{T}_{2j} \uplus \mathfrak{T}_{3..n} .$$

The first term in this union contributes the sum

$$\sum_{\sigma\in\mathfrak{T}_{12}}\sigma(\triangle\otimes 1^{\otimes(n-2)}) = 2(n-2)!\cdot\triangle\otimes 1^{\otimes(n-2)} .$$

The second term contributes, for each j between 3 and n,

$$\begin{aligned}\sum_{\sigma\in\mathfrak{T}_{1j}}\sigma(\triangle\otimes 1^{\otimes(n-2)}) &= \frac{1}{8}\sum_{i=0}^{7}\frac{1}{N(e_i)}\sum_{\sigma\in\mathfrak{T}_{1j}}\sigma(e_i\otimes e_i\otimes 1^{\otimes(n-2)})\\ &= \frac{2(n-2)!}{8}\sum_{i=0}^{7}\frac{1}{N(e_i)}e_i\otimes 1^{\otimes(j-2)}\otimes e_i\otimes 1^{\otimes(n-j)} .\end{aligned}$$

Similarly, the third term contributes, for each j between 3 and n,

$$\sum_{\sigma\in\mathfrak{T}_{2j}}\sigma(\triangle\otimes 1^{\otimes(n-2)}) = \frac{2(n-2)!}{8}\sum_{i=0}^{7}\frac{1}{N(e_i)}1\otimes e_i\otimes 1^{\otimes(j-3)}\otimes e_i\otimes 1^{\otimes(n-j)} .$$

Finally, the fourth term contributes

$$\sum_{\sigma\in\mathfrak{T}_{3..n}}\sigma(\triangle\otimes 1^{\otimes(n-2)}) = \frac{2(n-2)!}{8}\sum_{i=0}^{7}\frac{1}{N(e_i)}\sum_{3\leq j<k\leq n}1^{\otimes(j-1)}e_i\otimes 1^{\otimes(k-j-1)}\otimes e_i\otimes 1^{\otimes(n-k)} .$$

Note that

$$\begin{aligned}\triangle^{(n-2)} &= \frac{1}{8}\sum_{i=0}^{7}\frac{1}{N(e_i)}\cdot\frac{1}{(n-2)!}\sum_{\sigma\in\mathfrak{S}_{n-2}}\sigma(e_i\otimes e_i\otimes 1^{\otimes(n-4)})\\ &= \frac{1}{(n-2)!}\frac{2(n-4)!}{8}\sum_{i=0}^{7}\frac{1}{N(e_i)}\sum_{3\leq j<k\leq n}1^{\otimes(j-3)}e_i\otimes 1^{\otimes(k-j-1)}\otimes e_i\otimes 1^{\otimes(n-k)} .\end{aligned}$$

Hence,

$$\sum_{\sigma\in\mathfrak{T}_{3..n}}\sigma(\triangle\otimes 1^{\otimes(n-2)}) = (n-2)!\cdot(n-2)(n-3)\cdot 1\otimes 1\otimes\triangle^{(n-2)} .$$

Thus, since $\triangle(e_i\otimes 1+1\otimes e_i)=\mathrm{tr}(e_i)\triangle=2\delta_{i0}\triangle$ (Lemma 2.2) and since $\triangle^2=\triangle$, by multiplying each of the partial sums by $\triangle\otimes 1^{\otimes(n-2)}$ and adding up, we get

$$\begin{aligned}(\triangle\otimes 1^{\otimes(n-2)})\cdot\triangle^{(n)}=&\frac{1}{n!}\cdot\Big(2(n-2)!\cdot\triangle\otimes 1^{\otimes(n-2)}\\ &+\frac{2(n-2)!}{8}\sum_{i=0}^{7}\frac{1}{N(e_i)}\sum_{j=3}^{n}\mathrm{tr}(e_i)\cdot\triangle\otimes 1^{\otimes(j-3)}\otimes e_i\otimes 1^{\otimes(n-j)}\\ &+(n-2)!(n-2)(n-3)\cdot\triangle\otimes\triangle^{(n-2)}\Big)\\ =&\frac{n+2}{2n(n-1)}\triangle\otimes 1^{\otimes(n-2)}+\frac{(n-2)(n-3)}{n(n-1)}\triangle\otimes\triangle^{(n-2)},\end{aligned}$$

as desired. □

For each integer m between $\lceil\frac{n}{2}\rceil$ (= rounding up $\frac{n}{2}$) and n we denote

$$\beta_m^{(n)}=\frac{(n-m)(m+3)}{2n(n-1)}.$$

PROPOSITION 4.3. $\prod_{\lceil\frac{n}{2}\rceil\le m\le n}\big(\triangle^{(n)}-\beta_m^{(n)}\cdot 1^{\otimes n}\big)=\mathbf{0}$.

PROOF. We prove the proposition by induction on n. For $n=2$ and $n=3$ we have by Lemma 4.2,

$$(\triangle^{(2)}-\beta_2^{(2)}\cdot 1\otimes 1)(\triangle^{(2)}-\beta_1^{(2)}\cdot 1\otimes 1)=\triangle(\triangle-1\otimes 1)=\mathbf{0}$$

and

$$(\triangle^{(3)}-\beta_3^{(3)}\cdot 1^{\otimes 3})(\triangle^{(3)}-\beta_2^{(3)}\cdot 1^{\otimes 3})=\Big((\triangle\otimes 1)(\triangle^{(3)}-\frac{5}{12}\cdot 1^{\otimes 3})\Big)^{\vee}=\mathbf{0}.$$

Suppose $n\ge 4$. For each integer m between $\lceil\frac{n}{2}\rceil$ and $n-1$ we get, by Lemma 4.2, the identity

$$(\triangle\otimes 1^{\otimes(n-2)})\big(\triangle^{(n)}-\beta_m^{(n)}\cdot 1^{\otimes n}\big)=\frac{(n-2)(n-3)}{n(n-1)}\cdot\triangle\otimes\big(\triangle^{(n-2)}-\beta_{m-1}^{(n-2)}\cdot 1^{\otimes(n-2)}\big).$$

Hence, by the induction hypothesis,

$$\begin{aligned}\prod_{\lceil\frac{n}{2}\rceil\le m\le n}\big(\triangle^{(n)}-\beta_m^{(n)}\cdot 1^{\otimes n}\big)&=\triangle^{(n)}\prod_{\lceil\frac{n}{2}\rceil\le m\le n-1}\big(\triangle^{(n)}-\beta_m^{(n)}\cdot 1^{\otimes n}\big)\\ &=\Big((\triangle\otimes 1^{\otimes(n-2)})\prod_{\lceil\frac{n}{2}\rceil\le m\le n-1}\big(\triangle^{(n)}-\beta_m^{(n)}\cdot 1^{\otimes n}\big)\Big)^{\vee}\\ &=\left(\frac{\binom{n-2}{2}}{\binom{n}{2}}\right)^{n-\lceil\frac{n}{2}\rceil}\Big(\triangle\otimes\prod_{\lceil\frac{n-2}{2}\rceil\le m-1\le n-2}\big(\triangle^{(n-2)}-\beta_{m-1}^{(n-2)}\cdot 1^{\otimes(n-2)}\big)\Big)^{\vee}=\mathbf{0},\end{aligned}$$

as required. □

We assume now that $\mathrm{char}(F)=0$ or $\mathrm{char}(F)>n+2$. We claim that $\beta_{m_1}^{(n)}\ne\beta_{m_2}^{(n)}$ for integers $m_1\ne m_2$ between $\lceil\frac{n}{2}\rceil$ and n.

Indeed, suppose on the contrary that $\beta_{m_1}^{(n)}=\beta_{m_2}^{(n)}$ with $m_1<m_2$. Then,

$$(n-m_1)(m_1+3)=(n-m_2)(m_2+3)$$

and therefore

$$n(m_1-m_2)=(m_1-m_2)(m_1+m_2+3).$$

Hence, since $m_1 < m_2$, $n = m_1 + m_2 + 3$. If $\mathrm{char}(F) = 0$, then this equality is impossible since

$$m_1 + m_2 + 3 \geq \left\lceil \frac{n}{2} \right\rceil + \left\lceil \frac{n}{2} \right\rceil + 1 + 3 \geq n + 4\,.$$

If $\mathrm{char}(F) = p > n+2$, the equality $n \equiv m_1 + m_2 + 3 \bmod p$ is impossible since

$$m_1 + m_2 + 3 \leq n - 1 + n + 3 = 2n + 2 < n + p\,.$$

Thus, $\beta_{m_1}^{(n)} \neq \beta_{m_2}^{(n)}$, as claimed.

A partition of n of length 2 is a pair (m_1, m_2) of two integers with $m_1 \geq m_2 \geq 0$ and $m_1 + m_2 = n$. We denote by $\Lambda_2^{(n)}$ the set of all partitions of n of length 2. For each m between $\lceil \frac{n}{2} \rceil$ and n, let $\lambda = (m, n-m)$ be the corresponding partition of n and denote

$$e_\lambda = e_m^{(n)} = \prod_{\substack{\lceil \frac{n}{2} \rceil \leq m' \leq n \\ m' \neq m}} \frac{\triangle^{(n)} - \beta_{m'}^{(n)} \cdot 1^{\otimes n}}{\beta_m^{(n)} - \beta_{m'}^{(n)}}\,.$$

COROLLARY 4.4. *$e_\lambda e_{\lambda'} = \mathbf{0}$ if $\lambda \neq \lambda'$ for all $\lambda, \lambda' \in \Lambda_2^{(n)}$. Moreover, $\sum_{\lambda \in \Lambda_2^{(n)}} e_\lambda = 1^{\otimes n}$. In particular, $e_\lambda^2 = e_\lambda$ for all $\lambda \in \Lambda_2^{(n)}$.*

PROOF. By Proposition 4.3, $e_m^{(n)} e_{m'}^{(n)} = \mathbf{0}$ if $m \neq m'$ are two integers between $\lceil \frac{n}{2} \rceil$ and n.

For every polynomial $f \in F[X]$ of degree d with distinct roots $\alpha_1, \ldots, \alpha_d$ in some algebraic closure of F, there is a polynomial identity

$$1 = \sum_{i=1}^{d} \frac{1}{f'(\alpha_i)} \frac{f(X)}{X - \alpha_i} = \sum_{i=1}^{d} \prod_{j \neq i} \frac{X - \alpha_j}{\alpha_i - \alpha_j}\,.$$

Hence, by Proposition 4.3, $\sum_{\lceil \frac{n}{2} \rceil \leq m \leq n} e_m^{(n)} = 1^{\otimes n}$, as desired. □

We denote $e_0^{(0)} = 1_F$ and $e_1^{(1)} = 1_C$.

LEMMA 4.5. *Let m be an integer between $\lceil \frac{n}{2} \rceil$ and $n-1$. Then*

$$e_m^{(n)} \in F^\times \cdot \left(\triangle \otimes e_{m-1}^{(n-2)}\right)^\vee.$$

PROOF. Since $e_1^{(2)} = \triangle$ and $e_2^{(3)} = \frac{12}{5}\triangle^{(3)} = \frac{12}{5}(\triangle \otimes 1)^\vee$, the lemma is true if $n = 2$ and $m = 1$ or if $n = 3$ and $m = 2$.

Suppose $n \geq 4$. As in the proof of Proposition 4.3,

$$\begin{aligned}
&\prod_{\substack{\lceil \frac{n}{2} \rceil \leq m' \leq n \\ m' \neq m}} \left(\triangle^{(n)} - \beta_{m'}^{(n)} \cdot 1^{\otimes n}\right) = \triangle^{(n)} \prod_{\substack{\lceil \frac{n}{2} \rceil \leq m' \leq n-1 \\ m' \neq m}} \left(\triangle^{(n)} - \beta_{m'}^{(n)} \cdot 1^{\otimes n}\right) \\
&= \Big((\triangle \otimes 1^{\otimes(n-2)}) \prod_{\substack{\lceil \frac{n}{2} \rceil \leq m' \leq n-1 \\ m' \neq m}} \left(\triangle^{(n)} - \beta_{m'}^{(n)} \cdot 1^{\otimes n}\right)\Big)^\vee \\
&= \left(\frac{\binom{n-2}{2}}{\binom{n}{2}}\right)^{n - \lceil \frac{n}{2} \rceil - 1} \Big(\triangle \otimes \prod_{\substack{\lceil \frac{n-2}{2} \rceil \leq m'-1 \leq n-2 \\ m'-1 \neq m-1}} \left(\triangle^{(n-2)} - \beta_{m'-1}^{(n-2)} \cdot 1^{\otimes(n-2)}\right)\Big)^\vee.
\end{aligned}$$

Hence, $e_m^{(n)} \in F^\times \cdot \left(\triangle \otimes e_{m-1}^{(n-2)}\right)^\vee$, as desired. □

LEMMA 4.6. *$e_n^{(n)}\triangle^{(n)} = \mathbf{0}$ and $e_n^{(n)} \equiv 1^{\otimes n} \mod F[\triangle^{(n)}]\triangle^{(n)}$.*

PROOF. Note that for each integer m between $\lceil \frac{n}{2} \rceil$ and $n-1$ we have

$$\frac{\triangle^{(n)} - \beta_m^{(n)} \cdot 1^{\otimes n}}{\beta_n^{(n)} - \beta_m^{(n)}} = -\frac{\triangle^{(n)}}{\beta_m^{(n)}} + 1^{\otimes n}.$$

Hence

$$e_n^{(n)} = \prod_{\lceil \frac{n}{2} \rceil \le m < n} \frac{\triangle^{(n)} - \beta_m^{(n)} \cdot 1^{\otimes n}}{\beta_n^{(n)} - \beta_m^{(n)}} \equiv 1^{\otimes n} \mod F[\triangle^{(n)}]\triangle^{(n)}.$$

Also, since $\beta_n^{(n)} = 0$, it follows from Proposition 4.3 that $e_n^{(n)}\triangle^{(n)} = \mathbf{0}$. □

For each partition $\lambda = (m, n-m)$ of n of length 2, we denote

$$\mathbb{S}_\lambda \mathcal{C} = \mathbb{S}_m^{(n)} \mathcal{C} = \operatorname{Sym}^n \mathcal{C} \cdot e_m^{(n)} = \operatorname{Sym}^n \mathcal{C} \cdot e_\lambda.$$

By Corollary 4.4, we have:

COROLLARY 4.7. $\operatorname{Sym}^n \mathcal{C} = \bigoplus_{\lambda \in \Lambda_2^{(n)}} \mathbb{S}_\lambda \mathcal{C}$.

We denote $T_n\mathcal{C} = \mathbb{S}_n^{(n)}\mathcal{C} = \operatorname{Sym}^n \mathcal{C} \cdot e_n^{(n)}$. By Lemma 4.6,

$$e_n^{(n)}\triangle^{(n)} = \mathbf{0} \text{ and } \operatorname{Sym}^n \mathcal{C} = T_n\mathcal{C} \oplus \operatorname{Sym}^n \mathcal{C} \cdot \triangle^{(n)}. \tag{4.1}$$

Let $\operatorname{Sym}^1 \mathcal{C} = \mathcal{C}$, $\operatorname{Sym}^0 \mathcal{C} = F$, and consider the epimorphism

$$\theta \colon \operatorname{Sym}^n \mathcal{C} \to \operatorname{Sym}^{n-2} \mathcal{C}$$

defined by

$$\theta(x^{\otimes n}) = N(x) x^{\otimes(n-2)}$$

for $x \in \mathcal{C}$ (use Lemma 3.1). Note that by Lemma 3.2, θ is well defined, since by Lemma 2.2,

$$\begin{aligned} \triangle^{(n)} x^{\otimes n} &= (\triangle \otimes 1^{\otimes(n-2)})^\vee x^{\otimes n} = \big((\triangle \otimes 1^{\otimes(n-2)}) x^{\otimes n}\big)^\vee \\ &= N(x)\big(\triangle \otimes x^{\otimes(n-2)}\big)^\vee = \big(\triangle \otimes \theta(x^{\otimes n})\big)^\vee \end{aligned}$$

for each $x \in \mathcal{C}$. Hence, $\triangle^{(n)} r = \big(\triangle \otimes \theta(r)\big)^\vee$ for each $r \in \operatorname{Sym}^n \mathcal{C}$. Thus, by (4.1), $T_n\mathcal{C} = \operatorname{Ker}(\theta)$.

COROLLARY 4.8. $\dim_F T_n\mathcal{C} = \binom{n+7}{n} - \binom{n+5}{n-2} = \binom{n+5}{n-1} + \binom{n+6}{n}$.

PROOF. By (3.4),

$$\dim T_n\mathcal{C} = \dim \operatorname{Ker}(\theta) = \dim \operatorname{Sym}^n \mathcal{C} - \dim \operatorname{Sym}^{n-2} \mathcal{C} = \binom{n+7}{n} - \binom{n+5}{n-2}.$$

The second equality is a known identity. □

REMARK 4.9. Assuming that $\operatorname{char}(F) = 0$ and F is algebraically closed, [**MPP01**, §4] shows, using methods that rely on the representation theory of some Lie algebras, that $T_n\mathcal{C}$ is a unital simple algebra.

In the next section we prove that $T_n\mathcal{C}$ is a central simple algebra without the assumption that F is algebraically closed and with mild assumption on the characteristic of F.

We reduce the trace map $\operatorname{tr}_{\mathcal{C}^{\otimes n}}$ of $\mathcal{C}^{\otimes n}$ to a trace map $\operatorname{tr}_{T_n\mathcal{C}}$ on $T_n\mathcal{C}$.

LEMMA 4.10. *$\mathrm{tr}_{T_n\mathcal{C}}$ is non-degenerate.*

PROOF. Let $x \in T_n\mathcal{C}$ be such that $\mathrm{tr}_{T_n\mathcal{C}}(xx') = 0$ for each $x' \in T_n\mathcal{C}$. Then, by (4.1), $\mathrm{tr}_{\mathcal{C}^{\otimes n}}(xr) = 0$ for each $r \in \mathrm{Sym}^n \mathcal{C}$. Therefore

$$\mathrm{tr}_{\mathcal{C}^{\otimes n}}(xa) = \mathrm{tr}_{\mathcal{C}^{\otimes n}}((xa)^\vee) = \mathrm{tr}_{\mathcal{C}^{\otimes n}}(xa^\vee) = 0$$

for each $a \in \mathcal{C}^{\otimes n}$. Hence, since $\mathrm{tr}_{\mathcal{C}^{\otimes n}}$ is non-degenerate, $x = 0$. Thus, $\mathrm{tr}_{T_n\mathcal{C}}$ is non-degenerate. □

We denote $T_0\mathcal{C} = \mathbb{S}_0^{(0)}\mathcal{C} = F$ and $T_1\mathcal{C} = \mathbb{S}_1^{(1)}\mathcal{C} = \mathcal{C}$.

PROPOSITION 4.11. *Let m be an integer between $\lceil \frac{n}{2} \rceil$ and $n-1$. Then*

$$\mathbb{S}_m^{(n)}\mathcal{C} \cong \mathbb{S}_{m-1}^{(n-2)}\mathcal{C}\,.$$

PROOF. By Lemma 4.5, $e_m^{(n)} = \alpha\big(\triangle \otimes e_{m-1}^{(n-2)}\big)^\vee$ with $\alpha \in F^\times$. Hence, by Lemma 2.2,

$$x^{\otimes n}e_m^{(n)} = \alpha\Big(N(x)\triangle \otimes x^{\otimes(n-2)}e_{m-1}^{(n-2)}\Big)^\vee\,. \tag{4.2}$$

Consider the epimorphism

$$\varphi\colon \mathbb{S}_m^{(n)}\mathcal{C} = \mathrm{Sym}^n \mathcal{C}\cdot e_m^{(n)} \to \mathrm{Sym}^{n-2} \mathcal{C}\cdot e_{m-1}^{(n-2)} = \mathbb{S}_{m-1}^{(n-2)}\mathcal{C}$$

defined, using Lemma 3.1, by

$$\varphi\big(x^{\otimes n}e_m^{(n)}\big) = N(x)x^{\otimes(n-2)}e_{m-1}^{(n-2)}\,.$$

By Lemma 3.2 and (4.2), for each $r \in \mathrm{Sym}^n \mathcal{C}$, $re_m^{(n)} = \mathbf{0}$ if and only if $\varphi(re_m^{(n)}) = \mathbf{0}$. Hence, φ is well defined. Moreover, $\mathrm{Ker}(\varphi) = \mathbf{0}$, so φ is an isomorphism. □

Let m be an integer between $\lceil \frac{n}{2} \rceil$ and $n-1$. Since $n - 2k = m - k$ if and only if $k = n - m$, we get by Proposition 4.11,

$$\mathbb{S}_m^{(n)}\mathcal{C} \cong \mathbb{S}_{m-1}^{(n-2)}\mathcal{C} \cong \cdots \cong \mathbb{S}_{n-m}^{(n-m)}\mathcal{C}\,.$$

Thus, we have:

COROLLARY 4.12. *Let m be an integer between $\lceil \frac{n}{2} \rceil$ and $n-1$. Then*

$$\mathbb{S}_m^{(n)}\mathcal{C} \cong T_{n-m}\mathcal{C}\,.$$

5. Central simple algebras

Let $n \geq 2$ be an integer. Assume that $\mathrm{char}(F) = 0$ or $\mathrm{char}(F) = p > n + 5$. We prove in this section that $\mathbb{S}_\lambda\mathcal{C}$ is a central simple algebra for each partition λ of n of length 2. By Corollary 4.12, it suffices to prove that $T_n\mathcal{C}$ is a central simple algebra.

LEMMA 5.1. $\big(\sum_{i\in\mathcal{I}} e_i^{-1}x \otimes ye_i\big)^\vee \equiv 2(1 \otimes yx)^\vee \bmod F\cdot\triangle$ *for all $x, y \in \mathcal{C}$.*

PROOF. For each $j,k \in \mathcal{I}$, we have by (2.5) and Lemma 2.2,

$$\begin{aligned}\Big(\sum_{i\in\mathcal{I}} e_i^{-1}e_j \otimes e_k e_i\Big)^\vee &= \Big(\Big(\sum_{i\in\mathcal{I}}(2\delta_{ik} - \beta_i\beta_k)e_i^{-1}\otimes e_i\Big)(e_j\otimes e_k)\Big)^\vee \\ &= 2\big((e_k^{-1}\otimes e_k)(e_j\otimes e_k)\big)^\vee - \beta_k\Big(\Big(\sum_{i\in\mathcal{I}}\beta_i e_i^{-1}\otimes e_i\Big)(e_j\otimes e_k)\Big)^\vee \\ &= 2(e_ke_j\otimes 1)^\vee - 8\beta_k\big(\triangle(e_j\otimes e_k)\big)^\vee = 2(1\otimes e_ke_j)^\vee - 8\beta_k\triangle(e_j\otimes e_k)^\vee \\ &\equiv 2(1\otimes e_ke_j)^\vee \bmod F\cdot\triangle\,.\end{aligned}$$

Hence, the lemma follows by linearity. □

LEMMA 5.2. $\sum_{i\in\mathcal{I}}\big(1^{\otimes(n-1)}\otimes e_i^{-1}\big)^\vee e_n^{(n)}\big(1^{\otimes(n-1)}\otimes e_i\big)^\vee = \frac{2n+6}{n}\cdot e_n^{(n)}$.

PROOF. by Lemma 5.1 and (4.1),

$$\begin{aligned}&\sum_{i\in\mathcal{I}}\big(1^{\otimes(n-1)}\otimes e_i^{-1}\big)^\vee e_n^{(n)}\big(1^{\otimes(n-1)}\otimes e_i\big)^\vee \\ &\quad= \sum_{i\in\mathcal{I}}\Big(\big(1^{\otimes(n-1)}\otimes e_i^{-1}\big)^\vee\big(1^{\otimes(n-1)}\otimes e_i\big)\Big)^\vee e_n^{(n)} \\ &\quad= \frac{1}{n}\Big(\sum_{i\in\mathcal{I}} e_i^{-1}\otimes 1^{\otimes(n-2)}\otimes e_i + \cdots + \sum_{i\in\mathcal{I}} 1^{\otimes(n-2)}\otimes e_i^{-1}\otimes e_i + 8\cdot 1^{\otimes n}\Big)^\vee e_n^{(n)} \\ &\quad= \frac{1}{n}\big(2(n-1)+8\big)\cdot e_n^{(n)} = \frac{2n+6}{n}\cdot e_n^{(n)}\,.\end{aligned}$$

as required. □

For each integer m between 0 and n, we define the F-vector space

$$T_{n,m}\mathcal{C} = \{(1^{\otimes m}\otimes a)^\vee e_n^{(n)} \in T_n\mathcal{C} \mid a \in \mathcal{C}^{\otimes(n-m)}\}\,.$$

Note that $T_{n,0}\mathcal{C} = T_n\mathcal{C}$ and $T_{n,n}\mathcal{C} = F\cdot e_n^{(n)}$.

PROPOSITION 5.3. *Let m be an integer between 0 and $n-1$, let $x \in T_{n,m}\mathcal{C}$, and let*

$$y = \sum_{i\in\mathcal{I}} \frac{n^2 + (4n - 4mn - 10m + 2m^2)\delta_{i0}}{2(n-m)(n-m+5)}\big(1^{\otimes(n-1)}\otimes e_i^{-1}\big)^\vee x\big(1^{\otimes(n-1)}\otimes e_i\big)^\vee\,.$$

Then, $y \in T_{n,m+1}\mathcal{C}$ and $\operatorname{tr}_{\mathcal{C}^{\otimes n}}(x) = \operatorname{tr}_{\mathcal{C}^{\otimes n}}(y)$. We call y the **derived element** *of x.*

PROOF. First note that by Artin's theorem, there is no issue with associativity. Also, by the assumption on the characteristic of F, the denominator does not vanish.

By linearity, we may assume that $x = (1^{\otimes m}\otimes x_{m+1}\otimes\cdots\otimes x_n)^\vee e_n^{(n)}$ with $x_{m+1},\dots,x_n \in \mathcal{C}$. For each integer j between 1 and n and for each $a \in \mathcal{C}$, we denote

$$a_{(j)} = 1^{\otimes(j-1)}\otimes a\otimes 1^{\otimes(n-j)}\,.$$

In particular, $(1^{\otimes(n-1)}\otimes a)^\vee = \frac{1}{n}\sum_{j=1}^n a_{(j)}$. Let

$$z = \sum_{i\in\mathcal{I}}\big(1^{\otimes(n-1)}\otimes e_i^{-1}\big)^\vee x\big(1^{\otimes(n-1)}\otimes e_i\big)^\vee\,.$$

Then

$$z=\Big(\sum_{i\in\mathcal{I}}\big(1^{\otimes(n-1)}\otimes e_i^{-1}\big)^{\vee}\big(1^{\otimes m}\otimes x_{m+1}\otimes\cdots\otimes x_n\big)\big(1^{\otimes(n-1)}\otimes e_i\big)^{\vee}\Big)^{\vee}e_n^{(n)}$$
$$=\frac{1}{n^2}\sum_{j,k=1}^{n}\Big(\sum_{i\in\mathcal{I}}(e_i^{-1})_{(j)}\big(1^{\otimes m}\otimes x_{m+1}\otimes\cdots\otimes x_n\big)(e_i)_{(k)}\Big)^{\vee}e_n^{(n)}.$$

Let

$$z_{jk}=\sum_{i\in\mathcal{I}}(e_i^{-1})_{(j)}\big(1^{\otimes m}\otimes x_{m+1}\otimes\cdots\otimes x_n\big)(e_i)_{(k)}.$$

If $1\le j=k\le m$, then $z_{jk}^{\vee}e_n^{(n)}=8x$. By (2.7), $\sum_{i\in\mathcal{I}}e_i^{-1}ae_i=-4a+6\,\mathrm{tr}(a)1$, for each $a\in\mathcal{C}$. Hence, if $m+1\le j=k\le n$, then $z_{jk}^{\vee}e_n^{(n)}\equiv-4x \bmod T_{n,m+1}\mathcal{C}$. If $1\le j\le m$ and $k\ne j$ or if $1\le k\le m$ and $j\ne k$, then by Lemma 5.1 and since $\triangle^{(n)}e_n^{(n)}=\mathbf{0}$ (by (4.1)), $z_{jk}^{\vee}e_n^{(n)}=2x$. Finally, if $m+1\le j\ne k\le n$, then by Lemma 5.1 and by (4.1), $z_{jk}^{\vee}e_n^{(n)}\in T_{n,m+1}\mathcal{C}$. Putting all this together, we get

$$z=\frac{1}{n^2}\sum_{j,k=1}^{n}z_{jk}^{\vee}e_n^{(n)}\equiv\frac{1}{n^2}\big(8m-4(n-m)+2\cdot(2m(n-1)-m(m-1))\big)x \bmod T_{n,m+1}\mathcal{C}.$$

Thus,

$$z\equiv\frac{4mn-4n+10m-2m^2}{n^2}x \mod T_{n,m+1}\mathcal{C}.$$

Therefore

$$y_0=\sum_{i\in\mathcal{I}}\Big(1-\frac{4mn-4n+10m-2m^2}{n^2}\delta_{i0}\Big)\big(1^{\otimes(n-1)}\otimes e_i^{-1}\big)^{\vee}x\big(1^{\otimes(n-1)}\otimes e_i\big)^{\vee}$$
$$=z-\frac{4mn-4n+10m-2m^2}{n^2}x\in T_{n,m+1}\mathcal{C}.$$

By Lemma 5.2,

$$\sum_{i\in\mathcal{I}}\Big(1-\frac{4mn-4n+10m-2m^2}{n^2}\delta_{i0}\Big)\big(1^{\otimes(n-1)}\otimes e_i^{-1}\big)^{\vee}e_n^{(n)}\big(1^{\otimes(n-1)}\otimes e_i\big)^{\vee}$$
$$=\Big(\frac{2n+6}{n}-\frac{4mn-4n+10m-2m^2}{n^2}\Big)\cdot e_n^{(n)}=\frac{2(n-m)(n-m+5)}{n^2}\cdot e_n^{(n)}.$$

Note that

$$y=\frac{n^2}{2(n-m)(n-m+5)}y_0.$$

Hence, since

$$\mathrm{tr}_{\mathcal{C}^{\otimes n}}\Big(\big(1^{\otimes(n-1)}\otimes e_i^{-1}\big)^{\vee}x\big(1^{\otimes(n-1)}\otimes e_i\big)^{\vee}\Big)$$
$$=\mathrm{tr}_{\mathcal{C}^{\otimes n}}\Big(x\big(1^{\otimes(n-1)}\otimes e_i\big)^{\vee}\big(1^{\otimes(n-1)}\otimes e_i^{-1}\big)^{\vee}\Big)$$
$$=\mathrm{tr}_{\mathcal{C}^{\otimes n}}\Big(x\big(1^{\otimes(n-1)}\otimes e_i\big)^{\vee}e_n^{(n)}\big(1^{\otimes(n-1)}\otimes e_i^{-1}\big)^{\vee}\Big),$$

we have $\mathrm{tr}_{\mathcal{C}^{\otimes n}}(y)=\mathrm{tr}_{\mathcal{C}^{\otimes n}}(x)$. □

THEOREM 5.4. *$T_n\mathcal{C}$ is a central simple algebra.*

PROOF. Let $x = x^{(0)} \in T_n\mathcal{C}$. For each integer m between 0 and $n-1$, define inductively $x^{(m+1)} \in T_{n,m+1}\mathcal{C}$ as the derived element of $x^{(m)}$ that Proposition 5.3 gives. Then $x^{(n)} = \alpha_x \cdot e_n^{(n)}$, with $\alpha_x \in F$, and $\mathrm{tr}_{\mathcal{C}^{\otimes n}}(x) = \mathrm{tr}_{\mathcal{C}^{\otimes n}}(\alpha_x \cdot e_n^{(n)})$. Since $\mathrm{tr}_{T_n\mathcal{C}}$ is non-degenerate (Lemma 4.10), $\mathrm{tr}_{T_n\mathcal{C}}(e_n^{(n)}) \neq 0$. Hence

$$x^{(n)} = \frac{\mathrm{tr}_{\mathcal{C}^{\otimes n}}(x)}{\mathrm{tr}_{\mathcal{C}^{\otimes n}}(e_n^{(n)})} e_n^{(n)}. \tag{5.1}$$

If $x \neq 0$, then since $\mathrm{tr}_{T_n\mathcal{C}}$ is non-degenerate, there exists $x' \in T_n\mathcal{C}$ such that $\mathrm{tr}_{\mathcal{C}^{\otimes n}}(xx') \neq 0$. By (5.1), $(xx')^{(n)} \neq 0$, and by its definition, $(xx')^{(n)}$ belongs to the (two-sided) ideal in $T_n\mathcal{C}$ generated by xx'. Thus, $e_n^{(n)}$ belongs to the ideal generated by x. It follows that $T_n\mathcal{C}$ is simple, since $e_n^{(n)}$ is the unit element in $T_n\mathcal{C}$.

Suppose now that x lies in the center of $T_n\mathcal{C}$. In particular, it commutes with each $\left(1^{\otimes(n-1)} \otimes e_i\right)^{\vee} e_n^{(n)}$. Hence, by the definition of the $x^{(m)}$'s, $x^{(n)} = x \cdot \left(e_n^{(n)}\right)^{(n)}$. By (5.1), $\left(e_n^{(n)}\right)^{(n)} = e_n^{(n)}$ and $x^{(n)} \in F \cdot e_n^{(n)}$. Thus, $x \in F \cdot e_n^{(n)}$. It follows that $T_n\mathcal{C}$ is central. □

By Corollary 4.7, Corollary 4.12, and Theorem 5.4, we reach our main result:

COROLLARY 5.5. *$\mathrm{Sym}^n \mathcal{C}$ is a direct sum of central simple algebras over F.*

If A is a finite dimensional central simple alternative algebra over F, then A is either associative or an octonion algebra, by a result of M. Zorn [**Sch43**]. Thus, by Corollary 1.6 and Corollary 5.5, we get:

THEOREM 5.6. *Let A be a finite dimensional central simple alternative algebra over a field F of characteristic 0 or of positive characteristic p with*

- *$p > n$ if A is associative, or*
- *$p > n + 5$ if A is not associative.*

Then $\mathrm{Sym}^n A$ is a direct sum of central simple algebras over F.

References

[Ba02] John C. Baez, *The octonions*, Bull. Amer. Math. Soc. (N.S.) **39** (2002), no. 2, 145–205, DOI 10.1090/S0273-0979-01-00934-X. MR1886087

[FuH04] W. Fulton and J. Harris, *Representation Theory*, Springer, 2004.

[GG73] Murat Günaydin and Feza Gürsey, *Quark structure and octonions*, J. Mathematical Phys. **14** (1973), 1651–1667, DOI 10.1063/1.1666240. MR329463

[GG74] M. Günaydin and F. Gürsey, *Quark statistics and octonions*, Physical Review D, **9** (1974), no. 12, 3387–3391.

[Jac85] Nathan Jacobson, *Basic algebra. I*, 2nd ed., W. H. Freeman and Company, New York, 1985. MR780184

[Jac89] Nathan Jacobson, *Basic algebra. II*, 2nd ed., W. H. Freeman and Company, New York, 1989. MR1009787

[MPP01] Patrick J. Morandi, José M. Pérez-Izquierdo, and S. Pumplün, *On the tensor product of composition algebras*, J. Algebra **243** (2001), no. 1, 41–68, DOI 10.1006/jabr.2001.8773. MR1851653

[Row88] Louis H. Rowen, *Ring theory. Vol. II*, Pure and Applied Mathematics, vol. 128, Academic Press, Inc., Boston, MA, 1988. MR945718

[Sal99] David J. Saltman, *Lectures on division algebras*, CBMS Regional Conference Series in Mathematics, vol. 94, Published by American Mathematical Society, Providence, RI; on behalf of Conference Board of the Mathematical Sciences, Washington, DC, 1999. MR1692654

[Sch43] R. D. Schafer, *Alternative algebras over an arbitrary field*, Bull. Amer. Math. Soc. **49** (1943), 549–555, DOI 10.1090/S0002-9904-1943-07967-0. MR8610

[Sch66] Richard D. Schafer, *An introduction to nonassociative algebras*, Pure and Applied Mathematics, Vol. 22, Academic Press, New York-London, 1966. MR0210757

[Wey46] Hermann Weyl, *The Classical Groups. Their Invariants and Representations*, Princeton University Press, Princeton, N.J., 1939. MR0000255

Elta Systems Ltd., Ashdod, Israel
Email address: razona@elta.co.il

Contemporary Mathematics
Volume **767**, 2021
https://doi.org/10.1090/conm/767/15406

Representations of étale fundamental groups and specialization of algebraic cycles

Anna Cadoret

In honor of Gerhard Frey's 75th birthday

ABSTRACT. We describe an elementary strategy to study the locus where a finite family of linearly independent 1-cohomology classes for the étale fundamental group remains linearly independent under specialization. We apply this strategy to the injectivity of the specialization morphism on the second graded piece of the ℓ-adic Abel-Jacobi filtration on Chow groups varying in the fibers of a smooth projective morphism. When the base scheme is a curve this provides a generalization (to Chow groups and arbitrary finitely generated fields of characteristic 0) of a theorem of Silverman about the sparsity of the jumping locus of the rank in the fibers of an abelian scheme.

1. Introduction

1.1. Let k be a field of characteristic $p \geq 0$. A variety over k means a separated scheme of finite type over k. Let S be a smooth, geometrically connected variety over k. Set $\overline{S} := S \times_k \overline{k}$. Let η denote the generic point of S, $|S|$ the set of closed points of S. For a subset $\Sigma \subset |S|$ and an integer $d \geq 1$, let $\Sigma^{\leq d}$ denote the set of all $s \in \Sigma$ with $[k(s) : k] \leq d$ (in particular, $|S|^{\leq 1} = S(k)$). For every $s \in S$, fix a geometric point $\overline{s}$ over s and an étale path $\overline{\eta} \tilde{\to} \overline{s}$, which induces an isomorphism $\pi_1(s, \overline{s}) \simeq \pi_1(s, \overline{\eta})$. By functoriality of the étale fundamental group, $s \in S$ regarded as a morphism $s : \mathrm{spec}(k(s)) \to S$ induces a continuous group morphism $\sigma_s : \pi_1(s, \overline{s}) \simeq \pi_1(s, \overline{\eta}) \to \pi_1(S, \overline{\eta})$, which is injective if $s \in |S|$. As the choice of base points will basically play no part in the following, we will omit them from the notation unless necessary. For a field K and an (any) algebraic closure $\overline{K}$ of K, write

$$\pi_1(K) := \pi_1(\mathrm{spec}(K), \mathrm{spec}(\overline{K}))$$

for the absolute Galois group of K.

1.2. Fix a prime $\ell \neq p$ and V a finite-dimensional $\mathbb{Q}_\ell$-vector space endowed with a continuous action of $\pi_1(S)$. Write Π_V, $\overline{\Pi}_V$ and $\Pi_{s,V}$ for the image of $\pi_1(S)$, $\pi_1(\overline{S})$ and $\pi_1(s)$ acting on V respectively. Let $S(V) \subset |S|$ denote the set of all $s \in |S|$ such that $\Pi_{s,V} \subset \Pi_V$ is open. For a finite-dimensional $\mathbb{Q}_\ell$-subvector space $E \hookrightarrow \mathrm{H}^1(\pi_1(S), V)$, let $S(E, V) \subset |S|$ denote the set of all $s \in |S|$ such that

$$res_s|_E : E \hookrightarrow \mathrm{H}^1(\pi_1(S), V) \stackrel{res_s}{\to} \mathrm{H}^1(\pi_1(s), V)$$

2020 *Mathematics Subject Classification.* Primary 14H30, 14C25; Secondary 14K15.

is injective, where res_s denotes the restriction morphism with respect to $\sigma_s : \pi_1(s) \to \pi_1(S)$.

The problem addressed in this note is to describe the arithmetico-geometric structure of $S(E,V)$ when k is 'arithmetically rich' that is, essentially, with a "large absolute Galois group $\pi_1(k)$".

This is motivated by studying the variation of certain type of arithmetico-geometric invariants in families of smooth, proper varieties. Here is an example. Let $A \to S$ be an abelian scheme. The Kummer map and the Néron extension property for abelian schemes give rise to a canonical commutative diagram

$$\begin{array}{ccc} A_\eta(k(\eta)) \otimes \mathbb{Q} & \overset{\kappa_\ell}{\hookrightarrow} & \mathrm{H}^1(\pi_1(\eta), V_\ell(A_{\overline{\eta}})) \\ \uparrow \simeq & & \uparrow \simeq \, infl \\ A(S) \otimes \mathbb{Q} & \hookrightarrow & \mathrm{H}^1(\pi_1(S), V_\ell(A_{\overline{\eta}})) \\ \downarrow sp_s & & \downarrow res_s \\ A_s(k(s))) \otimes \mathbb{Q} & \overset{\kappa_\ell}{\hookrightarrow} & \mathrm{H}^1(\pi_1(s), V_\ell(A_{\overline{s}})), \end{array}$$

where $V_\ell(-) := (\lim_{\leftarrow} - [\ell^n]) \otimes \mathbb{Q}_\ell$ denotes the $\mathbb{Q}_\ell$-Tate module of $-$. Taking $V = V_\ell(A_{\overline{\eta}})$ and $E = A_\eta(k(\eta)) \otimes \mathbb{Q}$, the locus $S(E,V) \subset |S|$ is the set of all $s \in |S|$ where the specialization map $sp_s : A_\eta(k(\eta)) \otimes \mathbb{Q} \to A(S) \otimes \mathbb{Q}$ is injective and, in particular, it is contained in the set of all $s \in |S|$ where $\mathrm{rank}(A_\eta(k(\eta))) \leq \mathrm{rank}(A_s(k(s)))$.

1.3. In general, one expects $S(E,V)$ to be 'huge'. This relies on the observation that one can attach to (E,V) a "universal extension" $\tilde{E}$, which is a finite-dimensional $\mathbb{Q}_\ell$-vector space endowed with a continuous action of $\pi_1(S)$ fitting into a $\pi_1(S)$-equivariant short exact sequence $0 \to V \to \tilde{E} \to E \to 0$ (with E endowed with the trivial $\pi_1(S)$-action) and with the property that $S(\tilde{E}) \subset S(E,V)$; see Subsection 2.1 and Lemma 2.1.3 for details. The point is that, under mild assumptions on V, one expects $S(\tilde{E})$ to be 'huge' (see 3). More precisely, if k is Hilbertian, there always exists an integer $d \geq 1$ such that $S(\tilde{E})^{\leq d}$ is infinite (Fact 3.1.1). This is already enough for applications which only require the existence of a single closed point in $S(E,V)$ (see *e.g* Subsection 6.1). Under mild assumptions on V (satisfied by representations V arising from geometry as in the example of 1.2) and when k is finitely generated over $\mathbb{Q}$, one expects stronger abundance results for $S(E,V)$. For higher dimensional varieties S such results are still highly conjectural but when S is a curve, one can ensure that if $\overline{\Pi}_V$ has perfect Lie algebra then for every integer $d \geq 1$, $(|S| \setminus S(E,V))^{\leq d}$ is finite (Fact 5). To apply our strategy, we have to find conditions on V which can be checked in practice and which ensure that $\overline{\Pi}_{\tilde{E}}$ has perfect Lie algebra. This is basically the content of Lemma 2.2.

We also provide a variant of our results for families $\underline{V} = V_\ell$, $\ell \in \mathcal{L}$ of finite dimensional $\mathbb{F}_\ell$-vector spaces of bounded rank and endowed with a continuous action of $\pi_1(S)$ for ℓ varying in an infinite set $\mathcal{L}$ of primes.

1.4. The group-theoretical results described in Subsection 1.3 are stated and proved in Section 2. The remaining sections of the paper (Sections 4 - 6) are devoted to applications of these result to motivic representations (that is those arising from the étale cohomology of smooth proper morphisms $X \to S$). We discuss in particular the injectivity of the specialization map for the second graded piece of the ℓ-adic Abel-Jacobi filtration on Chow groups in the fibers of a smooth projective morphism $X \to S$. This provides a generalization (to Chow groups and arbitrary finitely generated fields) of a theorem of Silverman about the sparsity of the dropping locus of the rank in the fibers of an abelian scheme; the theorem of Silverman corresponds to the example of Subsection 1.2 (see Corollary 5.2.2 and, for the case of Abelian schemes, 6.2). We also show that the locus where the unipotent part of the Mumford-Tate group of a 1-motive degenerates is sparse (See Subsection 6.2.3). The reader who is mostly interested in these applications can jump directly to Sections 4 - 6 and browse the formal Section 2 only when needed.

1.5. This note does not intend to exhaustivity but rather to present a method that could hopefully be applied to other situations. In particular, we only deal with $\mathbb{Q}_\ell$ or $\mathbb{F}_\ell$-coefficients but the method extend to more general coefficients. It would also be interesting to investigate what could be said for higher cohomology groups (for i smaller than the cohomological dimension of the fields $k(s)$, $s \in |S|$). The method itself (see Section 2) relies on elementary group-theoretic observations. What is *not* elementary and involves subtle arithmetico-geometric inputs are the specialization results (see Subsection 3) we inject in the method and the fact that the method can be applied to motivic representations (see Section 4). It seems unlikely that our sparsity results could be enhanced for arbitrary $E \subset \mathrm{H}^1(\pi_1(S), V)$. But when $E \subset \mathrm{H}^1(\pi_1(S), V)$ parametrizes cocycles of geometric origin (as it is the case in our application to Chow groups), one expects stronger sparsity results (algebraicity, finiteness, bounded height - see Subsection 6.2.3 for discussion). This illustrates both the generality of the method (which applies to arbitrary cohomology classes) and its limits (since it cannot see the difference between transcendental and geometric cohomology classes).

1.6. In a first version of this note, we also included additional applications to the case where the base field k has characteristic $p > 0$ (and for primes $\ell \neq p$). These applications were relying on specialization results in [**CT19**], [**A18**], which in turn were relying on [**EEHK09**, Thm. 2] - a tentative positive characteristic analogue of the Mordell conjecture. But just before resubmitting the revised version of this note, Akio Tamagawa exhibited a counter-example to [**EEHK09**, Thm. 2], showing that the statement of [**EEHK09**, Thm. 2] has to be refined (to include some isotriviality phenomena). The results of [**CT19**], [**A18**] should be adjusted consequently. These could then be (re)injected in the strategy described in this note to obtain similar applications when the base field k has characteristic $p > 0$.

2. Representation-theoretic results

Let Λ denote $\mathbb{Q}_\ell$, $\mathbb{Z}_\ell$ or $\mathbb{F}_\ell$ for some prime ℓ and let Π be a profinite group. Write $\mathrm{Mod}_\Lambda(\Pi)$ for the (abelian) category of finitely generated Λ-modules equipped with a continuous Λ-linear action of Π. In particular, $\mathrm{Mod}_\Lambda := \mathrm{Mod}_\Lambda(1)$ is just the category of finitely generated Λ-modules. For $V \in \mathrm{Mod}_\Lambda(\Pi)$ let $\Pi_V \subset \mathrm{Aut}_\Lambda(V)$, $\Pi^V \subset \Pi'$ denote respectively the image and kernel of the corresponding morphism $\Pi \to \mathrm{Aut}_\Lambda(V)$ so that one has a short exact sequence of profinite groups

$$1 \to \Pi^V \to \Pi \to \Pi_V \to 1.$$

For a continuous morphism of profinite groups $\Gamma \to \Pi$, write $-|_\Gamma : \mathrm{Mod}_\Lambda(\Pi) \to \mathrm{Mod}_\Lambda(\Gamma)$ for the obvious restriction functor. Let $\mathrm{H}^i(\Pi, -) : \mathrm{Mod}_\Lambda(\Pi) \to \mathrm{Mod}_\Lambda$

denote the ith continuous cohomology group functor of Π (defined by means of the continuous i-cochains $Z^i(\Pi, V)$- see [**T76**]).

2.1. The universal extension $\tilde{E}$. Let Λ denote $\mathbb{Q}_\ell$ or $\mathbb{F}_\ell$ for some prime ℓ and let Π be a profinite group. Fix $V \in \mathrm{Mod}_\Lambda(\Pi)$. Recall that if $\Lambda = \mathbb{Q}_\ell$, Π_V being a closed subgroup of $\mathrm{Aut}_\Lambda(V)$ is a (compact) ℓ-adic Lie group; we write $\mathrm{Lie}(\Pi_V)$ for its Lie algebra (as a ℓ-adic Lie group).

2.1.1. *Construction.* Let $\mathrm{Ext}^1_\Pi(-, V) : \mathrm{Mod}_\Lambda \to Ab$ denote the functor sending $E \in \mathrm{Mod}_\Lambda$ to the abelian group $\mathrm{Ext}^1_\Pi(E, V)$ of equivalence classes of extensions

$$0 \to V \to \tilde{E} \to E \to 0$$

in $\mathrm{Mod}_\Lambda(\Pi)$. There is a canonical isomorphism of functors $\mathrm{Mod}_\Lambda \to \mathrm{Mod}_\mathbb{Z}$

$$\phi := \phi(\Pi, V) : \mathrm{Hom}(-, \mathrm{H}^1(\Pi, V)) \tilde{\to} \mathrm{Ext}^1_\Pi(-, V)$$

defined as follows. Let $\widetilde{-} : \mathrm{H}^1(\Pi, V) \hookrightarrow Z^1(\Pi, V)$ be a section in Mod_Λ of the canonical projection $Z^1(\Pi, V) \twoheadrightarrow \mathrm{H}^1(\Pi, V)$. For $E \in \mathrm{Mod}_\Lambda$ and a morphism $f : E \to \mathrm{H}^1(\Pi, V)$ in Mod_Λ, a representative of the isomorphism class $\phi(E)(f) \in \mathrm{Ext}^1_\Pi(E, V)$ is given by $\tilde{E} = V \oplus E$ endowed with the Λ-linear action of Π given by $\pi \cdot (m \oplus e) = (\pi m + \widetilde{f(e)}(\pi)) \oplus e$, $\pi \in \Pi$. One easily checks that $\phi(E)(f)$ does not depend on the choice of the section $\widetilde{-} : \mathrm{H}^1(\Pi, V) \hookrightarrow Z^1(\Pi, V)$ and that the construction defines a morphism of functors $\mathrm{Mod}_\Lambda \to Ab$. To show that ϕ is an isomorphism, one can exhibit an explicit inverse $\psi := \psi(\Pi, V) : \mathrm{Ext}^1_\Pi(-, V) \to \mathrm{Hom}_{\mathrm{Mod}_\Lambda}(-, \mathrm{H}^1(\Pi, V))$ as follows. For the isomorphism class $[\tilde{E}] \in \mathrm{Ext}^1_\Pi(E, V)$ of an extension $\tilde{E}$, the morphism $\psi(E)([\tilde{E}]) : E \to \mathrm{H}^1(\Pi, V)$ sends $e \in E$ to the cohomology class of the 1-cocycle

$$\begin{array}{ccc} \Pi & \to & V \\ \pi & \to & \pi\tilde{e} - \tilde{e}, \end{array}$$

where $\widetilde{-} : E \to \tilde{E}$ is a section in Mod_Λ of the projection $\tilde{E} \twoheadrightarrow E$. Again, one easily checks that $\psi(E)([\tilde{E}])$ does not depend on the choice of $\widetilde{-} : E \to \tilde{E}$ and that $\psi(E) \circ \phi(E) = Id$, $\phi(E) \circ \psi(E) = Id$.

2.1.2. *Functoriality.* The construction of $\phi := \phi(\Pi, V)$ is functorial in Π, V. In particular, for short exact sequence of profinite groups $1 \to N \to \Pi \to \Pi/N \to 1$, one has a canonical inflation-restriction commutative diagram with exact lines

$$\begin{array}{ccccccc} 0 & \longrightarrow & \mathrm{Hom}(-, \mathrm{H}^1(\Pi/N, V^N)) & \xrightarrow{infl} & \mathrm{Hom}(-, \mathrm{H}^1(\Pi, V)) & \xrightarrow{res} & \mathrm{Hom}(-, \mathrm{H}^1(N, V)) \\ & & \Big\downarrow {\scriptstyle \phi(\Pi/N, V^N)} & & \Big\downarrow {\scriptstyle \phi(\Pi, V)} & & \Big\downarrow {\scriptstyle \phi(N, V)} \\ 0 & \longrightarrow & \mathrm{Ext}^1_{\Pi/N}(-, V^N) & \xrightarrow[(V^N \hookrightarrow V)_*]{} & \mathrm{Ext}^1_\Pi(-, V) & \xrightarrow[|_N]{} & \mathrm{Ext}^1_N(-, V) \end{array}$$

2.1.3. *Injectivity criterion for the restriction morphism on H^1.* Let $\Gamma \to \Pi$ be a morphism of profinite groups, let $V \in \mathrm{Mod}_\Lambda(\Pi)$ and $\iota : E \hookrightarrow \mathrm{H}^1(\Pi, V)$ a finitely generated Λ-submodule. Let $\tilde{E}$ be a representative of $\phi(E)(\iota) \in \mathrm{Ext}^1_\Pi(E, V)$. Recall that $\Pi_{\tilde{E}} \subset \mathrm{Aut}(\tilde{E})$ denote the image of Π acting on $\Pi_{\tilde{E}}$ and that $\Pi^{\tilde{E}} = \ker(\Pi \to \Pi_{\tilde{E}})$.

LEMMA 1. *Fix a closed normal subgroup* $N \subset \Pi_{\tilde{E}}$. *Assume* $\Lambda = \mathbb{Q}_\ell$ *(resp.* $\Lambda = \mathbb{F}_\ell$*) and*

(SS) $\Pi_{\tilde{E}}/N$ *acts semisimply on* V^N *(resp. and* $\ell \gg 0$ *compared with the* Λ*-rank of* V^N*).*

(O) $\Gamma_{\tilde{E}} \cap N \subset N$ *is open of index* $[N : \Gamma_{\tilde{E}} \cap N]$ *invertible in* Λ.

Then $E \overset{\iota}{\hookrightarrow} \mathrm{H}^1(\Pi, V) \overset{res}{\to} \mathrm{H}^1(\Gamma, V)$ *is injective.*

PROOF. By construction, $\phi(E)(\iota)|_{\Pi^{\tilde{E}}} = 0$ hence by 2.1.2, $\iota : E \hookrightarrow \mathrm{H}^1(\Pi, V)$ factors through

$$(*)\qquad \begin{array}{ccccccc}
 & & & & E & & \\
 & & & \overset{\iota}{\swarrow} & \downarrow \iota & \overset{0}{\searrow} & \\
0 \longrightarrow & \mathrm{H}^1(\Pi_{\tilde{E}}, V) & \xrightarrow{infl} & \mathrm{H}^1(\Pi, V) & \xrightarrow{res} & \mathrm{H}^1(\Pi^{\tilde{E}}, V) \\
 & \downarrow res & & \downarrow res & & \downarrow res \\
0 \longrightarrow & \mathrm{H}^1(\Gamma_{\tilde{E}}, V) & \xrightarrow{infl} & \mathrm{H}^1(\Gamma, V) & \xrightarrow{res} & \mathrm{H}^1(\Gamma \cap \Pi^{\tilde{E}}, V)
\end{array}$$

So it is enough to prove that the left vertical arrow of $(*)$ is injective. From the inflation-restriction diagram

$$(**)\qquad \begin{array}{cccccc}
0 \longrightarrow & \mathrm{H}^1(\Pi_{\tilde{E}}/N, V^N) & \xrightarrow{infl} & \mathrm{H}^1(\Pi_{\tilde{E}}, V) & \xrightarrow{res} & \mathrm{H}^1(N, V) \\
 & \downarrow res & & \downarrow res & & \downarrow res \\
0 \longrightarrow & \mathrm{H}^1(\Gamma_{\tilde{E}} \cap N/N, V^N) & \xrightarrow{infl} & \mathrm{H}^1(\Gamma_{\tilde{E}}, V) & \xrightarrow{res} & \mathrm{H}^1(\Gamma_{\tilde{E}} \cap N, V)
\end{array}$$

to show that $res : \mathrm{H}^1(\Pi_{\tilde{E}}, V) \to \mathrm{H}^1(\Gamma_{\tilde{E}}, V)$ is injective it is enough to show that $\mathrm{H}^1(\Pi_{\tilde{E}}/N, V^N) = 0$ and $res : \mathrm{H}^1(N, V) \to \mathrm{H}^1(\Gamma_{\tilde{E}} \cap N, V)$ is injective. The injectivity of $res : \mathrm{H}^1(N, V) \to \mathrm{H}^1(\Gamma_{\tilde{E}} \cap N, V)$ follows from (O) by corestriction while $\mathrm{H}^1(\Pi_{\tilde{E}}/N, V^N) = 0$ follows from (SS). More precisely, if $\Lambda = \mathbb{Q}_\ell$ then $\Pi_{\tilde{E}}/N$ is a compact ℓ-adic Lie group hence its ℓ-adic Lie algebra $\mathrm{Lie}(\Pi_{\tilde{E}}/N)$ is reductive. This implies $\mathrm{H}^1(\mathrm{Lie}(\Pi_{\tilde{E}}/N), V) = 0$ and the assertion follows from the injectivity of the canonical morphism $\mathrm{H}^1(\Pi_{\tilde{E}}/N, V^N) \to \mathrm{H}^1(\mathrm{Lie}(\Pi_{\tilde{E}}/N), V^N)$ [**Se64**, Prop. 12]. If $\Lambda = \mathbb{F}_\ell$, this is [**N87**, Thm. E]. □

In applications, we will consider $N = \Pi_{\tilde{E}}$ (in which case Condition (SS) is empty) or $N = \ker(\Pi_{\tilde{E}} \twoheadrightarrow \Pi_V)$.

2.2. Transfer of perfectness. We retain the notation and conventions of Subsection 2.1. If $\Lambda = \mathbb{Q}_\ell$ (resp. $\Lambda = \mathbb{F}_\ell$) we say that $V \in \mathrm{Mod}_\Lambda(\Pi)$ has no Π-Lie-coinvariants (resp. has no Π-coinvariants) if $W_{\mathrm{Lie}(\Pi_V)} = 0$ for every subobject $W \subset V$ in $\mathrm{Mod}_\Lambda(\Pi)$ (resp. if $W_\Pi = 0$ for every subobject $W \subset V$ in $\mathrm{Mod}_\Lambda(\Pi)$). In practice, to ensure Condition (O) of Lemma 2.1.3 holds for the groups $\Gamma(:= \pi_1(s)) \subset \Pi(:= \pi_1(S))$ we are interested in, we will use the following observation.

LEMMA 2. *Let* $E \in \mathrm{Mod}_\Lambda$ *and* $\tilde{E} \in \mathrm{Ext}^1_\Pi(E, V)$. *Assume* $\Lambda = \mathbb{Q}_\ell$ *(resp.* $\Lambda = \mathbb{F}_\ell$*) and* V *has no* Π*-Lie-coinvariants (resp. no* Π*-coinvariants). Then* $\mathrm{Lie}(\Pi_{\tilde{E}})^{ab} = \mathrm{Lie}(\Pi_V)^{ab}$ *(resp.* $\Pi_{\tilde{E}}^{ab} = \Pi_V^{ab}$*), where* $(-)^{ab}$ *denote the abelianization functor for Lie algebras (resp. finite groups).*

PROOF. We only give the proof for $\Lambda = \mathbb{Q}_\ell$; the proof for $\Lambda = \mathbb{F}_\ell$ is exactly similar, working in the category of groups instead of Lie algebras. Applying the Lie functor to the short exact sequence of ℓ-adic Lie groups $1 \to N \to \Pi_{\tilde{E}} \to \Pi_V \to 1$ we get the short exact sequence of $\mathbb{Q}_\ell$-Lie algebras

$$0 \to \mathrm{Lie}(N) \to \mathrm{Lie}(\Pi_{\tilde{E}}) \to \mathrm{Lie}(\Pi_V) \to 0$$

and, applying the abelianization functor, the short exact sequence

$$0 \to \mathrm{Lie}(N) \cap [\mathrm{Lie}(\Pi_{\tilde{E}}), \mathrm{Lie}(\Pi_{\tilde{E}})] \to \mathrm{Lie}(N) \to \mathrm{Lie}(\Pi_{\tilde{E}})^{ab} \to \mathrm{Lie}(\Pi_V)^{ab} \to 0.$$

To show $\mathrm{Lie}(N) \subset [\mathrm{Lie}(\Pi_{\tilde{E}}), \mathrm{Lie}(\Pi_{\tilde{E}})]$, it is enough to show $\mathrm{Lie}(N_E) = [\mathrm{Lie}(\Pi_{\tilde{E}}), \mathrm{Lie}(N)]$, which follows from

$$\mathrm{Lie}(N)/[\mathrm{Lie}(\Pi_{\tilde{E}}), \mathrm{Lie}(N)] = \mathrm{Lie}(N)_{\mathrm{Lie}(\Pi_{\tilde{E}})}$$

and $\mathrm{Lie}(N)_{\mathrm{Lie}(\Pi_{\tilde{E}})} = 0$ since $\mathrm{Lie}(N) \subset E^\vee \otimes V \simeq V^{\dim(E^\vee)}$ is a Π-invariant submodule. □

3. Specialization of representations of the étale fundamental group and injectivity of the restriction morphism on H^1

Let S be a smooth, geometrically connected variety over k. We are now going to apply the results of Section 2 to the case where $\Pi := \pi_1(S)$ and $\Gamma = \Pi_s := \pi_1(s)$ for $s \in |S|$.

3.1. $\Lambda = \mathbb{Q}_\ell$**.** Let $H \in \mathrm{Mod}_{\mathbb{Q}_\ell}(\Pi)$. Recall that $S(H) \subset |S|$ denote the set of all $s \in |S|$ such that $\Pi_{s,H} \subset \Pi_H$ is open. By definition, when $H = \tilde{E}$ is the universal extension attached to a pair (E, V) (Subsection 2.1) Condition (O) of Lemma 2.1.3 is satisfied for $N = \Pi_{\tilde{E}}$, $\Gamma_{\tilde{E}} := \Pi_{s,\tilde{E}}$, $s \in S(\tilde{E})$. In particular, $S(\tilde{E}) \subset S(E, V)$.

When k is 'arithmetically rich' in the sense that it has a huge non-abelian absolute Galois group, one expect $S(H)^{\leq d}$ to be 'huge'.

3.1.1. The first result in this direction is the following elementary group-theoretical and very general observation.

FACT 3. [1] ([**Se89**, 10.6]) Assume k is Hilbertian of characteristic $p \geq 0$ and let $\mathcal{L}$ be a finite set of primes $\neq p$. For each $\ell \in \mathcal{L}$ fix $H_\ell \in \mathrm{Mod}_{\mathbb{Q}_\ell}(\Pi)$. Then there exists an integer $d \geq 1$ such that for infinitely many $s \in |S|^{\leq d}$, the images of Π_s and Π acting on $\oplus_{\ell \in \mathcal{L}} H_\ell$ coincide. In particular, $\cap_{\ell \in \mathcal{L}} S(H_\ell)^{\leq d}$ is infinite.

Fact 3 applies in particular to fields k which are finitely generated over their prime field [**Se89**, 9.5 - Rem. 4), 5), 9.6]. Combining Fact 3.1.1 and Lemma 2.1.3, one gets:

COROLLARY 4. *Assume k is Hilbertian of characteristic $p \geq 0$ and let $\mathcal{L}$ be a finite set of primes $\neq p$. For each $\ell \in \mathcal{L}$, let $V_\ell \in \mathrm{Mod}_{\mathbb{Q}_\ell}(\Pi)$ and $E_\ell \subset H^1(\Pi, V_\ell)$ a finite-dimensional $\mathbb{Q}_\ell$-vector subspace. Then there exists an integer $d \geq 1$ such that $\cap_{\ell \in \mathcal{L}} S(E_\ell, V_\ell)^{\leq d}$ is infinite.*

---

[1] More generally, if $\Pi \twoheadrightarrow \Pi''$ is any continuous quotient such that the Frattini subgroup of Π'' is open in Π then there exists an integer $d \geq 1$ such that for infinitely many $s \in |S|^{\leq d}$, the images of Π_s and Π in Π'' coincide. The condition that the Frattini subgroup of Π'' is open in Π is equivalent to the fact that the set of primes dividing the order of Π'' is finite and that the p-Sylow of Π'' are topologically finitely generated. In particular, it is satisfied if Π'' is a closed subgroup of a finite product of compact ℓ-adic Lie groups. See [**Se89**, 10.6, Prop. p. 148].

3.1.2. When k is finitely generated and provided Π_H satisfies some mild assumption, one expects $S(H)^{\leq d}$ satisfies much stronger abundance results. For instance, if H is motivic, $d = 1$ and k is a number field, it should follow from the Bombieri-Lang conjecture and some (more tractable) conjectures on the geometric properties of some projective systems of étale covers of S attached to V that $(|S| \setminus S(H))(k)$ is not Zariski-dense in S (See *e.g.* the brief discussion in [**CCh20**, §2.2]). In this direction, one has the following unconditional result, which ultimately relies on Mordell and Mordell-Lang conjectures.

FACT 5. (([**CT12**], [**CT13**]) Assume k is finitely generated over $\mathbb{Q}$ and S is a curve. Suppose (P) $\mathrm{Lie}(\overline{\Pi}_H)$ is perfect. Then for every integer $d \geq 1$, $(|S|\setminus S(H))^{\leq d}$ is finite.

REMARK. (1) Furthermore, $\sup\{[\Pi_V : \Pi_{s,V}] \mid s \in (|S| \setminus S(V))^{\leq d}\} < +\infty$ but we will not need this uniformity result in our applications.
(2) For an analogue of Fact 5 to the char $p > 0$ case when $d = 1$ see [**A18**].

From Lemma 2.2, the condition that V has no $\overline{\Pi}$-Lie-coinvariants ensures Condition (P) of Fact 5 is satisfied by the universal extension $\tilde{E}$ attached to any pair (E, V). Combining this with Lemma 2.1.3 and Fact 5, one gets:

COROLLARY 6. *Assume k is finitely generated of characteristic* 0 *and S is a curve. Let $V \in \mathrm{Mod}_{\mathbb{Q}_\ell}(\Pi)$. Assume*

(P) $\mathrm{Lie}(\overline{\Pi}_V)$ *is perfect;*
(cI) V *has no $\overline{\Pi}$-Lie-coinvariants.*

Then, for every finite-dimensional $\mathbb{Q}_\ell$-vector subspace $E \subset H^1(\Pi, V)$ and integer $d \geq 1$, $(|S| \setminus S(E, V))^{\leq d}$ is finite.

3.2. $\Lambda = \mathbb{F}_\ell$. Let $H \in \mathrm{Mod}_{\mathbb{F}_\ell}(\Pi)$. Let $S(H) \subset |S|$ denote the set of all $s \in |S|$ such that $\Pi_{s,H} \subset \Pi_H$ is of index prime-to-ℓ. By definition, when $H = \tilde{E}$ is the universal extension attached to a pair (E, V) (Subsection 2.1) Condition (O) of Lemma 2.1.3 is satisfied for $N = \Pi_{\tilde{E}}$, $\Gamma_{\tilde{E}} := \Pi_{s,\tilde{E}}$, $s \in S(\tilde{E})$. In particular, $S(\tilde{E}) \subset S(E, V)$.

Let $\mathcal{L}$ be an infinite set of primes and let $\underline{H} = H_\ell$, $\ell \in \mathcal{L}$ be a family of elements in $\mathrm{Mod}_{\mathbb{F}_\ell}$ of uniformly bounded Λ_ℓ-rank $r_\ell \leq r$. Since $\mathcal{L}$ is infinite, Fact 3.1.1 (under its general form stated in footnote 1) fails. Still, one has the following analogue of Fact 3.1.2. For a finite subgroup $G \subset \mathrm{GL}(V_\ell)$, let $G^+ \subset G$ denote the (characteristic) subgroup of G generated by its order ℓ elements.

FACT 7. ([**EHK12**], [**CT19**, §7]) Assume k is finitely generated over $\mathbb{Q}$ and S is a curve. Suppose $\underline{H}$ satisfies

(U) There exists an open subgroup $U \subset \overline{\Pi}$ such that $U_{H_\ell} = U^+_{H_\ell}$, $\ell \in \mathcal{L}$;
(P) $\overline{\Pi}^+_{H_\ell}$ is perfect for $\ell \gg 0$.

Then for every integer $d \geq 1$ there exists an integer $B_d \geq 1$ such that for $\ell \gg 0$ the set of all $s \in |S|^{\leq d}$ with $[\Pi_{H_\ell} : \Pi_{s,H_\ell}] > B_d$ is finite. In particular, $(|S| \setminus S(H_\ell))^{\leq d}$ is finite.

REMARK. For an analogue of the above Fact to the char $p > 0$ case see [**CT19**], [**CT20**].

Again, from Lemma 2.2, the condition that V_ℓ has no $\overline{\Pi}$-coinvariants ensures Condition (P) of Fact 3.2 is satisfied by the universal extension $\tilde{E}_\ell$ attached to any pair (E_ℓ, V_ℓ). Combining this with Lemma 2.1.3 and Fact 3.2, one gets:

COROLLARY 8. *Assume k is finitely generated of characteristic* 0 *and S is a curve. Let $\mathcal{L}$ be an infinite set of primes and let $\underline{V} = V_\ell$, $\ell \in \mathcal{L}$ be a family of elements in* $\mathrm{Mod}_{\mathbb{F}_\ell}(\Pi)$. *Assume*

(U) There exists an open subgroup $U \subset \overline{\Pi}$ such that $U_{V_\ell} = U^+_{V_\ell}$, $\ell \in \mathcal{L}$;

(P) $\overline{\Pi}^+_{V_\ell}$ is perfect for $\ell \gg 0$;

(cI) $\underline{V}$ has no $\overline{\Pi}$-quasi-coinvariants,

Then, for every family of finitely generated $\mathbb{F}_\ell$-submodules $E_\ell \subset \mathrm{H}^1(\Pi, V_\ell)$, $\ell \in \mathcal{L}$ of uniformly bounded $\mathbb{F}_\ell$-dimension $r_\ell \leq r$ and integer $d \geq 1$, $(|S| \setminus S(V_\ell, E_\ell))^{\leq d})$ is finite, $\ell \gg 0$.

3.3. Base change. Let $\Lambda = \mathbb{Q}_\ell$ or $\mathbb{F}_\ell$ and $V \in \mathrm{Mod}_\Lambda(\Pi)$. Let $f : S' \to S$ be a Galois cover and write $\Pi' := \pi_1(S') \subset \Pi$. Let $s \in |S|$ and $s' \in |S'|$ lying over s. The restriction morphisms induce a canonical commutative diagram

$$\begin{array}{ccc} \mathrm{H}^1(\Pi, V) & \xrightarrow{res} & \mathrm{H}^1(\Pi', V) \\ \Big\downarrow{\scriptstyle res_s} & & \Big\downarrow{\scriptstyle res_{s'}} \\ \mathrm{H}^1(\Pi_s, V) & \xrightarrow{res} & \mathrm{H}^1(\Pi_{s'}, V) \end{array}$$

and the inflation-restriction exact sequence shows that the horizontal arrows are injective provided $S' \to S$ has degree invertible in Λ. In particular, $f^{-1}(S(E, V)) = S(res(E), V|_{\Pi'})$.

This shows that the following are equivalent:

- Corollary 4 for V_ℓ, $\ell \in \mathcal{L}$ (resp. Corollary 6 for V);
- Corollary 4 for $V_\ell|_{\Pi'}$, $\ell \in \mathcal{L}$ (resp. Corollary 6 for $V|_{\Pi'}$) for some Galois cover $S' \to S$;
- Corollary 4 for $V_\ell|_{\Pi'}$, $\ell \in \mathcal{L}$ (resp. Corollary 6 for $V|_{\Pi'}$) for every Galois cover $S' \to S$.

and, similarly, that the following are equivalent:

- Corollary 8 for $\underline{V}$;
- Corollary 8 for $\underline{V}$ for some Galois cover $S' \to S$;
- Corollary 8 for $\underline{V}$ for every Galois cover $S' \to S$.

In particular, to apply Corollaries 4, 6, 8 we may freely replace S by a connected étale cover hence, in the setting[2] of Corollary 8 , replace (U) with the seemingly stronger assumption

$$(\mathrm{U}+)\ \overline{\Pi}_{V_\ell} = \overline{\Pi}^+_{V_\ell} \text{ for } \ell \gg 0.$$

3.4. Remarks about assumptions (P), (SS), (cI), (I).

3.4.1. (P) in Corollary 6 (resp. in Corollary 8) can be ensured by the stronger (resp. - see [**CT17**, Cor. 3.3]) assumption (SS) $Lie(\overline{\Pi}_V)$ is semisimple. (resp. (SS) $\overline{\Pi}$ acts semisimply on V_ℓ, $\ell \gg 0$). But (SS), contrary to (P), does not transfer in general from V to $\tilde{E}$ (Lemma 2.2).

3.4.2. In the setting of Corollary 6 (resp. of Corollary 8), consider the condition (I) $V^{\mathrm{Lie}(\overline{\Pi}_V)} = 0$ (resp. (I) $V_\ell^{\overline{\Pi}} = 0$, $\ell \gg 0$). Then, under (SS), (I) and (cI) are equivalent.

3.4.3. The restriction that $E \subset \mathrm{H}^1(\Pi, V)$ be a finite dimensional $\mathbb{Q}_\ell$-vector subspace in Corollaries 4, 6 (resp. that $E_\ell \subset \mathrm{H}^1(\Pi, V_\ell)$, $\ell \in \mathcal{L}$ be of uniformly bounded $\mathbb{F}_\ell$-dimension $r_\ell \leq r$ in Corollary 8) becomes tautological under (I) (resp. provided the V_ℓ, $\ell \in \mathcal{L}$ are of uniformly bounded $\mathbb{F}_\ell$-dimension, $\ell \in \mathcal{L}$). Indeed, if

(I) holds, the inflation-restriction exact sequence implies that the restriction morphism $H^1(\Pi, V) \to H^1(\overline{\Pi}, V)$ is injective (resp. that the restriction morphism $H^1(\Pi, V_\ell) \to H^1(\overline{\Pi}, V_\ell)$ is injective, $\ell \gg 0$). Thus it is enough to prove that $\dim_{\mathbb{Q}_\ell}(H^1(\overline{\Pi}, V)) < +\infty$ (resp. that $\dim_{\mathbb{F}_\ell}(H^1(\overline{\Pi}, V_\ell))$ is uniformly bounded, $\ell \in \mathcal{L}$). In the non resp. situation, this follows from the fact that $H^1(\overline{\Pi}, V) \tilde{\to} H^1_{et}(S_{\overline{k}}, V)$. In the resp. situation, the fact that $H^1(\overline{\Pi}, V_\ell) \tilde{\to} H^1_{et}(S_{\overline{k}}, V)$ already shows that $E_\ell := H^1(\overline{\Pi}, V_\ell)$ has finite $\mathbb{F}_\ell$-dimension. Let $\tilde{E}_\ell$ denote the universal extension attached to (E_ℓ, V_ℓ).

- (i) By construction, $N_\ell := \ker(\overline{\Pi}_{\tilde{E}_\ell} \twoheadrightarrow \overline{\Pi}_{V_\ell}) \subset \mathrm{Hom}_{\mathbb{F}_\ell}(E_\ell, V_\ell)$ is an elementary ℓ-group.
- (ii) By construction and the inflation-restriction exact sequence, we have a canonical isomorphism $H^1(\overline{\Pi}_{\tilde{E}_\ell}, V_\ell) \tilde{\to} H^1(\overline{\Pi}, V_\ell)$.

Now, recall $\overline{\Pi}$ is topologically finitely generated; let s denote the minimal number of topological generators of $\overline{\Pi}$. Then $\overline{\Pi}_{\tilde{E}_\ell}$ is generated by $\leq s$ elements. As $H^1(\overline{\Pi}_{\tilde{E}_\ell}, V_\ell)$ is a quotient of the $\mathbb{F}_\ell$-module of 1-cocycles $Z^1(\overline{\Pi}_{\tilde{E}_\ell}, V_\ell)$ and as 1-cocyles are entirely determined by their image on generators of $\overline{\Pi}_{\tilde{E}_\ell}$, one has

$$\dim_{\mathbb{F}_\ell}(H^1(\overline{\Pi}_{\tilde{E}_\ell}, V_\ell)) \leq \dim_{\mathbb{F}_\ell}(Z^1(\overline{\Pi}_{\tilde{E}_\ell}, V_\ell)) \leq s\dim_{\mathbb{F}_\ell}(V_\ell).$$

3.4.4. In the setting of Corollary 6 (resp. Corollary 8), (P) implies (I) for $V/V^{\mathrm{Lie}(\overline{\Pi}_V)}$ (resp. (U+) and (P) imply (I) for $V_\ell/V_\ell^{\overline{\Pi}}$, $\ell \in \mathcal{L}$). Indeed, if for every $v \in V$ the 1-cocycle $\mathrm{Lie}(\overline{\Pi}_V) \to V$, $g \to gv$ takes its value in $V^{\mathrm{Lie}(\overline{\Pi}_V)}$, it factors through $\mathrm{Lie}(\overline{\Pi}_V)^{ab}$, which is 0 by (P) (resp. if for every $v \in V_\ell$, $\ell \in \mathcal{L}$ the 1-cocycle $\overline{\Pi}_{V_\ell} \to V_\ell$, $\pi \to \pi v - v$ takes its value in $V_\ell^{\overline{\Pi}}$, it factors through ${\overline{\Pi}_{V_\ell}}^{ab}$, which is 0 by (U+) and (P)).

3.4.5. (cI) does not hold in general. However by (3.4.4), in the setting of Corollary 6, (SS) implies (cI) (=(I)) for $V/V^{\mathrm{Lie}(\overline{\Pi}_V)}$ and, in the setting of Corollary 8, (U+), (SS) imply (cI) (=(I)) for $V_\ell/V_\ell^{\overline{\Pi}}$, $\ell \in \mathcal{L}$.

4. Motivic representations

The formulation of the group-theoretical Corollaries 6, 8 is motivated by its possible applications to motivic representations that is (subquotients cut out by algebraic correspondance of) those of the form

- $V = H^u(X_{\overline{\eta}}, \mathbb{Q}_\ell(v)) \in \mathrm{Mod}_{\mathbb{Q}_\ell}(\Pi)$;
- $V_\ell = H^u(X_{\overline{\eta}}, \mu_\ell^{\otimes v}) \in \mathrm{Mod}_{\mathbb{F}_\ell}(\Pi)$, $\ell \neq p$.

for some smooth proper morphism $X \to S$ (smooth-proper base change - see [**SGA4-III**, Exp. XVI]).

We recall briefly the arguments ensuring assumptions (SS) (hence (P)), (U) and (I) (hence (cI)) and give references for more details.

- In Corollary 6, (P) is a consequence of (SS), which follows from comparison between Betti and étale cohomology and Deligne's semisimplicity theorem for variation of polarizable Hodge structures ([**D71**]).
- In Corollary 8,
  – The fact that the $\mathbb{F}_\ell$-rank of V_ℓ is uniformly bounded follows from the comparison between Betti and étale cohomology and the fact that Betti-cohomology with $\mathbb{Z}$-coefficients in finitely generated.
  – For (U), see [**CT17**].

– (P) is a consequence of (SS) which follows from comparison between Betti and étale cohomology, the fact that Betti-cohomology with $\mathbb{Z}$-coefficients in finitely generated and the constructibility of semisimplicity - see [**CT11**, 2.2].
- (I) does not hold in general; this is closely related to isotriviality. For instance, if $u = 1$, by the geometric Lang-Néron theorem ([**LN59**]), (I) holds if and only if $\mathrm{Pic}^{\circ}(X_{\overline{\eta}})$ contains no non-trivial $\overline{k}$-isotrivial abelian subvariety.

5. Specialization of the ℓ-adic Abel-Jacobi filtration on Chow groups

For a noetherian scheme $\mathcal{X}$ and an integer $i \geq 0$, let $\mathrm{Z}^i(\mathcal{X})$ denote the group of codimension i algebraic cycles and $\mathrm{CH}^i(\mathcal{X})$ denote the Chow group of codimension i algebraic cycles modulo rational equivalence. Let X be a smooth projective geometrically connected scheme of dimension g over a field K of characteristic $p \geq 0$. Determining the structure of the groups $\mathrm{CH}^i(X)$, $i \geq 0$ is a difficult still widely open problem. In particular, these groups are not finitely generated in general and they are very sensitive to extensions of the base field K. If K is finitely generated over its prime field, one technics is to endow $\mathrm{CH}^i(X)$ with the so-called ℓ-adic Abel-Jacobi filtrations, $\ell \neq p$ and study the graded pieces of these, which embed in continuous Galois cohomology groups. To simplify the exposition, we work with $\mathbb{Q}_\ell$-coefficients. Our arguments can be basically transposed as they are to $\mathbb{F}_\ell$-coefficients provided $\ell \gg 0$. We refer to [**J88**] for the technical setting of continuous cohomology and to [**R95**] for a nice introduction to the formalism of ℓ-adic Abel-Jacobi filtrations and related motivic conjectures (see also [**J90**]).

5.1. Absolute setting. Let X be a smooth projective geometrically connected scheme of dimension g over a finitely generated field K of characteristic $p \geq 0$.

5.1.1. *Construction.* Consider the Hochschild-Serre spectral sequence for continuous étale cohomology

$$E_2^{j,2i-j} = \mathrm{H}^j(K, \mathrm{H}^{2i-j}(X_{\overline{K}}, \mathbb{Q}_\ell)(i)) \Rightarrow \mathrm{H}^{2i}(X, \mathbb{Q}_\ell(i)),$$

and pull-back the induced filtration

$$F_{H,\ell}^{i,0}(X) = \mathrm{H}^{2i}(X, \mathbb{Q}_\ell(i)) \supset F_{H,\ell}^{i,1}(X) \supset \cdots \supset F_{H,\ell}^{i,2i}(X) \supset 0$$

to $\mathrm{CH}^i(X) \otimes \mathbb{Q}$ *via* the cycle class map $cl_\ell^i : \mathrm{CH}^i(X) \otimes \mathbb{Q} \to \mathrm{H}^{2i}(X, \mathbb{Q}_\ell(i))$. The resulting filtration

$$F_\ell^{i,0}(X) = \mathrm{CH}^i(X) \otimes \mathbb{Q} \supset F_\ell^{i,1}(X) \supset \cdots \supset F_\ell^{i,2i}(X) \supset 0$$

is called the ℓ-adic Abel-Jacobi filtration and the canonical morphisms

$$a_\ell^{i,j} : F_\ell^{i,j}(X) \to F_{H,\ell}^{i,j}(X)/F_{H,\ell}^{i,j+1}(X) =: Gr_{H,\ell}^{i,j}(X),\ j \geq 0$$

the ℓ-adic Abel-Jacobi maps. Write $Gr_\ell^{i,j}(X) := F_\ell^{i,j}(X)/F_\ell^{i,j+1}(X)$. The ℓ-adic Abel-Jacobi maps induce an injective graded morphism $Gr_\ell^{i,\bullet}(X) \hookrightarrow Gr_{H,\ell}^{i,*}(X)$. It is known that the Hochschild-Serre spectral sequence degenerates at E_2 ([**R95**, Th. 1]) so, what we get, actually, is a morphism of graded groups

$$\bigoplus_{0 \leq j \leq 2i} Gr_\ell^{i,j}(X) \to \bigoplus_{0 \leq j \leq 2i} \mathrm{H}^j(K, \mathrm{H}^{2i-j}(X_{\overline{K}}, \mathbb{Q}_\ell)(i))$$

which is injective in degree $\leq 2i - 1$.

WARNING. A priori, the canonical morphism $Gr_\ell^{i,\bullet}(X)\otimes\mathbb{Q}_\ell\to Gr_{H,\ell}^{i,\bullet}(X)$ is neither injective nor surjective in general. In particular, the $\mathbb{Q}$-rank of $Gr_\ell^{i,\bullet}(X)$ might be infinite even though the $\mathbb{Q}_\ell$-rank of $Gr_{H,\ell}^{i,\bullet}(X)$ is finite.

5.1.2. *Example.* For codimension 1 and dimension 0-cycles, the filtration is rather explicit. For an abelian variety A over a field K and an integer $n\geq 1$, let $A[n]$ denote the kernel of the multiplication-by-n morphism on $A(\overline{K})$. For a prime ℓ, write $T_\ell(A):=\lim_{\longleftarrow}A[\ell^n]$, $V_\ell(A):=T_\ell(A)\otimes\mathbb{Q}_\ell$.

Let $\alpha:X\to\mathrm{Alb}_X$ denote the Albanese variety and Pic_X the Picard variety of X; let $\mathrm{Pic}_X^\circ\subset\mathrm{Pic}_X$ denote the connected component of Pic_X. One has canonical isomorphisms $\mathrm{H}^1(X_{\overline{K}},\mathbb{Q}_\ell(1))\tilde{\to}V_\ell(\mathrm{Pic}_X^\circ)$ and $\mathrm{H}^{2g-1}(X_{\overline{K}},\mathbb{Q}_\ell(g))\tilde{\to}V_\ell(\mathrm{Alb}_X)$. These isomorphisms induce commutative diagrams (see *e.g.* [**R95**, Appendix]), where κ_ℓ is the ℓ-adic Kummer morphism (induced by taking cohomology, projective limit and $-\otimes_{\mathbb{Z}_\ell}\mathbb{Q}_\ell$ from the projective system of Kummer short exact sequences $0\to\mathrm{Alb}_X[\ell^n]\to\mathrm{Alb}_X\stackrel{\ell^n}{\to}\mathrm{Alb}_X\to 0$, $n\geq 1$)

$$\begin{array}{ccc}
 & \stackrel{a_\ell^{1,1}=\kappa_\ell}{\nearrow} & Gr_{H,\ell}^{1,1}(X)=\mathrm{H}^1(K,V_\ell(\mathrm{Pic}_X^\circ)) \\
 & & \uparrow \\
F_\ell^{1,1}(X)=\mathrm{Pic}^\circ(X)\otimes\mathbb{Q} & \longrightarrow & F_{H,\ell}^{1,1}(X) \\
\downarrow & \square & \downarrow \\
F_\ell^{1,0}(X)=\mathrm{Pic}(X)\otimes\mathbb{Q} & \longrightarrow & F_{H,\ell}^{1,0}(X) \\
 & \stackrel{cl_\ell^1}{\searrow} & \downarrow \\
 & & Gr_{H,\ell}^{1,0}(X)=\mathrm{H}^2(X_{\overline{K}},\mathbb{Q}_\ell(1))^{\pi_1(K)}
\end{array}$$

$$\begin{array}{ccc}
F_\ell^{g,1}(X) & \stackrel{a_\ell^{g,1}}{\longrightarrow} & Gr_{H,\ell}^{g,1}(X)=\mathrm{H}^1(K,V_\ell(\mathrm{Alb}_X)) \\
\alpha\downarrow & \stackrel{\kappa_\ell}{\nearrow} & \\
\mathrm{Alb}_X(K)\otimes\mathbb{Q} & &
\end{array}$$

5.2. Relative setting. Let S be a smooth, geometrically connected curve (to simplify) over a field k of characteristic $p\geq 0$. Let $f:X\to S$ be a smooth, projective morphism with geometrically connected fibers. Fix an integer $i\geq 1$ and a prime $\ell\neq p$. We first compare the ℓ-adic Abel-Jacobi filtrations on $\mathrm{CH}^i(X_\eta)$ and $\mathrm{CH}^i(X_s)$. This can be done using the arguments in [**R95**, 2], which we recall briefly.

5.2.1. For $s\in|S|$, let $\mathcal{O}_s$ denote the local ring of S at s and $\mathcal{O}_s^h$ its henselisation. Write $X(s)\to S(s)$ and $X(s)^h\to S(s)^h$ for the base-change of $X\to S$ *via* $S(s):=\mathrm{spec}(\mathcal{O}_s)\to S$ and $S(s)^h:=\mathrm{spec}(\mathcal{O}_s^h)\to S$ respectively. Let η_s denote the generic point of $S(s)^h$. Then one has a canonical commutative specialization diagram

([**Fu75**, 4.4], [**Fu84**, 20.2, 20.3])

$$\begin{array}{ccc} & \mathrm{CH}^i(X(s)) & \\ \swarrow^{\eta^*} & & \searrow^{s^*} \\ \mathrm{CH}^i(X_\eta) & \xrightarrow[sp_s]{} & \mathrm{CH}^i(X_s), \end{array}$$

which is compatible with the cycle maps, that is such that the following diagram commutes

$$\begin{array}{ccccc} & & \overset{sp_s}{\frown\!\!\!\longrightarrow} & & \\ \mathrm{CH}^i(X_\eta) & \xleftarrow[\eta^*]{} & \mathrm{CH}^i(X(s)) & \xrightarrow[i^*]{} & \mathrm{CH}^i(X_s) \\ \downarrow & & \downarrow & & \downarrow \\ \mathrm{H}^{2i}(X_\eta, \mathbb{Q}_\ell(i)) & \xleftarrow[\eta^*]{} & \mathrm{H}^{2i}(X(s), \mathbb{Q}_\ell(i)) & \xrightarrow[i^*]{} & \mathrm{H}^{2i}(X_s, \mathbb{Q}_\ell(i)) \\ \downarrow & & \downarrow & \nearrow_{i^*}^{\simeq} & \\ \mathrm{H}^{2i}(X_{\eta_s}, \mathbb{Q}_\ell(i)) & \xleftarrow[\eta_s^*]{} & \mathrm{H}^{2i}(X(s)^h, \mathbb{Q}_\ell(i)) & & \end{array}$$

One also has the Leray spectral sequence $E_2^{j,2i-j} = \mathrm{H}^j(S(s)^h, \mathrm{H}^{2i-j}(X_{\overline{\eta}_s}, \mathbb{Q}_\ell(i)) \Rightarrow \mathrm{H}^{2i}(X(s)^h, \mathbb{Q}_\ell(i))$, which is compatible with the Hoschshild-Serre spectral sequences of 5.1.1 for X_s over $k(s)$ and X_{η_s} over $k(\eta_s)$. The key point is that the morphisms

$$\mathrm{H}^j(S(s)^h, \mathrm{H}^{2i-j}(X_{\overline{\eta}_s}, \mathbb{Q}_\ell)(i)) \to \mathrm{H}^j(\eta_s, \mathrm{H}^{2i-j}(X_{\overline{\eta}_s}, \mathbb{Q}_\ell)(i))$$

are injective ([**R95**, Lemma 2.13]; see also [**CTHK97**, Thm. B.2.1]). This shows that the Leray spectral sequence degenerates at E_2 (since the Hoschshild-Serre spectral sequence does) and then, by a straightforward induction on j, that

$$sp_s(F_\ell^{i,j}(X_\eta)) \subset F_\ell^{i,j}(X_s), \quad j \geq 0$$

As a result, for every $j \geq 0$ we obtain a commutative specialization diagram

$$\begin{array}{ccccc} Gr_\ell^{i,j}(X_\eta) & \longrightarrow & \mathrm{H}^j(\Pi_\eta, \mathrm{H}^{2i-j}(X_{\overline{\eta}}, \mathbb{Q}_\ell(i))) & \longrightarrow & \mathrm{H}^j(\Pi_{\eta_s}, \mathrm{H}^{2i-j}(X_{\overline{\eta}_s}, \mathbb{Q}_\ell(i))) \\ \Big| & & \uparrow {\scriptstyle infl} & & \uparrow \\ \Big| {\scriptstyle sp_s} & & \mathrm{H}^j(\Pi, \mathrm{H}^{2i-j}(X_{\overline{\eta}_s}, \mathbb{Q}_\ell(i))) & \longrightarrow & \mathrm{H}^j(S(s)^h, R^{2i-j}f_*\mathbb{Q}_\ell(i)) \\ \downarrow & & \downarrow {\scriptstyle res_s} & \swarrow^{\simeq} & \\ Gr_\ell^{i,j}(X_s) & \longrightarrow & \mathrm{H}^j(\Pi_s, \mathrm{H}^{2i-j}(X_{\overline{s}}, \mathbb{Q}_\ell(i))) & & \end{array}$$

In the above diagram, the central upper vertical arrow is the inflation map attached to the epimorphism $\Pi_\eta \twoheadrightarrow \Pi$ (recall that S is smooth hence normal) and the right vertical arrow is the one induced by the morphisms of sites $S(s)_{et}^h \to S_{et} \to \mathcal{B}\Pi$. Write $V := \mathrm{H}^{2i-1}(X_{\overline{\eta}_s}, \mathbb{Q}_\ell(i))$.

LEMMA 9. *The inflation morphism* $infl : \mathrm{H}^1(\Pi, V) \hookrightarrow \mathrm{H}^1(\Pi_\eta, V)$ *is an isomorphism.*

PROOF. Let $S \hookrightarrow S^{cpt}$ denote the smooth compactification of S and set $\partial S := S^{cpt} \setminus S$. The kernel N of $\Pi_\eta \twoheadrightarrow \Pi$ is generated by the inertia groups at $s \in \partial S$. By the Hochschild-Serre spectral sequence, it is enough to show that $\mathrm{H}^1(N, V)^\Pi = 0$. By the smooth-proper base change theorem N acts trivially on V so that $\mathrm{H}^1(N, V)^\Pi = \mathrm{Hom}_\Pi(N, V)$. Hence it is enough to show that for every inertia group $I_s \subset N$ at $s \in \partial S$ one has $\mathrm{Hom}_{\Pi_s}(I_s^{(\ell)}, V) = 0$, where $(-)^{(\ell)}$ denotes pro-ℓ completion. But, as Π_s-modules, $I_s^{(\ell)} \simeq \mathbb{Z}_\ell(1)$ hence has Weil weight -2 whereas V has Weil weights -1. □

5.2.2. Write $\widetilde{V} := V/V^{\mathrm{Lie}(\overline{\Pi}_V)}$. Since V is a semisimple $\overline{\Pi}$-module (see 4) that is V satisfies (SS), $\widetilde{V}^{\mathrm{Lie}(\overline{\Pi}_V)} = 0$. Equivalently, $\widetilde{V}$ satisfies (I). Since $\widetilde{V}$ also satisfies (SS), it satisfies (cI). Eventually, from 3.4.3, $\dim_{\mathbb{Q}_\ell}(\mathrm{H}^1(\Pi_\eta, \widetilde{V})) < +\infty$.

For $s \in S$, let $\widetilde{Gr}_\ell^{i,1}(X_s)$ denote the image of

$$F_\ell^{i,1}(X_s) \overset{a_\ell^{i,1}}{\to} \mathrm{H}^1(\Pi_s, V) \to \mathrm{H}^1(\Pi_s, \widetilde{V}).$$

For $s \in |S|$ the specialization morphism $sp_s : Gr_\ell^{i,1}(X_\eta) \to Gr_\ell^{i,1}(X_s)$ induces a commutative square

$$\begin{array}{ccc} \widetilde{Gr}_\ell^{i,1}(X_\eta) & \hookrightarrow & \mathrm{H}^1(\Pi_\eta, \widetilde{V}) \\ \downarrow{\scriptstyle sp_s} & & \downarrow{\scriptstyle res_s} \\ \widetilde{Gr}_\ell^{i,1}(X_s) & \hookrightarrow & \mathrm{H}^1(\Pi_s, \widetilde{V}) \end{array}$$

If k is finitely generated over its prime field, the image of $Gr_\ell^{i,1}(X_\eta)$ in $\mathrm{H}^1(\Pi, V)$ always generates a $\mathbb{Q}_\ell$-submodule of finite $\mathbb{Q}_\ell$-rank ([**R95**, Prop. 2.5]). Combining these observations and Lemma 9, Corollary 3.1.1 and Corollary 6 yield the following results about specialization of $Gr_\ell^{i,1}$.

COROLLARY 10. *Assume k is Hilbertian. Let $\mathcal{L}$ be a finite set of primes not containing p. Set $V_\ell := \mathrm{H}^{2i-1}(X_{\overline{\eta}_s}, \mathbb{Q}_\ell(i))$, $\ell \in \mathcal{L}$ and let $E_\ell \subset Gr_\ell^{i,1}(X_\eta)$ whose image in $\mathrm{H}^1(\Pi, V_\ell)$ generates a finite-dimensional $\mathbb{Q}_\ell$-subvector space, $\ell \in \mathcal{L}$. Then there exists an integer $d \geq 1$ such that the specialization morphisms*

$$sp_s^1 : E_\ell \hookrightarrow Gr_\ell^{i,1}(X_\eta) \to Gr_\ell^{i,1}(X_s),\ \ell \in \mathcal{L}$$

are injective for infinitely many $s \in |S|^{\leq d}$.

In particular, if k is finitely generated over its prime field, there always exists $s \in |S|$ such that $\mathrm{rank}(Gr_\ell^{i,1}(X_\eta)) \leq \mathrm{rank}(Gr_\ell^{i,1}(X_s))$, $\ell \in \Sigma$ (where we allow infinite $\mathbb{Q}$-rank).

COROLLARY 11. *Assume k is finitely generated over its prime field and $p = 0$. Then for every integer $d \geq 1$ and all but finitely many $s \in |S|^{\leq d}$, the specialization morphism $\widetilde{sp}_s^1 : \widetilde{Gr}_\ell^{i,1}(X_\eta) \to \widetilde{Gr}_\ell^{i,1}(X_s)$ is injective.*

REMARK. Since the specialization morphism $Gr_\ell^{i,0}(X_\eta) \to Gr_\ell^{i,0}(X_s)$ is injective for every $s \in |S|$, we may replace $Gr_\ell^{i,1}$ (resp. $\widetilde{Gr}_\ell^{i,1}$) with $Gr_\ell^{i,0} \oplus Gr_\ell^{i,1}$ (resp. $Gr_\ell^{i,0} \oplus \widetilde{Gr}_\ell^{i,1}$) in Corollaries 10, 11.

5.3. Note that our method only allows to tackle specialization issues for families over a 'geometric' basis that is schemes of finite type over a field k. Very little seems to be known for specialization over 'arithmetic' bases that is scheme of finite type over $\mathbb{Z}$. See however [**Sc06**] for a discussion about ℓ-primary torsion classes in Chow groups.

6. Applications, examples

6.1. About the ℓ-independence of the second step of the ℓ-adic Abel-Jacobi filtration in characteristic 0. Jannsen's injectivity conjecture ([**J90**, Conj. 9.15]; see also [**BlK90**, Conj. 5.3]) predicts that for a smooth, projective scheme X over a global field K the morphisms induced by the the ℓ-adic Abel-Jacobi maps

$$a_\ell^{i,1} : F_\ell^{i,1}(X) \otimes \mathbb{Q}_\ell \to \mathrm{H}^1(K, \mathrm{H}^{2i-1}(X_{\overline{K}}, \mathbb{Q}_\ell)(i))$$

are injective, $i \geq 0$. On the other hand, Beilinson's conjecture on the existence of filtrations on Chow groups predicts that, at least in characteristic 0, the ℓ-adic Abel-Jacobi filtration should be independent of ℓ ([**J94**, Lemma 2.7]). The following is a typical application of a specialization result as Corollary 10.

COROLLARY 12. *Assume Jannsen's injectivity conjecture for number fields. Then for every smooth, projective scheme X over a field K finitely generated over $\mathbb{Q}$, the second piece $F_\ell^{i,2}(X) \subset \mathrm{CH}^j(X) \otimes \mathbb{Q}$ of the ℓ-adic Abel-Jacobi filtration is independent of ℓ.*

PROOF. By comparison between Betti and ℓ-adic cohomologies[2], $F_\ell^{i,1}(X) \subset \mathrm{CH}^i(X) \otimes \mathbb{Q}$ is independent of ℓ. We proceed by induction on the transcendence degree d of K. If $d = 0$, this follows from Jannsen's conjecture [**J90**, Conj. 9.15]. If $d \geq 1$, $K = k(\eta)$ for k a field finitely generated over $\mathbb{Q}$ of transcendence degree $d-1$ and η the generic point of a smooth, geometrically connected curve S over k. Up to replacing S by a non-empty open subscheme, one may assume X extends to a smooth projective scheme $\mathcal{X} \to S$. From Corollary 10, there exists $s \in |S|$ such that the restriction morphisms $res_{s,\ell} : \mathrm{H}^1(\Pi, V_\ell) \to \mathrm{H}^1(\Pi_s, V_\ell)$ and $res_{s,\ell'} : \mathrm{H}^1(\Pi, V_{\ell'}) \to \mathrm{H}^1(\Pi_s, V_{\ell'})$ are injective on the image of the corresponding first higher ℓ-adic Abel-Jacobi maps (here we write $V_\ell = \mathrm{H}^{2i-1}(X_{\overline{K}}, \mathbb{Q}_\ell(i))$). But, then

$$F_\ell^{i,2}(X) = sp_s^{-1}(F_\ell^{i,2}(\mathcal{X}_s) \cap F_\ell^{i,1}(X)) \overset{(*)}{=} sp_s^{-1}(F_{\ell'}^{i,2}(\mathcal{X}_s) \cap F_{\ell'}^{i,1}(X)) = F_{\ell'}^{i,2}(X),$$

where (*) follows from the ℓ-independence of $F_\ell^{i,1}(X) \subset \mathrm{CH}^i(X) \otimes \mathbb{Q}$ and the induction hypothesis applied to $\mathcal{X}_s$. □

6.2. Abelian schemes (5.1.2 cont.) Assume $X = A \to S$ is an abelian scheme.

[2]This is the only place where we use the assumption that K is of characteristic 0; it does not seem that for $p > 0$ one knows that $F_\ell^{i,1}(X) \subset \mathrm{CH}^i(X) \otimes \mathbb{Q}$ is independent of ℓ.

6.2.1. The specialization diagram of 5.2.2 reads as

$$\begin{array}{ccc} A_\eta(k(\eta))\otimes\mathbb{Q} & \overset{\kappa_\ell}{\hookrightarrow} & \mathrm{H}^1(\Pi_\eta, V_\ell(A_\eta)) \\ \simeq\uparrow & & \simeq\uparrow infl \\ A(S)\otimes\mathbb{Q} & \hookrightarrow & \mathrm{H}^1(S,V_\ell(1))\simeq \mathrm{H}^1(\Pi, V_\ell(A_\eta)) \\ \downarrow sp_s & & \downarrow res_s \\ A_s(k(s)))\otimes\mathbb{Q} & \overset{\kappa_\ell}{\hookrightarrow} & \mathrm{H}^1(\Pi_s, V_\ell(A_s)), \end{array}$$

where the upper left isomorphism is the Néron extension property for abelian schemes. In particular, Corollary 10 and Corollary 11 together with the Lang-Néron theorem [**LN59**] imply respectively the following.

COROLLARY 13. *Assume k is finitely generated over its prime field. Then there exists an integer $d\geq 1$ such that the specialization morphism $sp_s : A_\eta(k(\eta))\otimes\mathbb{Q}\to A_s(k(s))\otimes\mathbb{Q}$ is injective for infinitely many $s\in|S|^{\leq d}$. In particular,* $\mathrm{rank}(A_\eta(k(\eta)))\leq\mathrm{rank}(A_s(k(s)))$ *for infinitely many $s\in|S|^{\leq d}$.*

COROLLARY 14. *Assume k is finitely generated over $\mathbb{Q}$ and X_η contains no non-zero $\overline{k}$-isotrivial abelian subvariety. Then for every integer $d\geq 1$ and all but finitely many $s\in|S|^{\leq d}$, the specialization map $sp_s : A_\eta(k(\eta))\otimes\mathbb{Q}\to A_s(k(s))\otimes\mathbb{Q}$ is injective. In particular,* $\mathrm{rank}(A_\eta(k(\eta)))\leq\mathrm{rank}(A_s(k(s)))$.

6.2.2. *Comparison with Silverman's theorem.* When k is a number field, one recovers Silverman's specialization theorem [**Si83**, Thm. C]. More precisely, in this setting, our method shows, more generally, that for all but finitely many $s\in|S|^{\leq d}$, the restriction morphism $res_s : \mathrm{H}^1(\Pi, V_\ell(A_\eta))\to\mathrm{H}^1(\Pi_s, V_\ell(A_s))$ is injective and Corollary 14 appears as a special case of a more general specialization theorem about cohomology classes. On the other hand, Silverman's specialization theorem says more about algebraic classes, namely that the set of all $s\in|S|$ where the specialization map $sp_s : A_\eta(k(\eta))\otimes\mathbb{Q}\to A_s(k(s))\otimes\mathbb{Q}$ is non-injective is of bounded height.

Silverman's theorem follows from the non-degeneracy of the Néron-Tate height pairing $\langle -,-\rangle_{A_\eta} : A_\eta(k(\overline{\eta}))_{\mathbb{Q}}\times A_\eta(k(\overline{\eta}))_{\mathbb{Q}}\to\mathbb{R}$ at the generic fiber, the limit formula: for every $P,Q\in A(S)$

$$\lim_{s\in|S|,\ h_S(s)\to+\infty}\frac{\langle P_s,Q_s\rangle_{A_s}}{h_S(s)}=\langle P_\eta,Q_\eta\rangle_{A_\eta}.$$

and the Northcott property for heights over number fields.

So provided a good theory of heights is available for Chow groups, Silverman's argument should extend to $Gr_\ell^{1,i}(A_\eta)$. Several constructions of height pairings on Chow groups generalizing the Néron-Tate height pairing have been proposed ([**Bei87**], [**Bl84**], [**GiS84**]) but the constructions and properties (such as the non-degeneracy or limit formula) of these height pairings are still partly conjectural. However, for cycles geometrically algebraically equivalent to zero modulo incidence equivalence, which are parametrized by abelian varieties - the so-called higher Picard varieties (see [**S79**] for details) - one can essentially reduces to Silverman's theorem. More precisely, let

$$F^{i,1}_{inc}(A_\eta)\subset F^{i,1}_{alg}(A_\eta)\subset F^{i,1}_\ell(A_\eta)$$

denote the $\mathbb{Q}$-subvector spaces of cycles geometrically incidence equivalent to zero and geometrically algebraically equivalent to zero respectively. Set

$$F^{i,2}_{alg,\ell}(A_\eta) := F^{i,1}_{alg}(A_\eta) \cap F^{i,2}_\ell(A_\eta) \subset F^{i,1}_{alg}(A_\eta).$$

Then $F^{i,2}_{alg,\ell}(A_\eta) \subset F^{i,1}_{inc}(A_\eta)$ that is,

$$F^{i,1}_{alg}(A_\eta)/F^{i,1}_{inc}(A_\eta) \twoheadleftarrow F^{i,1}_{alg}(A_\eta)/F^{i,1}_{alg,\ell}(A_\eta) \hookrightarrow Gr^{i,1}_\ell(A_\eta).$$

Let $I \subset F^{i,1}_{alg}(A_\eta)/F^{i,1}_{alg,\ell}(A_\eta)$ be any $\mathbb{Q}$-vector subspace mapping injectively into $F^{i,1}_{alg}(A_\eta)/F^{i,1}_{inc}(A_\eta)$. Then, if the higher Picard variety $\mathrm{Pic}^i_{X_{\overline{\eta}}}$ contains no non-zero $\overline{k}$-isotrivial abelian subvariety, Silverman's theorem shows that the set of all $s \in |S|$ such that the induced morphism

$$I \subset Gr^{i,1}_\ell(A_\eta) \stackrel{sp_s}{\to} Gr^{i,1}_\ell(A_s)$$

is non-injective is of bounded height.

Remark. The geometric meaning of the condition that $\mathrm{Pic}^i_{A_{\overline{\eta}}}$ contains no non-zero $\overline{k}$-isotrivial abelian subvariety is unclear to us. We do not know if there is a simple criterion that ensures it.

6.2.3. *The set $S^1(E,V)$.* When k is finitely generated, $V := V_\ell(A_\eta)$ is a semi-simple $\pi_1(S)$-module hence for every finite-dimensional $\mathbb{Q}_\ell$-vector subspace $E \subset \mathrm{H}^1(\Pi, V)$ one can apply Lemma 2.1.3 with $N = \ker(\Pi_{\tilde{E}} \to \Pi_V)$. Define $\Pi := \Pi_E : |S| \to \mathbb{Z}_{\geq 0}$ by $\Pi(s) = \dim(N) - \dim(N \cap \Pi_{s,\tilde{E}})$. Let $S^1(E,V) \subset |S|$ denote the set of all $s \in |S|$ such that $\Pi(s) = 0$. Then one has

$$S^0(E,V) := S(E,V) \supset S^1(E,V) \supset S^2(E,V) := S(\tilde{E}) = S^1(E,V) \cap S(V) \subset S(V)$$

Let A be an abelian variety over a field K of characteristic $p \geq 0$. Fix a prime $\ell \neq p$ and consider $V := V_\ell(A_{\overline{K}}) \in \mathrm{Mod}_{\mathbb{Q}_\ell}(\Pi)$. Fix a free $\mathbb{Z}$-module $E_0 \subset A(K)$ of finite rank and set $E =: E_0 \otimes \mathbb{Q}_\ell$, which we identify with its image in $\mathrm{H}^1(\pi_1(K), V)$ *via* the ℓ-adic Kummer morphism. Then the 'universal' extension $\tilde{E}$ of 2.1.1 is the Tate module $V_\ell([\underline{E} \to A])$ of the 1-motive $[\underline{E} \to A]$ defined by E (see [**Jo14**]). Pick a $\mathbb{Z}$-basis $\underline{P} = (P_1, \dots, P_r) \in E_0$. Let $A(\underline{P}) \subset X^r$ denote the connected component of the Zariski closure of $\mathbb{Z}\underline{P} \subset A^r$ and set $V(\underline{P}) := V_\ell(A(\underline{P})_{\overline{K}})$. Up to non-canonical isomorphism, $A(\underline{P})$ and $V(\underline{P})$ only depend on E and not on $\underline{P}$. By construction, $N \hookrightarrow \mathrm{Hom}_{\mathbb{Q}_\ell}(E,V) \simeq V^r$ and we have the following description of N ([**Jo14**, (Comment after) Thm. 2]),

Fact 15. (Unipotent part of the Mumford-Tate conjecture for 1-motives) The inclusion $N \hookrightarrow V^r$ factors through $V(\underline{P})$ and induces an (a non-canonical) isomorphism of ℓ-adic Lie groups $N \otimes \mathbb{Q}_\ell \simeq V(\underline{P})$.

We come back to our abelian scheme $A \to S$. If $E_0 \subset A(S)$ is a free $\mathbb{Z}$-module of finite rank and $\underline{P} = (P_1, \dots, P_r) \in E_0$ a $\mathbb{Z}$-basis, for $s \in |S|$, one has

$$\Pi(s) = \dim(V(\underline{P}_\eta)) - \dim(V(\underline{P}_s)) = \dim(A(\underline{P}_\eta)) - \dim(A(\underline{P}_s)).$$

On the one hand, this shows that $S^1(E,V)$ is independent of ℓ and, on the other hand that

Corollary 16. *Assume k is finitely generated over $\mathbb{Q}$. Then the set of all $s \in S(k)$ (resp. for every integer $d \geq 1$ the set of all $s \in |S|^{\leq d}$) where the unipotent part $A(\underline{P}_s)$ of the motivic Galois group of $[\underline{E}_s \to A_s]$ degenerates is finite.*

6.2.4. *An example with $\mathbb{F}_\ell$-coefficients.* We conclude by a refinement of Corollary 14, using Corollary 8. For an abelian group M and a prime ℓ, let M_{tors} and $M[\ell^\infty]$ denote the torsion subgroup of M and the ℓ-Sylow of M_{tors} respectively. With the notation of 6.2, one has

COROLLARY 17. *Assume k is finitely generated over $\mathbb{Q}$ and A_η contains no non-zero $\overline{k}$-isotrivial abelian subvariety. Then for every integer $d \geq 1$ and all but finitely many $s \in |S|^{\leq d}$, the specialization morphism $sp_s : A_\eta(k(\eta)) \to A_s(k(s))$ is injective and satisfies* $\mathrm{coker}(sp_s)[\ell] = 0$.

PROOF. For $s \in |S|$, the specialization map induces a morphism

$$sp_s : A_\eta(k(\eta))_{tors} \to A_s(k(s))_{tors}$$

which is injective. Since $A_\eta(k(\eta))$ is a finitely generated abelian group, $A_\eta(k(\eta))[\ell] = 0$ for $\ell \gg 0$. By the geometric Lang-Néron theorem ([**LN59**]), the assumption that $A_{\overline{\eta}}$ contains no non-zero $\overline{k}$-isotrivial abelian subvariety ensures (I). This, in turn, implies that for every integer $B \geq 1$, for $\ell \gg 0$ and every $0 \neq v_\ell \in A_\eta[\ell]$ one has $[\overline{\Pi}_{V_\ell} : \mathrm{Stab}_{\overline{\Pi}_{V_\ell}}(v_\ell)] > B^2$ ([**CT16a**, Lemma 2.8]). Then, by Corollary 8, for every integer $d \geq 1$, there exists an integer $B \geq 1$ such that for $\ell \gg 0$ and all but finitely many $s \in |S|^{\leq d}$, $[\Pi_{V_\ell} : \Pi_{s,V_\ell}] \leq B$. Hence

$$\begin{aligned} B[\Pi_{s,V_\ell} : \mathrm{Stab}_{\Pi_{s,V_\ell}}(v_\ell)] \quad &\geq [\Pi_{V_\ell} : \mathrm{Stab}_{\Pi_{s,V_\ell}}(v_\ell)] \quad \geq [\Pi_{V_\ell} : \mathrm{Stab}_{\Pi_{V_\ell}}(v_\ell)] \\ &\geq [\overline{\Pi}_{V_\ell} : \mathrm{Stab}_{\overline{\Pi}_{V_\ell}}(v_\ell)] > B^2, \end{aligned}$$

which shows that $A_s(k(s))[\ell] = 0$. On the other hand, by Corollary 8, for every integer $d \geq 1$, for $\ell \gg 0$ and for all but finitely many $s \in |S|^{\leq d}$, the specialization map $sp_s : A_\eta(k(\eta))/\ell \to A_s(k(s))/\ell$ is injective. The conclusion thus follows from the formal Lemma 18. □

LEMMA 18. *Let R be a principal ideal domain, $\varphi : N \to M$ a morphism in* Mod_R *and let $0 \neq r \in R \setminus R^\times$ such that the induced morphism $\overline{\varphi} : N/r \to M/r$ is injective. If* $\ker(\varphi)_{tors} = M[r] = 0$ *then $\varphi : N \to M$ is injective and* $\mathrm{coker}(\varphi)[r] = 0$.

PROOF. Apply the snake lemma to

$$\begin{array}{ccccccccc} 0 & \longrightarrow & M & \xrightarrow{r} & M & \longrightarrow & M/r & \longrightarrow & 0 \\ & & \downarrow{\scriptstyle\varphi} & & \downarrow{\scriptstyle\varphi} & & \downarrow{\scriptstyle\overline{\varphi}} & & \\ 0 & \longrightarrow & N & \xrightarrow{r} & N & \longrightarrow & N/r & \longrightarrow & 0 \end{array}$$

□

REMARK. For ℓ fixed, increasing k produces lots of $s \in S(k)$ such that $\mathrm{coker}(sp_s)[\ell] \neq 0$. More precisely, if A_η contains no non-zero $\overline{k}$-isotrivial abelian subvariety, $A(S_{\overline{k}})$ is a finitely generated $\mathbb{Z}$-module. Up to replacing S by a finite cover, we may assume $A(S_{\overline{k}})$ has rank ≥ 1 and up to replacing k with a finite field extension, we may assume $X(S) = X(S_{\overline{k}})$. Then, $X(S)$ remains unchanged when k is replaced with a finite field extension. Pick $x \in A(S)$ a non-torsion, non-divisible section and $s \in S(k)$ such that $sp_s : A(S) \hookrightarrow A_s(k(s))$ is injective (after replacing k by a finite field extension, such an s always exists by Corollary 17). But then, up to replacing k by a finite field extension, we may assume $sp_s(x) =: x_s \in k(s)$ is ℓ-divisible that is there exists $a \in A_s(k(s))$ such that $x_s = \ell a$.

Then $0 \neq a \in \mathrm{coker}(sp_s)[\ell]$ since, otherwise, there would exist $x' \in A(S)$ such that $a = sp_s(x')$ hence, by injectivity of $sp_s : A(S) \otimes \mathbb{Q} \hookrightarrow A_s(k(s)) \otimes \mathbb{Q}$, $x = \ell x'$. This would contradict our assumption on x.

Acknowledgments

The author thanks Jean-Louis Colliot-Thélène and Ofer Gabber for their technical explanations about the specialization of the ℓ-adic Abel Jacobi filtration. The author is very grateful to Bruno Kahn for his interest, accurate comments and the time he devoted to explain some aspects of the problems about Chow groups. The author also thanks Cédric Pépin for our discussion on the last remark in this paper.

References

[A18] E. Ambrosi, *A uniform open image theorem for ℓ-adic representations in positive characteristic*, Preprint 2018.

[Bei87] A. Beĭlinson, *Height pairing between algebraic cycles*, Current trends in arithmetical algebraic geometry (Arcata, Calif., 1985), Contemp. Math., vol. 67, Amer. Math. Soc., Providence, RI, 1987, pp. 1–24, DOI 10.1090/conm/067/902590. MR902590

[Bl84] Spencer Bloch, *Height pairings for algebraic cycles*, Proceedings of the Luminy conference on algebraic K-theory (Luminy, 1983), J. Pure Appl. Algebra **34** (1984), no. 2-3, 119–145, DOI 10.1016/0022-4049(84)90032-X. MR772054

[BlK90] Spencer Bloch and Kazuya Kato, *L-functions and Tamagawa numbers of motives*, The Grothendieck Festschrift, Vol. I, Progr. Math., vol. 86, Birkhäuser Boston, Boston, MA, 1990, pp. 333–400. MR1086888

[CTHK97] Jean-Louis Colliot-Thélène, Raymond T. Hoobler, and Bruno Kahn, *The Bloch-Ogus-Gabber theorem*, Algebraic K-theory (Toronto, ON, 1996), Fields Inst. Commun., vol. 16, Amer. Math. Soc., Providence, RI, 1997, pp. 31–94. MR1466971

[CCh20] Anna Cadoret and François Charles, *A remark on uniform boundedness for Brauer groups*, Algebr. Geom. **7** (2020), no. 5, 512–522, DOI 10.14231/ag-2020-017. MR4156415

[CHT17] Anna Cadoret, Chun-Yin Hui, and Akio Tamagawa, *Geometric monodromy—semisimplicity and maximality*, Ann. of Math. (2) **186** (2017), no. 1, 205–236, DOI 10.4007/annals.2017.186.1.5. MR3665003

[CT11] Anna Cadoret and Akio Tamagawa, *On a weak variant of the geometric torsion conjecture*, J. Algebra **346** (2011), 227–247, DOI 10.1016/j.jalgebra.2011.09.002. MR2842079

[CT12] Anna Cadoret and Akio Tamagawa, *A uniform open image theorem for ℓ-adic representations, I*, Duke Math. J. **161** (2012), no. 13, 2605–2634, DOI 10.1215/00127094-1812954. MR2988904

[CT13] Anna Cadoret and Akio Tamagawa, *A uniform open image theorem for ℓ-adic representations, II*, Duke Math. J. **162** (2013), no. 12, 2301–2344, DOI 10.1215/00127094-2323013. MR3102481

[CT16a] Anna Cadoret and Akio Tamagawa, *Gonality of abstract modular curves in positive characteristic*, Compos. Math. **152** (2016), no. 11, 2405–2442, DOI 10.1112/S0010437X16007612. MR3577899

[CT17] Anna Cadoret and Akio Tamagawa, *On the geometric image of $\mathbb{F}_\ell$-linear representations of étale fundamental groups*, Int. Math. Res. Not. IMRN **9** (2019), 2735–2762, DOI 10.1093/imrn/rnx193. MR3947637

[CT19] Anna Cadoret and Akio Tamagawa, *Genus of abstract modular curves with level-ℓ structures*, J. Reine Angew. Math. **752** (2019), 25–61, DOI 10.1515/crelle-2016-0057. MR3975637

[CT20] A. Cadoret and A. Tamagawa, *Erratum to Genus of abstract modular curves with level-ℓ structures*, 2020.

[D71] Pierre Deligne, *Théorie de Hodge. II* (French), Inst. Hautes Études Sci. Publ. Math. **40** (1971), 5–57. MR498551

[EEHK09] Jordan S. Ellenberg, Christian Elsholtz, Chris Hall, and Emmanuel Kowalski, *Non-simple abelian varieties in a family: geometric and analytic approaches*, J. Lond. Math. Soc. (2) **80** (2009), no. 1, 135–154, DOI 10.1112/jlms/jdp021. MR2520382

[EHK12] Jordan S. Ellenberg, Chris Hall, and Emmanuel Kowalski, *Expander graphs, gonality, and variation of Galois representations*, Duke Math. J. **161** (2012), no. 7, 1233–1275, DOI 10.1215/00127094-1593272. MR2922374

[Fu75] William Fulton, *Rational equivalence on singular varieties*, Inst. Hautes Études Sci. Publ. Math. **45** (1975), 147–167. MR404257

[Fu84] William Fulton, *Intersection theory*, Ergebnisse der Mathematik und ihrer Grenzgebiete (3) [Results in Mathematics and Related Areas (3)], vol. 2, Springer-Verlag, Berlin, 1984. MR732620

[GiS84] Henri Gillet and Christophe Soulé, *Intersection sur les variétés d'Arakelov* (French, with English summary), C. R. Acad. Sci. Paris Sér. I Math. **299** (1984), no. 12, 563–566. MR770447

[SGA1] A. Grothendieck, *SGAI: Revêtements étales et groupe fondamental*, L.N.M. **224**, Springer-Verlag, 1971.

[SGA7-I] A. Grothendieck et al., *SGA VII: Groupes de monodromie en géométrie algébrique - tome 1*, L.N.M. **288**, Springer, 1972.

[SGA4-III] A. Grothendieck et al., *SGA IV: Théorie des topos et cohomologie étale des schémas - tome 3*, L.N.M. **305**, Springer, 1973.

[J88] Uwe Jannsen, *Continuous étale cohomology*, Math. Ann. **280** (1988), no. 2, 207–245, DOI 10.1007/BF01456052. MR929536

[J90] Uwe Jannsen, *Mixed motives and algebraic K-theory*, Lecture Notes in Mathematics, vol. 1400, Springer-Verlag, Berlin, 1990. With appendices by S. Bloch and C. Schoen. MR1043451

[J94] Uwe Jannsen, *Motivic sheaves and filtrations on Chow groups*, Motives (Seattle, WA, 1991), Proc. Sympos. Pure Math., vol. 55, Amer. Math. Soc., Providence, RI, 1994, pp. 245–302. MR1265533

[Jo14] Peter Jossen, *On the Mumford-Tate conjecture for 1-motives*, Invent. Math. **195** (2014), no. 2, 393–439, DOI 10.1007/s00222-013-0459-y. MR3152752

[LN59] S. Lang and A. Néron, *Rational points of abelian varieties over function fields*, Amer. J. Math. **81** (1959), 95–118, DOI 10.2307/2372851. MR102520

[LaP95] M. Larsen and R. Pink, *Abelian varieties, l-adic representations, and l-independence*, Math. Ann. **302** (1995), no. 3, 561–579, DOI 10.1007/BF01444508. MR1339927

[N87] Madhav V. Nori, *On subgroups of* $\mathrm{GL}_n(\mathbf{F}_p)$, Invent. Math. **88** (1987), no. 2, 257–275, DOI 10.1007/BF01388909. MR880952

[O03] Fabrice Orgogozo, *Altérations et groupe fondamental premier à p* (French, with English and French summaries), Bull. Soc. Math. France **131** (2003), no. 1, 123–147, DOI 10.24033/bsmf.2438. MR1975807

[R95] Wayne Raskind, *Higher l-adic Abel-Jacobi mappings and filtrations on Chow groups*, Duke Math. J. **78** (1995), no. 1, 33–57, DOI 10.1215/S0012-7094-95-07803-X. MR1328751

[S79] Hiroshi Saito, *Abelian varieties attached to cycles of intermediate dimension*, Nagoya Math. J. **75** (1979), 95–119. MR542191

[Sc06] Chad Schoen, *Specialization of the torsion subgroup of the Chow group*, Math. Z. **252** (2006), no. 1, 11–17, DOI 10.1007/s00209-005-0837-2. MR2209148

[Se64] Jean-Pierre Serre, *Sur les groupes de congruence des variétés abéliennes* (French, with Russian summary), Izv. Akad. Nauk SSSR Ser. Mat. **28** (1964), 3–20. MR0160783

[Se89] Jean-Pierre Serre, *Lectures on the Mordell-Weil theorem*, Aspects of Mathematics, E15, Friedr. Vieweg & Sohn, Braunschweig, 1989. Translated from the French and edited by Martin Brown from notes by Michel Waldschmidt. MR1002324

[Si83] Joseph H. Silverman, *Heights and the specialization map for families of abelian varieties*, J. Reine Angew. Math. **342** (1983), 197–211, DOI 10.1515/crll.1983.342.197. MR703488

[T76] John Tate, *Relations between K_2 and Galois cohomology*, Invent. Math. **36** (1976), 257–274, DOI 10.1007/BF01390012. MR429837

IMJ-PRG, Sorbonne Université, 4, place Jussieu, 75252 PARIS Cedex 05, France
Email address: `anna.cadoret@imj-prg.fr`

Selected Published Titles in This Series

767 **Moshe Jarden and Tony Shaska, Editors,** Abelian Varieties and Number Theory, 2021

765 **Michael Aschbacher,** Quaternion Fusion Packets, 2021

764 **Gabriel Cunningham, Mark Mixer, and Egon Schulte, Editors,** Polytopes and Discrete Geometry, 2021

763 **Tyler J. Jarvis and Nathan Priddis, Editors,** Singularities, Mirror Symmetry, and the Gauged Linear Sigma Model, 2021

762 **Atsushi Ichino and Kartik Prasanna,** Periods of Quaternionic Shimura Varieties. I., 2021

761 **Ibrahim Assem, Christof Geiß, and Sonia Trepode, Editors,** Advances in Representation Theory of Algebras, 2021

760 **Olivier Collin, Stefan Friedl, Cameron Gordon, Stephan Tillmann, and Liam Watson, Editors,** Characters in Low-Dimensional Topology, 2020

759 **Omayra Ortega, Emille Davie Lawrence, and Edray Herber Goins, Editors,** The Golden Anniversary Celebration of the National Association of Mathematicians, 2020

758 **Jan Šťovíček and Jan Trlifaj, Editors,** Representation Theory and Beyond, 2020

757 **Kaïs Ammari and Stéphane Gerbi, Editors,** Identification and Control: Some New Challenges, 2020

756 **Joeri Van der Veken, Alfonso Carriazo, Ivko Dimitrić, Yun Myung Oh, Bogdan D. Suceavă, and Luc Vrancken, Editors,** Geometry of Submanifolds, 2020

755 **Marion Scheepers and Ondřej Zindulka, Editors,** Centenary of the Borel Conjecture, 2020

754 **Susanne C. Brenner, Igor Shparlinski, Chi-Wang Shu, and Daniel B. Szyld, Editors,** 75 Years of Mathematics of Computation, 2020

753 **Matthew Krauel, Michael Tuite, and Gaywalee Yamskulna, Editors,** Vertex Operator Algebras, Number Theory and Related Topics, 2020

752 **Samuel Coskey and Grigor Sargsyan, Editors,** Trends in Set Theory, 2020

751 **Ashish K. Srivastava, André Leroy, Ivo Herzog, and Pedro A. Guil Asensio, Editors,** Categorical, Homological and Combinatorial Methods in Algebra, 2020

750 **A. Bourhim, J. Mashreghi, L. Oubbi, and Z. Abdelali, Editors,** Linear and Multilinear Algebra and Function Spaces, 2020

749 **Guillermo Cortiñas and Charles A. Weibel, Editors,** K-theory in Algebra, Analysis and Topology, 2020

748 **Donatella Danielli and Irina Mitrea, Editors,** Advances in Harmonic Analysis and Partial Differential Equations, 2020

747 **Paul Bruillard, Carlos Ortiz Marrero, and Julia Plavnik, Editors,** Topological Phases of Matter and Quantum Computation, 2020

746 **Erica Flapan and Helen Wong, Editors,** Topology and Geometry of Biopolymers, 2020

745 **Federico Binda, Marc Levine, Manh Toan Nguyen, and Oliver Röndigs, Editors,** Motivic Homotopy Theory and Refined Enumerative Geometry, 2020

744 **Pieter Moree, Anke Pohl, L'ubomír Snoha, and Tom Ward, Editors,** Dynamics: Topology and Numbers, 2020

743 **H. Garth Dales, Dmitry Khavinson, and Javad Mashreghi, Editors,** Complex Analysis and Spectral Theory, 2020

742 **Francisco-Jesús Castro-Jiménez, David Bradley Massey, Bernard Teissier, and Meral Tosun, Editors,** A Panorama of Singularities, 2020

For a complete list of titles in this series, visit the AMS Bookstore at **www.ams.org/bookstore/conmseries/**.